3rd International Workshop & Summer School on Plasma Physics 2008

AA002641

Journal of Physics: Conference Series Volume 207

Kiten, Bulgaria
30 June - 5 July 2008

ISBN: 978-1-61738-270-3

Printed from e-media with permission by:

Curran Associates, Inc.
57 Morehouse Lane
Red Hook, NY 12571

Some format issues inherent in the e-media version may also appear in this print version.

Copyright© (2008) by the Institute of Physics
All rights reserved.

Printed by Curran Associates, Inc. (2010)

For permission requests, please contact the Institute of Physics
at the address below.

Institute of Physics
Dirac House, Temple Back
Bristol BS1 6BE UK

Phone: 44 1 17 929 7481
Fax: 44 1 17 920 0979

techtracking@iop.org

Additional copies of this publication are available from:

Curran Associates, Inc.
57 Morehouse Lane
Red Hook, NY 12571 USA
Phone: 845-758-0400
Fax: 845-758-2634
Email: curran@proceedings.com
Web: www.proceedings.com

3rd International Workshop & Summer School on Plasma Physics 2008

Journal of Physics: Conference Series Volume 207

Kiten, Bulgaria
30 June - 5 July 2008

ISBN: 978-1-61738-270-3

IOP Institute of Physics

TABLE OF CONTENTS

PAPERS

Kinetic Models of Nonequilibrium Nitrogen and Hydrogen Plasma for Diagnostics of Gas Discharges 1
Yu A Lebedev, V A Shakhatov, I L Epstein

Electrode Microwave Discharge: Areas of Application and Recent Results of Discharge Physics 11
Yu A Lebedev, I L Epstein, A V Tatarinov, V A Shakhatov

Data Processing of Electric Probe Data Acquired Under Locally Close-to-coherent Noise Conditions 21
F M Dias

Resonance Phenomena in Microwave Nonmagnetized Plasma Source 30
I A Kossyi

Diagnostics of Plasma Plume Produced by Laser Ablation Using ICCD Imaging and Transient Electrical Probe Technique 40
I Mihaila, C Ursu, A Gegiuc, G Popa

Active Species in N_2 and N_2-O_2 Afterglows for Surface Treatments 47
A Ricard, A M Pointu, S Villeger, C Canal

Study of Dielectric Barrier Discharges in Nitrogen and Afterglows at Atmospheric Pressure 56
F Clément, E Panousis, A Ricard, E Lecoq, J-F Loiseau, B Held

Parametric Study of Afterglows Issued from Dielectric Barrier Discharges in Nitrogen and Air and Applied on Aureobasidium Fungi 62
E Lecoq, C Leclaire, F Clément, G Orial, E Panousis, A Ricard

Environmentally Friendly Plasma-based Surface Engineering Technologies 69
A Drenik, M Mozetic, A Vesel, U Cvelbar

Diaphragm Discharge in Liquids: Fundamentals and Applications 75
Frantisek Krcma, Zdenka Stara, Jana Prochazkova

Plasmachemical Removal of Corrosion Layers from Iron in Pulsed RF Discharge 81
Vera Sazavska, Frantisek Krcma, Tereza Simsova, Nikola Zemanek

Effects of Diaphragm Discharge in Water Solutions Containing Humic Substances 85
Ivana Halamova, Zdenka Stara, Frantisek Krcma

Experimental Characterization of the Hβ-line Profiles in Microwave-produced Plasmas at Atmospheric Pressure 89
J M Palomares, J Torres, M A Gigosos, J J A M van der Mullen, A Gamero, A Sola

Study of Spatial Distributions of F, H and CF_2 Radicals in DF CCP Discharge in Ar/CF_4 and Ar/CHF_3 Mixtures 95
O V Braginsky, A S Kovalev, D V Lopaev, T V Rakhimova, A T Rakhimov, A N Vasilieva, S M Zyryanov

Hairpin-resonator Probe Design and Measurement Considerations 101
Plamen Dankov, Peter Stefanov, Valentin Gueorguiev, Tzvetan Ivanov

Ion-bombardment Modification of the Surface of Mirrors Fabricated of ZrTiCuNiBe Amorphous Alloys 107
V S Voitsenya, A F Bardamid, A I Belyaeva, V G Konovalov, K V Kovtun, D I Naidenkova, V I Ryzhkov, A F Shtan', K A Slatin, S I Solodovchenko, O V Trembach, A A Vasil'ev

Application of a Multi-pass Absorption Cell for Measurement of Sputtered Tungsten Atoms' Density 112
A Blagoev, A Pashov, A Stanev, I Bozhinova, I Rusinov

Ion Energy Distribution Analysis of the TVA Plasma Ignited in Carbon Vapours Using RFA 116
C C Surdu-Bob, M Badulescu, C Iacob, C Porosnicu, C P Lungu

AC Plasma Induced Modifications in Sb_2S_3 Thin Films 120
M Calixto-Rodriguez, F Castillo, H Martínez, Y Peña, A Sanchez-Juarez

Measurements of Electron Temperature, and Ion Density, in a Mixture of O_2 and Ar Plasma 124
P G Reyes, E Mendez, D Osorio-González, F Castillo, H Martínez

Pulsed Probe Diagnostics of a High-current Low-pressure Discharge in Crossed ExH Fields 128
I K Fetisov, I V Ivanov, G V Khodachenko, D V Mozgrin, D B Petrov, A M Snegirev, T V Shukshina

Electrical Properties of PMMA Ion-implanted with Low-energy Si^+ Beam 132
G B Hadjichristov, V K Gueorguiev, Tz E Ivanov, Y G Marinov, V G Ivanov, E Faulques

The Main Properties of Microwave Argon Plasma at Atmospheric Pressure 136
E Benova, M Pencheva

Similarities in Calculated Stark Broadening Parameters of Argon Spectral Lines 142
M Christova, M S Dimitrijevi?, A Kova?evi?

Stark Broadening Parameters of Ne I 837.8 Nm Spectral Line 148
M Christova, M S Dimitrijevi?, Z Simi?, S Sahal-Bréchot
Dynamic Effects in Dual-frequency Capacitively Coupled Discharges 152
D Voloshin, O Proshina, T Rakhimova
Electromagnetic Modeling of Axis-symmetric Microwave Devices 158
R Álvarez, L L Alves
Diffuse, Constricted-stratified and Constricted Modes of a DC Discharge in Argon. Simulation of Transitions Between These Modes 164
I A Shkurenkov, Yu A Mankelevich, T V Rakhimova
Kinetics of Ar-He Atmospheric Pressure Surface Wave Discharges 170
J Muñoz, M D Calzada
Numerical Investigations of Sheath Structure in Presence of Magnetic Field 174
E Havlí?ková, R Hrach
The Role of a Metal Screen on the Coaxial Discharge Wave Propagation Characteristics 180
T Bogdanov, E Benova
Pre- and Post-processing of Data for Simulation of Gyrotrons by the GYREOSS Software Package 184
M Damyanova, S Sabchevski, I Zhelyazkov
Model Calculation of Positron States in Tungsten Containing Hydrogen and Helium 190
T Troev, E Popov, N Nankov, T Yoshiie
Application of the Fokker Planck Approach for Modelling of Nano-structural Defects in Fusion Materials 201
Tzveta Apostolova, Troyo Troev
Numerical Simulations of Displacement Cascades in Irradiated with Intensive Neutron Flux Iron 207
N Nankov, L Petrov, E Popov, T Troev
The Distributions of Plasma Parameters in the "Magnetor" Device with Hot Cathode Discharge 213
I A Shchelkanov, D S Naumov, G V Krashevskaya, M M Tsventoukh, D E Vovchenko, G V Khodachenko, V A Kurnaev
Model Calculations of Edge Dislocation Defects and Vacancies in α-iron Lattice 219
L Petrov, T Troev, N Nankov, E Popov
Positron Lifetime Calculation of Defect in Palladium Containing Hydrogen and Helium 223
T Troev, E Popov, P Staykov, L Petrov, K Berovski, S Peneva, N Nankov
Author Index

PREFACE

The Third International Workshop & Summer School on Plasma Physics (IWSSPP'08) organized by St Kliment Ohridsky University of Sofia, with co-organizers TCPA Foundation, Association EURATOM/IRNRE, The Union of the Physicists in Bulgaria, and the Bulgarian Academy of Sciences was held in Kiten, Bulgaria, at the Black Sea Coast, from 30 June to 5 July 2008. A Special Session on Plasmas for Environmental Issues was co-organised by the Institute of Plasmas and Nuclear Fusion, Lisbon, Portugal and the Laboratory of Plasmas and Energy Conversion, University of Toulouse, France. That puts the beginning of a series in Workshops on Plasmas for Environmental Issues, now as a satellite meeting of the European Physical Society Conference on Plasma Physics.

As the previous issues of this scientific meeting (IWSSPP'05, *J. Phys.: Conf. Series* **44** (2006) and IWSSPP'06, *J. Phys.: Conf. Series* **63** (2007)), its aim was to stimulate the creation and support of a new generation of young scientists for further development of plasma physics fundamentals and applications, as well as to ensure an interdisciplinary exchange of views and initiate possible collaborations by bringing together scientists from various branches of plasma physics.

This volume of *Journal of Physics: Conference Series* includes 38 papers (invited lectures, contributed talks and posters) devoted to various branches of plasma physics, among them fusion plasma and materials, dc and microwave discharge modelling, transport phenomena in gas discharge plasmas, plasma diagnostics, cross sections and rate constants of elementary processes, material processing, plasma-chemistry and technology. Some of them have been presented by internationally known and recognized specialists in their fields; others are MSc or PhD students' first steps in science. In both cases, we believe they will raise readers' interest.

We would like to thank the members of both the International Advisory Committee and the Local Organizing Committee, the participants who sent their manuscripts and passed through the (sometimes heavy and troublesome) refereeing and editing procedure and our referees for their patience and considerable effort to improve the manuscripts. We greatly appreciate the financial support from the sponsors: the Department for Language Teaching and International Students at the University of Sofia, the Austrian Science and Research Liason Offices and the Bulgarian Nuclear Society. We would like to express our gratitude to the invited lecturers who were willing to pay the participation fee. In this way, in addition to the intellectual support they provided by means of their excellent lectures, they also supported the school financially.

E Benova, F M Dias and Yu Lebedev

All papers published in this volume of *Journal of Physics: Conference Series* have been peer reviewed through processes administered by the proceedings Editors. Reviews were conducted by expert referees to the professional and scientific standards expected of a proceedings journal published by IOP Publishing

Third International Workshop & Summer School on Plasma Physics 2008 IOP Publishing
Journal of Physics: Conference Series **207** (2010) 012001 doi:10.1088/1742-6596/207/1/012001

Kinetic models of nonequilibrium nitrogen and hydrogen plasma for diagnostics of gas discharges

Yu.A. Lebedev, V.A. Shakhatov, I.L. Epstein

Topchiev Institute of Petrochemical Synthesis RAS, Leninsky Prospect, 29, Moscow, 119991, Russia

lebedev@ips.ac.ru

Abstract. Computation-experimental methods for plasma diagnostics of nonequilibrium nitrogen and hydrogen gas discharges are presented. These methods are based on a combined use of collisional-radiative models and of experimental results on optical plasma diagnostics. Nonstationary self-consistent level-to-level collisional-radiative models include computer codes: for the processing and the simulation of the emission and the coherent anti-Stokes Raman scattering spectra; for the solution of the kinetic Boltzmann equation for plasma electrons, the thermal conductivity equation, and the balance equations for most of plasma neutral, excited, and charged particles. Banks of data were created for the processing and the simulation of the spectra, and the solution of the kinetic equations. The models were validated by the comparison of obtained results with numerous known results of experiments and with calculations for different discharges.

1. Introduction

Optical interferometry (OI), coherent anti–Stokes Raman scattering (CARS) spectroscopy, and optical emission spectroscopy (OES) have been widely used in diagnostics of nonequilibrium low temperature plasmas (NLTPs) [1]. The validation of these methods and the processing of measured data require the use of nonstationary semi-empirical self-consistent level-to-level collisional-radiative models (CRMs) of plasmas predicting population densities of the investigated diatomic molecules in the ground and electronically-excited states with resolution over vibrational and rotational levels. These models together with spectroscopic techniques allow the creation of computation-experimental methods gaining information on plasma parameters that could not be provided by any of these methods used separately [1-9]. Despite a wealth of publications [10-12] devoted to the development of the methods for describing collisional-radiative processes in NLTPs, the task remains topical up to this day. In the present work the nonstationary semi-empirical self-consistent level-to-level CRMs are developed for diagnostics of NLTPs of direct-current glow discharges, radio-frequency, and microwave discharges in N_2 and H_2. For plasmas in N_2 the discussion of the CRM is more detailed than for plasma in H_2, for which preliminary results will be given.

2. Collisional-radiative models of nonequilibrium low temperature plasma in N_2 and H_2

A comprehensive report on the semi-empirical self-consistent level-to-level CRM application to achieve plasma parameters for different types of discharges in N_2 from the spectral measurements can be found in [1, 4–6]. A detailed discussion and the development of CRMs in N_2, O_2 and H_2 are contained in [1–9, 13–15]. In brief, the CRM includes computing codes for modeling: the electron energy distribution function (EEDF) with the corresponding database on level cross sections of elementary processes involving electrons and heavy particles (molecules, atoms and their ions); the emission and CARS spectra with the corresponding spectroscopic database; the kinetic processes involving neutral molecules and molecular ions with resolution over vibrational levels of the ground and electronically-excited states, atoms and atomic ions in the ground and electronically-excited states with the corresponding database on level rate coefficients. Important components of the CRM are the database on the level cross sections and rate coefficients of elementary processes, the models for physical and chemical processes, and the radiative models of NLTP. These components of the CRM are formed and developed on the basis of the generalization of the accessible information in the literature on the spectroscopic data, rate coefficients, and cross sections of elementary processes, as well as on results of investigations on physical and chemical processes. A special attention is accented on the reproducibility of the results from measurement using the CRM. The reliability of the developed CRM is tested by comparing the calculated results with those from numerous data found in literature on plasma parameters of dc glow, radio-frequency, and microwave discharges. Thus the CRM includes a database on experiments devoted to the determination of EEDFs, vibrational and rotational functions of molecules and their ions in the ground and electronically-excited states, rate coefficients and cross sections of elementary processes, translational and rotational temperatures of molecules, specific power absorbed by plasma, etc.

In the following discussion we consider examples illustrating the application of the CRMs in N_2 and H_2 for the modeling of plasma parameters for different types of discharges: dc glow discharge (DCGD); capacitive and inductive-capacitive radio-frequency discharges (CRFD and ICRFD); electrode microwave discharge (EMD); microwave discharge excited by the surface wave; microwave discharges generated in the rectangular waveguide and in a prismatic resonator (MDRW and MDPR).

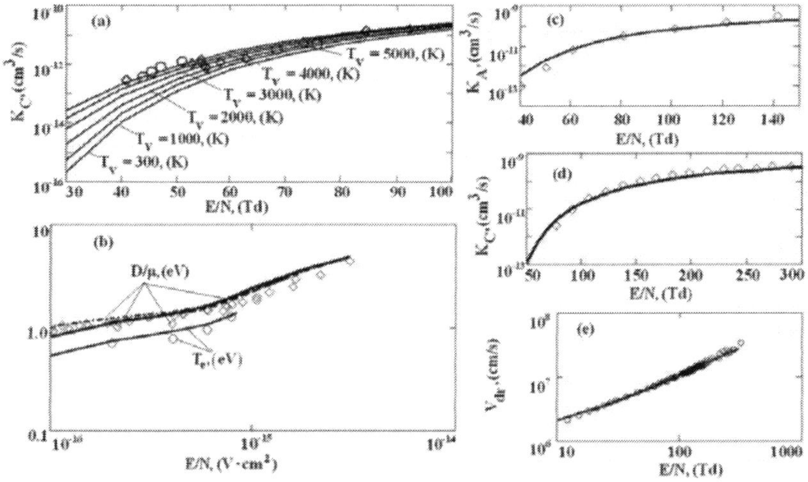

Figure 1. Measured (symbols) [10, 18–21] and calculated (solid curve by CRM, dotted line [22], and dash-dot [23]) for nitrogen: (a), (c), (d) values of excitation rate coefficients K_C and K_A of electronic states $C^3\Pi_u$ and $A^3\Sigma_u^+$, respectively (see text); (b) electron characteristic energy, D/μ, and temperature, T_e; (e) electron drift velocity, v_{dr}.

3. Application of CRMs for modeling of nonequilibrium low temperature plasma parameters in N_2 and H_2

The modeling and processing of the measured data are based on the self-consistent modeling of the kinetic processes in NLTP via the joint solution of the kinetic Boltzmann equation for the electron energy distribution function, the thermal conductivity equation, and the balance equations for heavy-particles in the ground and excited states considering heterogeneous processes on the surface of the reactor wall, as well as in the simulation of CARS and of the emission spectra.

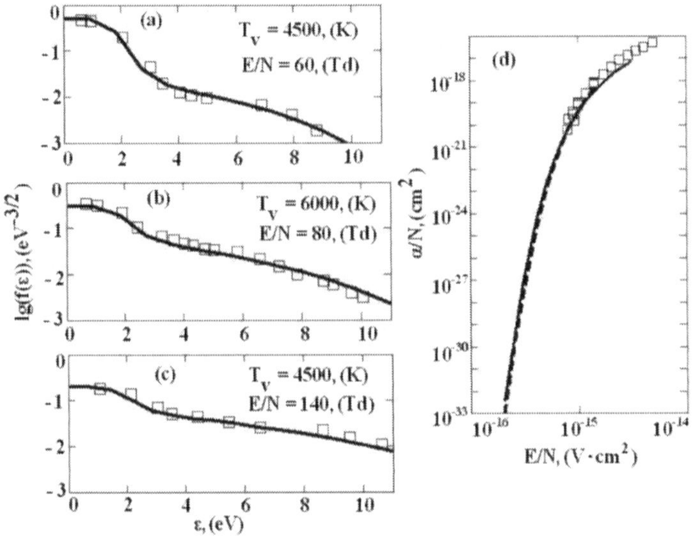

Figure 2. Measured (symbols) [10,17,18] and calculated (solid curve by CRM, and dotted line [22]) vs. T_v and the reduced electric field E/N: (a), (b), (c) EEDF in DCGD in N_2; (d) first Townsend ionization coefficient α/N in N_2.

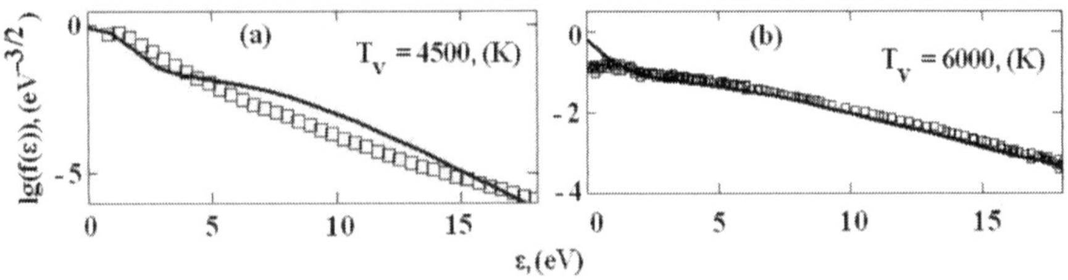

Figure 3. CRM calculated (solid curve) and measured (symbols) EEDF: a) in a nitrogen microwave discharge excited in a prismatic resonator [5]; b) in a nitrogen plasma excited by a microwave surface-wave [24].

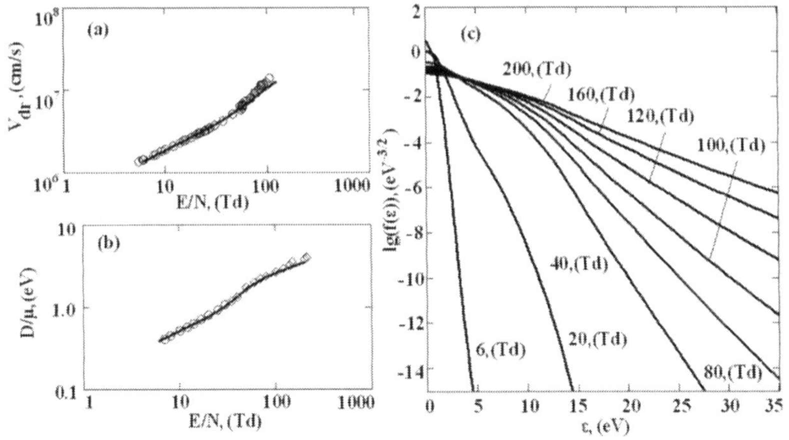

Figure 4. Measured (symbols) [18, 19] and calculated by the CRM (solid curve) in H_2: (a) electron drift velocity v_{dr}; (b) electron characteristic energy D/μ; (c) EEDF vs. E/N at T_g=T_v=325 K.

3.1. Electron energy distribution function

In an overwhelming number of the papers, which are reviewed in [10–13], on the solution of the kinetic Boltzmann equation for the EEDF are taken into account summary cross sections of excitation and deactivation for the total population densities of the electronically-excited states of diatomic molecules without resolution on vibrational levels. Preference is given to experimental data in the formation of the database. For the refinement of the absolute values of the cross sections the method of formation of the self-consistent set of cross sections is used [16]. Absolute values of the cross sections are determined via a comparison of the EEDF-based calculated data with the experimental one for the electron average energy, $<\varepsilon>$, the first Townsend ionization coefficient, α/N, the electron temperature, T_e, the drift velocity, v_{dr}, and the characteristic energy, D/μ (D is the electron diffusion coefficient, and μ is the electron mobility), as well as for the summary values of the electron-impact excitation rate coefficients, K_Y, of electronic states Y of molecules. However this method is ambiguous [17]. A more sequential approach for the database formation is presented in [3,5,8,13]. It implies that along with the data for $<\varepsilon>$, α/N, T_e, v_{dr}, D/μ, and K_Y additional experimental data on NLTP of different discharges is used: EEDF; vibrational distribution functions (VDFs) of molecules in the ground and electronically-excited states; translational, T_g, and vibrational, T_v, temperatures; specific power consumption, etc. In contrast to most of papers [10–13] (see also references therein) the level cross sections of excitation of the electronic excited states of the molecules are directly taken into account while solving the kinetic Boltzmann equation for the EEDF. This improves the quantitative description of experiments. The results of the database formation for the level cross sections and the comparison of their absolute values with the experimental data are given in [13].

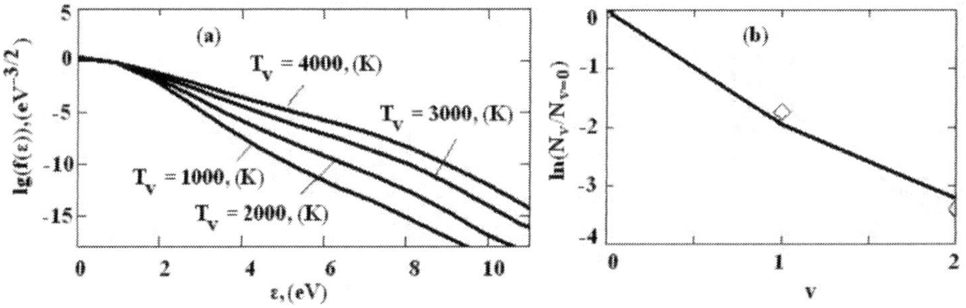

Figure 5. Measured (symbols) [1,7,25–27] and calculated by CRM (solid curve): (a) EEDF vs. T_v in DCGD in hydrogen at E/N = 20 Td; (b) VDF in inductive-capacitive RF discharge in hydrogen at a pressure 1.5 Torr.

Figures 1–5 show results from measurements [5,10,18-21,24] and calculations (in frame of CRM and from [22,23]) versus T_v and the reduced electric field E/N: EEDF in DCGD, radio-frequency and microwave discharges in H_2 and N_2; α/N and summary values of the excitation rate coefficients K_Y of the electronic state ($Y = A^3\Sigma_u^+$, $C^3\Pi_u$) of nitrogen in DCGD and for the conditions of electron swarm experiments in N_2; the characteristic energy D/μ, the temperature T_e, and drift velocity v_{dr} of electrons for the conditions of electron swarm experiments in N_2. As is seen from figures 1–5, there is a good agreement between theory and experiment for the formed self-consistent sets of level cross sections of elementary processes. The calculated values of α/N, T_e, and D/μ coincide also well with their calculated results by authors [22,23]. It is important to emphasize that the agreement between the measured and calculated EEDFs takes place at values of T_v in the range 4000–6000 K. The same values of T_v are obtained in calculations by the CRM and from experiments using Raman scattering technique [28], CARS spectroscopy [1–3, 29–37], and absorption spectroscopy [38] in DCGD and microwave discharges and their afterglows in N_2.

Journal of Physics: Conference Series **207** (2010) 012001 doi:10.1088/1742-6596/207/1/012001

Table 1. Computational and experimental results for T_g and T_v.

p (Torr)	R (cm)	N_e (cm^{-3})	t (ms)	E/N (Td)	Experiment		Theory		Ref.	Type of discharge
					T_v (K)	T_g (K)	T_v (K)	T_g (K)		
3.5	1.8	3.5×10^{9}	20	80	4000 ± 400^{e} 3790 ± 400^{s}	480 ± 40^{s} 460 ± 30^{e} $460\pm25^{\lambda}$	3475– 3700	470– 512	[1, 3]	DCGD in N_2
7.0	1.8	1.2×10^{10}	15	45	4100 ± 450^{s} 4320 ± 400^{s}	530 ± 40^{s} 540 ± 30^{e} $510\pm25^{\lambda}$ 570 ± 50^{i}	4200– 4255	530– 545	[1, 3]	DCGD in N_2
9.5	1.8	6×10^{9}	15	60	4150 ± 450^{e} 4270 ± 400^{s}	600 ± 40^{s} $580\pm30^{\lambda}$ 580 ± 35^{e} 570 ± 50^{i}	4240	605– 610	[1, 3]	DCGD in N_2
15	1.8	2×10^{10}	30	70		1000 ± 100^{s} 1100 ± 60^{B} 1140 ± 110^{i}		1135	[1, 3, 33]	DCGD in N_2
30	1.8	5×10^{10}	30	67		1350 ± 130^{s} 1300 ± 350^{i}		1300	[1, 3, 33]	DCGD in N_2
15	1.8	6.8×10^{9}	30	70		775 ± 60^{s} 760 ± 50^{B}			[1, 33]	DCGD in N_2
20	1.8	4×10^{10}	30	68		1200 ± 110^{s} 1230 ± 120^{i}		1230	[3, 33]	DCGD in N_2
2	1	2×10^{10}	11	80	5300 ± 400^{s}	540 ± 30^{s}	4250– 4960	420– 470	[34]	DCGD in N_2
12	0.7	$\approx1\times10^{9}$	30	>100	2850 ± 100^{s}	395 ± 12^{s}	2615– 2790	360– 400	[35]	DCGD in N_2
20	1	3.9×10^{10}	30	60		$1150–1200^{i}$		1170	[36]	DCGD in N_2
20	1.6	$\approx1\times10^{9}$	30		2400 ± 400^{b}	540 ± 40^{b}			[1, 37]	CIRFD in N_2
2	2^{d}	$\approx1\times10^{9}$	30		<2000	330 ± 20^{s}	<2000		[25]	CRFD in H_2
1.5	1.6	$\approx1\times10^{9}$	30		3400 ± 540^{b}	540 ± 120^{b}	3100		[26, 27]	CIRFD in H_2
8	1.6	$\approx1\times10^{9}$	30		2700 ± 420^{b}	750 ± 50^{b}	2700		[26, 27]	CIRFD in H_2
1	1	9.0×10^{10}	33		3500 ± 400^{e}	460 ± 50^{e}			[5]	MWRD in N_2
1.7	1.5	1.5×10^{11}	33		4500 ± 500^{e}	1300 ± 150^{e}			[5]	MWWD in N_2
1		1.1×10^{11}			4500 ± 500^{e}	550 ± 50^{e}			[9]	EMD in N_2

^{s}Scanning CARS spectroscopy with the collinear interaction of the laser beams.
^{b}Broadband CARS spectroscopy with the collinear interaction of the laser beams.
^{i}Optical interferometry.
^{e}Optical emission spectroscopy.
$^{\lambda}$Two-wavelength CARS spectroscopy with the collinear interaction of the laser beams.
^{B}Scanning CARS spectroscopy with sharp laser beam focusing (Planar BOXCARS).
^{d}RF discharge was created between two plane electrodes spaced by 2 cm.

3.2. Vibrational-translational nonequilibrium in nitrogen and hydrogen plasmas

The results of the database formation for the level rate coefficients, describing vibrational kinetics of N_2 and H_2 in the ground electronic state, and a comparison of their absolute values with the experimental data are given in [1,3]. The mechanisms causing gas heating and the formation of VDFs of these molecules in the ground electronic state as function of residence time in an active region of discharges are discussed in detail in [1,3].

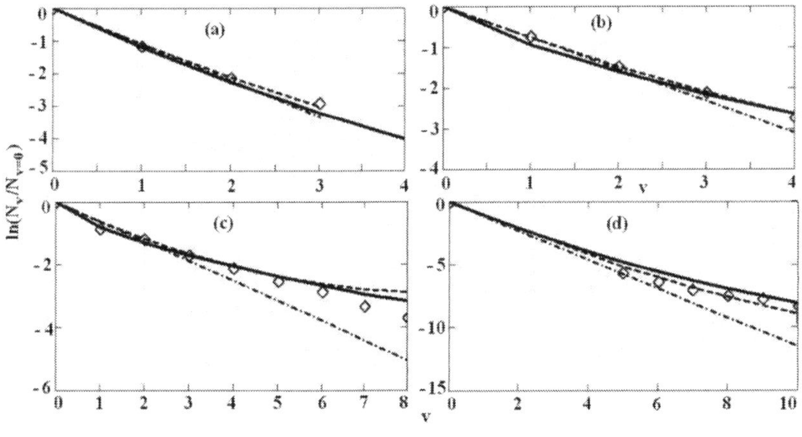

Figure 6. Calculated (solid curves by CRM, dash-dot lines by Boltzmann's, and dotted line by Treanor's formulas) and measured (symbols) (a) [29–32], (b) [35], (c) [34], and (d) [38] VDFs in the ground state $X^1\Sigma_g^+$ of nitrogen molecules.

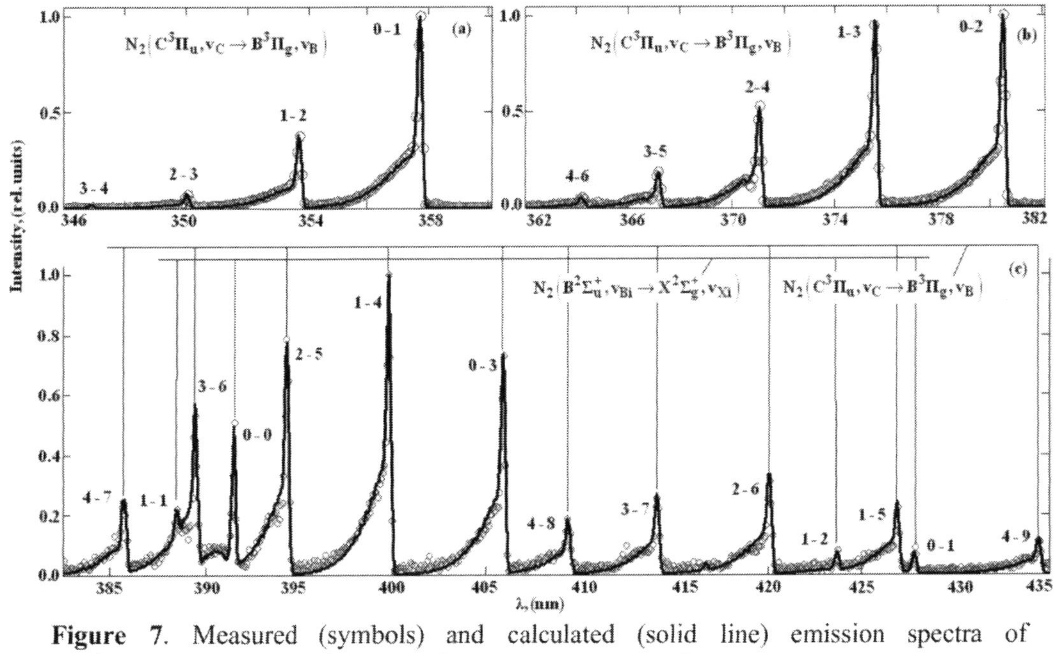

Figure 7. Measured (symbols) and calculated (solid line) emission spectra of $N_2\left(C^3\Pi_u, v_C \rightarrow B^3\Pi_g, v_B\right)$ and $N_2^+\left(B^2\Sigma_u^+, v_{Bi} \rightarrow X^2\Sigma_g^+, v_{Xi}\right)$ in nitrogen plasma of EMD at p = 1 Torr, electron density N_e=10^{11} cm^{-3} and electric field strength 139 V/cm [9] in the spectral range: (a) 340–360 nm; (b) 360–380 nm; (c) 380–435 nm.

Table 1 and figures 5(b) and 6 compare the computational results by CRMs for the relative population densities over vibrational levels, v, of nitrogen $N_2\left(X^1\Sigma_g^+, v\right)$ and hydrogen $H_2\left(X^1\Sigma_g^+, v\right)$ molecules, and the values of T_g and T_v with measurements performed by different techniques in plasmas of DCGD and contracted discharge, and radio-frequency and microwave discharges. The experimental data were obtained in our laboratory [1,3,5,9,29–33] and in a series of experiments published in [25–27,34,35,37]. The measured values of T_g and T_v listed in table 1 were obtained at different values of the pressure, p, the electron number density, N_e, the internal radius of discharge tube, R, and residence time, t, of the molecules in active region of discharges. The relative population densities in the $N_2\left(X^1\Sigma_g^+, v\right)$ and $H_2\left(X^1\Sigma_g^+, v\right)$ states and the values of T_g and T_v were determined using two approaches. In the first of them, the sought parameters are determined from the comparison of the amplitudes and line contours in the experimental and calculated broadband CARS spectra. In the second approach, the sought temperatures and distributions are determined from the processing of scanning CARS spectra by the amplitude technique. It is seen that measured and calculated parameters in frame of the CRM are in a good mutual agreement. The measured and calculated values of T_g and T_v in nitrogen plasmas of DCGD, contracted discharge and radio-frequency discharges lie in the ranges 395–1350 K and 2400–5300 K, respectively. In nitrogen plasmas of EMD the values of T_v and T_g, obtained from spectroscopic measurements [9], are equal to 4500±450 K and 550±50 K, respectively. For radio-frequency discharges in H_2 [25–27] the values of T_v and T_g, obtained from broadband CARS measurements, are not higher than 3400 K and 750 K, respectively.

3.3. Vibrational kinetics of electronically-excited states of nitrogen molecules

The results of the database formation for the level rate coefficients, describing vibrational kinetics of nitrogen molecules in the electronically-excited states, and a comparison of their absolute values with the experimental data are given in [8,13]. The mechanisms responsible for the formation of VDFs of molecules in the electronically-excited states as function residence time in an active region of the discharges are also discussed in detail in [8,13].

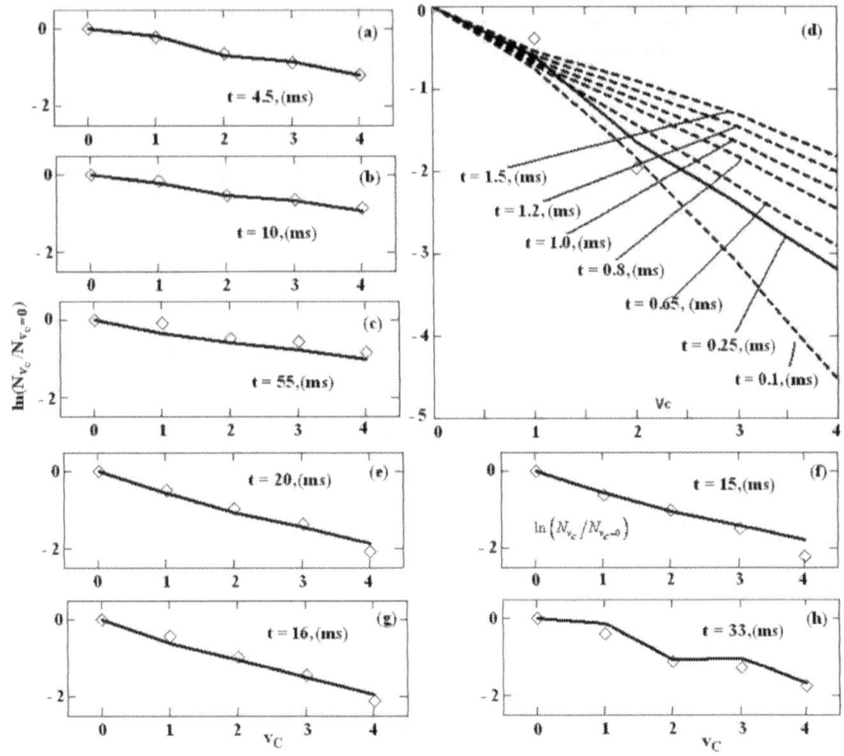

Figure 8. Calculated (solid and dotted line by CRM) and measured (symbols) VDF in the state $C^3\Pi_u$ of N_2 vs. t in the active discharge region: (a), (b), and (c) [39]; (d) [40]; (e), (f) and (g) [1,13]; (h) [1,13].

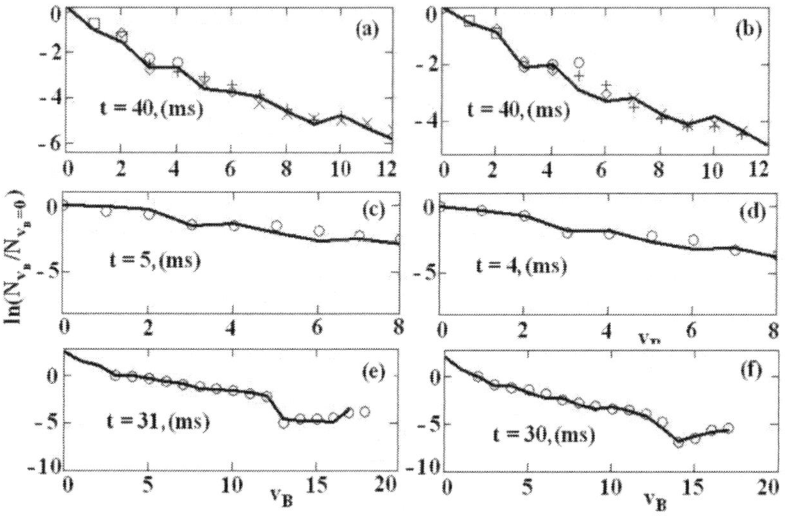

Figure 9. Calculated (solid line by CRM) and measured (symbols) VDF in the state $B^3\Pi_g$ of N_2 vs. t in the active discharge region: (a), (b) [41]; (c), (d) [42]; (e), (f) [43].

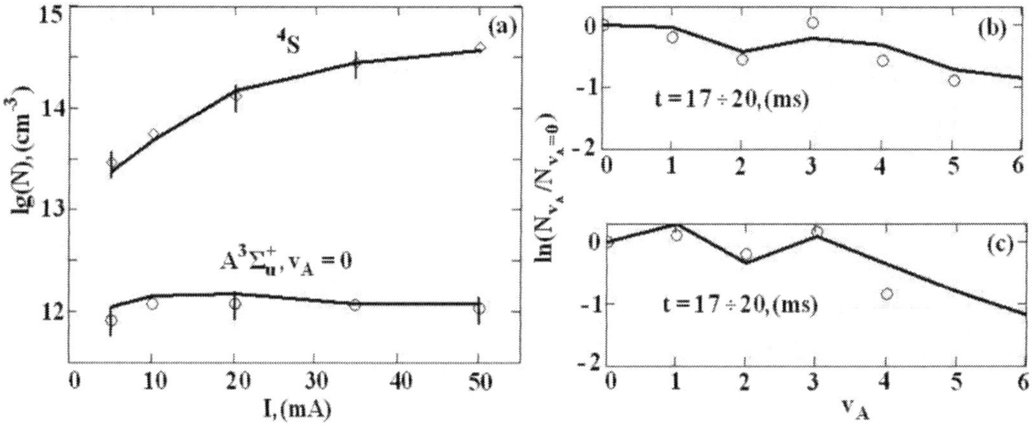

Figure 10. Calculated (solid line by CRM) and measured (symbols): (a) concentrations of atomic nitrogen in ^4S state and nitrogen molecules in $A^3\Sigma_u^+, v_A=0$ [45]; (c) and (d) VDF in $A^3\Sigma_u^+$ state [44].

During the modeling of the experiments, a special attention was given to the determination of populations on vibrational levels v_C = 0–4, v_B = 0–17, and v_A = 0–13 of the states $C^3\Pi_u$, $B^3\Pi_g$, and $A^3\Sigma_u^+$, of nitrogen molecules, respectively. These states are the upper ones for transitions in nitrogen: second positive system $N_2\left(C^3\Pi_u, v_C \rightarrow B^3\Pi_g, v_B\right)$, first positive system $N_2\left(B^3\Pi_g, v_B \rightarrow A^3\Sigma_u^+, v_A\right)$, and Vegard-Kaplan system $N_2\left(A^3\Sigma_u^+, v_A \rightarrow X^1\Sigma_g^+, v\right)$. The VDFs in electronically-excited states were determined from a comparison of the amplitude and contour of the vibronic bands of different sequences of measured and calculated emission spectra [6,8,9,13]. The comparison of spectra was performed at a constant value of the preliminary found rotational and translational temperatures. These temperatures were determined by an unresolved rotational structure technique. Figure 7, for example, illustrates a comparison of the calculated and experimentally recorded emission spectra of $N_2\left(C^3\Pi_u, v_C \rightarrow B^3\Pi_g, v_B\right)$, and first negative system $N_2^+\left(B^2\Sigma_u^+, v_{Bi} \rightarrow X^2\Sigma_g^+, v_{Xi}\right)$ of the molecular ion of nitrogen in an electrode microwave discharge. Values of T_g determined from bands, corresponding to the different sequences of $N_2\left(C^3\Pi_u,v_C \rightarrow B^3\Pi_g,v_B\right)$ and $N_2^+\left(B^2\Sigma_u^+,v_{Bi} \rightarrow X^2\Sigma_g^+,v_{Xi}\right)$, coincided. In nitrogen DCGD, the values of T_g determined from emission spectra are in good agreement with results of measurements performed by OI and CARS spectroscopy [4,8] (see table 1). Figure 8 shows the results of calculations and measurements [1,13,39,40] of populations on vibrational levels $v_C = 0–4$ of the $C^3\Pi_u$ state of N_2 as a dependence on the residence time of the molecules in the discharge region. For the states $B^3\Pi_g$ and $A^3\Sigma_u^+$ of N_2 a similar comparison of the calculation results and the experimental values [41–44] is shown in figure 9 and figure 10, respectively. The semi-empirical self-consistent level-to-level CRM in N_2 adequately reflects alternations of the relative populations on vibrational levels $v_C = 0–4$, $v_B = 0–17$, and $v_A = 0–13$ of the states $C^3\Pi_u$, $B^3\Pi_g$, and $A^3\Sigma_u^+$ measured in [1,13,40–44]. Figure 10(a) presents the results of calculations and measurements [45] of absolute concentrations of atomic nitrogen in the ground electronic state $N\left(^4S\right)$ and of nitrogen molecules in the metastable state $A^3\Sigma_u^+,v_A=0$. It was established that the calculated structure of the vibrational distributions and the concentrations of molecules and atoms in the ground and excited states depend on the kinetic scheme, experimental conditions, the database for level rate coefficients, and (which is especially important) on their dependence on the translational temperature, as well as on set of the level cross sections for elementary processes used in the semi-empirical self-consistent level-to-level CRM.

4. Conclusion

Computation-experimental methods of investigation of the nitrogen and hydrogen plasmas of different types of discharges are developed. These methods are based on processing of the emission and CARS spectra of the plasma by kinetic level-to-level semi-empirical collisional-radiative models. The reliability of the developed models is confirmed by a comparison of the results of simulation with a great body of the published data for different types of discharges. These models have been already effectively used in diagnostics of nonequilibrium plasma of DC and microwave discharges [1–9].

Acknowledgements

This study was partly supported by the Program # 9 for Fundamental Researches of the Presidium of the Russian Academy of Sciences and RFBR Grant 07-08-00020.

References

[1] *Plasma Diagnostics* Subject vol 1 part 1, ed Kolesnikov V N 2006 (Moscow: Yanus-K)
[2] Vereschagin K A, Smirnov V V and Shakhatov V A 1997 *Tech. Phys.* **42** 487
[3] Shakhatov V A and Gordeev O A 2005 *Tech. Phys.* **50** 1592
[4] Lebedev Yu A and Shakhatov V A 2006 *Plasma Phys. Rep.* **32** 56
[5] Lebedev Yu A and Shakhatov V A 2006 *High Temp.* **6** 795
[6] Lebedev Yu A, Solomakhin P V and Shakhatov V A 2007 *Plasma Phys. Rep.* **33** 157
[7] Shakhatov V A and Gordeev O A 2007 *Opt. Spectrosc.* **103** 468
[8] Shakhatov V A and Lebedev , Yu A 2008 *High Energ. Chem.* **42** 207
[9] Lebedev Yu A, Solomakhin P V and Shakhatov V A 2008 *Plasma Phys. Rep.* **34** 562
[10] Slovetskii D I 1980 *Mechanisms of Chemical Reactions in Nonequilibrium Plasma* (Moscow: Nauka)
[11] Capitelli M, Ferreira C M, Gordiets B F and Osipov A I 2000 *Plasma Kinetics in Atmospheric Gases* (Springer Series on Atomic, Optical, and Plasma Physics, Berlin: Springer)
[12] Guerra V, Sa P A and Loureiro J 2005 *Eur. Phys. J.: Appl. Phys.* **28** 56
[13] *Plasma Optics* vol XI, ed Ochkin V N 2008 (Moscow: Yanus-K)
[14] Lebedev Yu A and Shakhatov V A 2008 *Proc. V Int. Symp. on Theoretical and Applied Plasma Chemistry* ed Ivan G U (Ivanovo 2008) vol 1 p 121

[15] Lebedev Yu A, Tatarinov A V and Epstein I L 2008 *High Temp.* **6** 795
[16] Engelgardt A G, Phelps A V and Risk C G 1964 *Phys. Rev.* **A135** 1566.
[17] Aleksandrov N L and Son E E 1978 *Plasma Chemistry* vol 5, ed Smirnov B M (Moscow: Atomizdat)
[18] Huxley L G H and Crompton R W 1974 *The Diffusion and Drift of Electrons in Gases* (New York: Wiley J and Sons Inc – Intersience Publication)
[19] Dutton J J 1975 *Chem. Phys. Ref. Data* **4** 577
[20] Hyadon C S and Williams O M 1976 *J. Phys. D: Appl. Phys.* **9** 523
[21] McConkey J W and Simpson F R 1969 *J. Phys. B* **2** 923
[22] Brunet P V and Rocca S J 1983 *J. Appl. Phys.* **54** 4951
[23] Dyatko N A, Kochetov I V and Napartovich A P 1992 *Plasma Phys. Rep.* **18** 888
[24] Tatarova E, Dias F M *et al* 1997 *J. Phys. D: Appl. Phys.* **30** 2663
[25] Shakhatov V A, De Pascale O and Capitelli M 2004 *Eur. Phys. J. D* **29** 230
[26] Shakhatov V A, De Pascale O, Capitelli M *et al* 2005 *Phys. Plasmas* **12** 07331
[27] Hassouni K, Lombardi G, Gicquel A *et al* 2005 *Phys. Plasmas* **12** 023504
[28] Blois D, Supiot P, Barj M *et al* 1998 *J. Phys. D: Appl. Phys.* **31** 2521
[29] Bodronosov A V, Vereschagin K A, Gorshkov V A *et al* 1994 *Tech. Phys.* **39** 25
[30] Bodronosov A V, Vereschagin K A, Gorshkov V A *et al* 1994 *J. Quant. Electron.* **24** 832
[31] Gordeev O A and Shakhatov V A 1995 *Tech. Phys.* **40** 656
[32] Bodronosov A V, Vereschagin K A, Gordeev O A *et al* 1996 *High Temp.* **34** 656
[33] Shakhatov V A and Gordeev O A 2006 *High Temp.* **44** 206
[34] Massabieaux B, Gousset G, Lefebvre M *et al* 1987 *J. Physique* **48** 1939
[35] Smirnov V V and Fabelinskii V I 1978 *JETP Letters* **28** 424
[36] Golubovskii Yu B, Telezhko V M *et al* 1990 *Opt. Spectrosc.* **69** 322
[37] Shakhatov V A and Gordeev O A 2006 *High Temp.* **44** 12
[38] Mikheev I D and Badrautdinov O P 1988 *Plasma Phys. Rep.* **14** 1143
[39] Massabieaux B, Plain A, Ricard A *et al* 1983 *J. Phys. B.* **16** 1863
[40] Bruhl S P, Russell M W, Gomez B J *et al* 1997 *J. Phys. D: Appl. Phys.* **30** 2917
[41] Ferreira C M, Touzeau M *et al* 1984 *J. Phys. B: At. Mol. Phys.* **17** 4439
[42] Plain A, Gorse C *et al* 1985 *J. Phys. B: At. Mol. Phys.* **18** 843
[43] Polak L S, Slovetskii D I and Sokolov A I 1972 *Opt. Spectrosc.* **32** 472
[44] Cernogora G, Hochard L *et al* 1981 *J. Phys. B: At. Mol. Phys.* **14** 2977
[45] Cernogora G, Ferreira C M *et al* 1984 *J. Phys. B: At. Mol. Phys.* **17** 4429

Journal of Physics: Conference Series **207** (2010) 012002 doi:10.1088/1742-6596/207/1/012002

Electrode microwave discharge: areas of application and recent results of discharge physics

Yu. A. Lebedev, I. L. Epstein, A. V. Tatarinov, V. A. Shakhatov

Topchiev Institute of Petrochemical Synthesis RAS, Leninsky Prospect, 29, Moscow, 119991, Russia

lebedev@ips.ac.ru

Abstract. The first paper on the electrode microwave discharge (EMD) appeared in 1996. Presently many problems of EMD physics and applications have already been solved. Several examples of EMD application are discussed: diamond growth, deposition of CN_x films and nanotubes, deposition of metal films (Cu, Al), deposition of TiN and TiO_2 films, generation of $O_2(a^1\Delta)$, and EMD as a plasma cathode. Results of EMD experiments and modeling give rise to the assumption that an EMD consists of a self-sustained domain (near-electrode plasma region with overcritical plasma density) which is surrounded by a region of a non-self-sustained discharge (ball shaped region with undercritical plasma density). We assumed that the layer of charge separation and of induced electrostatic field originated at the outer EMD boundary was one of the reasons for the abrupt decrease of the plasma density which leads to the formation of a compact plasma structure. Recent modeling results of the strongly nonuniform electrode microwave plasma based on a quasi static, 1D spherically symmetric model showed that such a layer can be generated at the point where a sudden increase of the total ionization rate takes place.

1. Introduction

The intensive study of the electrode microwave discharge (EMD) started in 1996 [1] and now we have a lot of knowledge about the discharge physics and about the discharge applications [1–26]. The EMD is interesting due to its distinctive features and its promising applications. We call a discharge an EMD if the plasma is ignited near the tip of the powered electrode/antenna and the luminous plasma region is less than dimensions of the discharge chamber. The EMD turns into the ordinary discharge mode [2] at smaller distances between the electrodes.

The distinctive EMD features are: Extremely low maintaining discharge power (1–2 W) and a wide range of operating pressures; generation of compact plasma structures, of plasma active particles in large vessels and near the treated surface, and of plasmas in a chosen point of space (localization of plasma region; the area of plasma region being controlled by the electrode shape).

2. Experiment and modelling

A detailed description of the experimental set-up has been presented in [4,5,10,21,26]. Stainless steel cylindrical discharge chambers with diameters of discharge vessel of R_1 = 7 cm and R_2 = 4.2 cm have been used for the experiments. The plasmas in both arrangements revealed similar features, thus the properties of the described plasmas seem to be typical for electrode microwave discharge systems.

Plasma gases were Ar, Ne, H_2, N_2, O_2, CH_4, C_2H_2, air, and their mixtures at pressures 0.5–400 Torr. A gas flow system provides a total gas flow rate of less than 1000 sccm. A "gas feeding-pumping" system gives the possibility to operate with gas mixtures of 3 gases and introduces the input gas mixture through the channel inside the electrode or through the upper cover of the chamber. The latest modification of the setup (stand EMD-3) is shown in figure1.

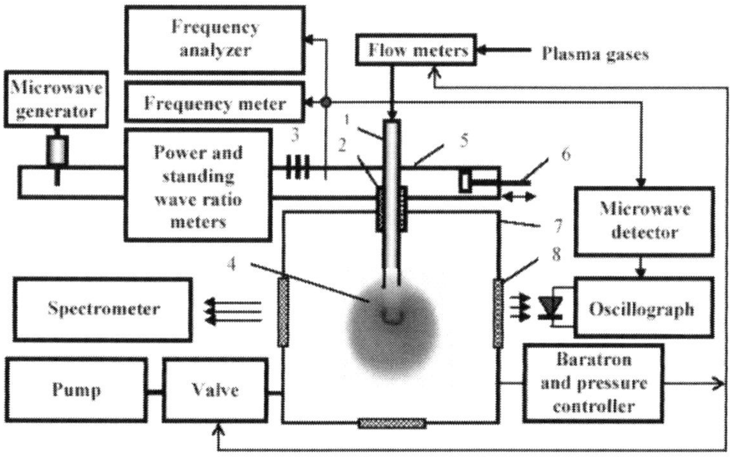

Figure 1a. Experimental set-up EMD-3. 1 – electrode (antenna), 2 – isolator, 3 – impedance transformer, 4 – plasma, 5 – coaxial-to-waveguide converter, 6 – shorting plunger, 7 – discharge chamber, 8 – optical windows

Figure 1b. View of the experimental set-up EMD-3.

Tubular and solid stainless steel and copper electrodes of different shapes (direct or bent rod and tube cylinder electrodes, trident electrode, spiral electrode, etc.) with diameters of 0.5–6 mm have been used as antennas. This set-up gives the possibility of breaking the galvanic coupling between antenna/electrode and the body of the discharge chamber. The latter property makes it possible to investigate the effect of external static fields on the properties of EMD plasma.

Most of results have been obtained when the discharge is ignited at the tip of a cylindrical electrode (antenna). The microwave power (2.45 GHz) was transmitted from magnetron generators with output powers of 2.5 kW or 150 W. The power absorbed in the system is measured by a directional coupler. For a given value of the input power the power absorbed in the plasma is obtained by subtracting the power losses in the empty chamber from the power losses in the chamber with plasma.

The discharge emission, which leaves the chamber through the lateral or bottom quartz windows, is focused by quartz lenses, collected by the optical fibre, and recorded with spectrometers (monochromators MDR-4 and MDR-23, spectrograph AvaSpec-2048, AvaSpec-3648, AvaSpec-2048-4-RM). The optical fibre can be moved both in the longitudinal and radial directions by means of a two-coordinate micrometric table with a spatial resolution of 150 μm. The discharge visualization is made by a video camera (the exposition time ranged between 1/20 and 1/8000 s), by a digital photo camera, and by electron-optic high time resolution digital cameras K008 and K011.

The set-up contains the devices for the spectral analysis of the microwave radiation (spectrum analyzer C4-27, stroboscopic oscilloscope C7-11) and for the registration of the time variation of the microwave power and of the integral plasma emission in the visible range (oscilloscope Tektronix TDS2012B).

The EMD had spherical symmetry at small flow rates. A few photos of the EMD are shown in figures 2, 3. The ball-like structure is deformed and the plasma expands along the electrode channel axis at flow rates higher than 50 sccm.

Methods used for the experimental study:

- Photometry of the discharge (study of EMD structure).
- Double probe method (determination of the local values of the electron temperature and density, electric field strength) [4,5,7–10].
- Emission spectroscopy in the visible range of the spectrum with spatial resolution (determination of plasma particles concentrations, gas temperature, electric fields, etc) [6,12–14,16,17,21,26].
- Electrodynamic measurements (determination of the absorbed power, frequency and microwave spectrum, time variation of the power).

Methods for EMD simulation:

- 0D kinetic numerical code [17,24].
- Self-consistent 1D modelling in quasi-static approximation [11,20,22,25].
- Self-consistent 2D modelling with the Maxwell equation [15,16,18,23].

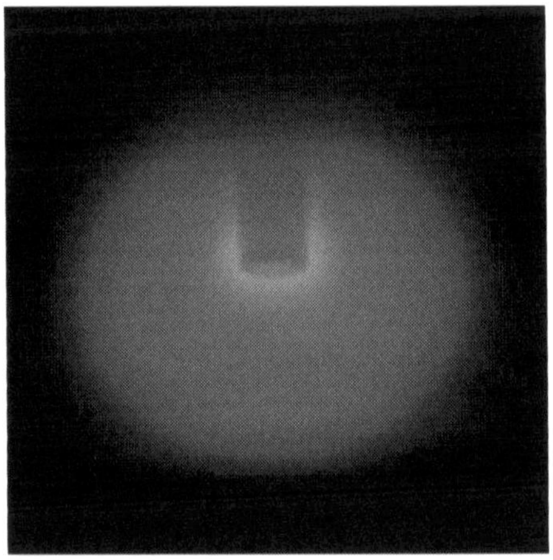

Figure 2. View of the air discharge at the tip of a 6 mm electrode in nitrogen (1 Torr)

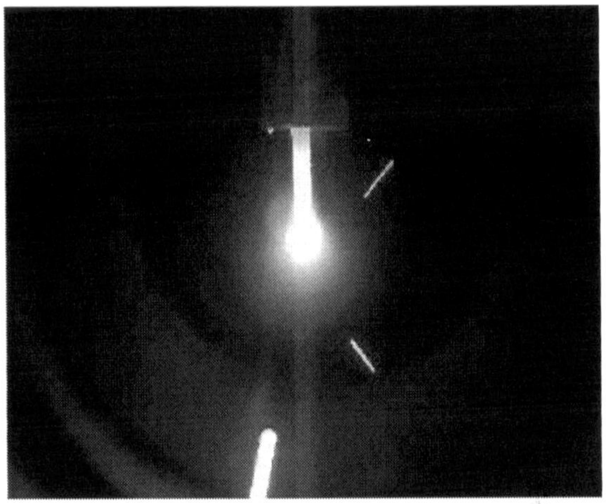

Figure 3. View of the air discharge with a needle electrode exploding (diameter – 1 mm, 50 Torr)

3. Brief summary on physics of the EMD in molecular gases and their mixtures with noble gases

3.1. General features and peculiarities of the EMD

I. General discharge properties (like spatial profiles of electron and ion densities, plasma emission) are independent of the diameter of the plasma chamber.

II. The discharge consists of (figure 2):

- a bright thin (1–1.5 mm) near electrode plasma region which covers the tip of the electrode;
- a ball-like emissive plasma region;
- a dark external conductive space.

III. The ball dimension increases with microwave power and decreases with pressure.

IV. There exists an upper limit for the microwave power. If the input power exceeds this value the discharge runs away from the tip of electrode towards the microwave generator.

IV. Thin electrodes can be destroyed due to melting and can even explode for high microwave powers and pressures (figure 3).

The structure of the EMD is clearly seen in figure 4.

3.2. Discharge boundary

The plasma ball has a sharp external boundary. Probe measurements showed an abrupt decrease of the electron concentration in a thin boundary layer even in diffusion controlled regime [7,8]. The only reason, why the spread of charged particles out of the plasma ball is suppressed, is the existence of a local electrostatic field. A double layer has the necessary property. This boundary reveals an elastic property in case of contact with an external body, e.g. with the probe. In case of two plasma balls ignited at two closely placed electrodes/antennas the increase of the microwave power at first leads to growing ball diameters. At higher microwave powers the shape of the plasma balls changes into ellipsoids, which are always separated by the dark space.

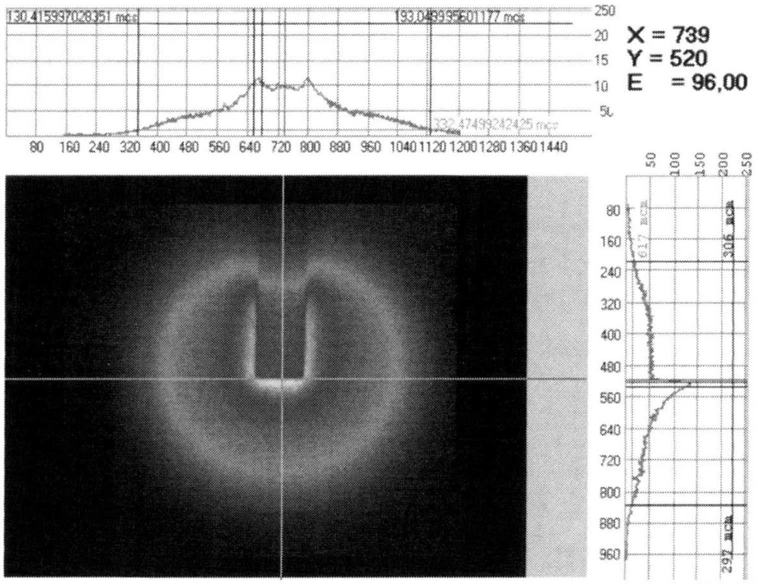

Figure 4. Color marked visualization of the total plasma emission of a nitrogen EMD together with the axial and radial distributions of the emission intensity (1.33 Torr).

3.3. EMD dimensions

What is the largest plasma dimension of an EMD and the highest power absorbed in the plasma? The answer to these questions was given by the 2D model, which included the Maxwell equations [18].

It was shown that at small levels of the input power the discharge is initiated in a thin region of high electric field at the tip of antenna. The size of the plasma region increases with the input power and the discharge grows in size. The discharge center is still fixed near the tip of the electrode. At a certain power value (the upper limit) the discharge gets detached from the tip of the antenna and runs towards the generator. In this case the stationary discharge is maintained at the entrance of the discharge chamber. This value of the input power determines the largest size of the EMD. It is shown that the largest size of the plasma ball can not exceed the quarter of the generator's wave length. Thus in order to increase the plasma size one needs to decrease the field frequency.

3.4. Gas temperature

The gas temperature in the hydrogen EMD plasma was determined from the relative intensities of rotational lines of the electron excited molecules $H_2(d^3\Pi_u)$. Calculations are based on the intensities of Q and R-branches emission for diagonal (v' = v'' = 0, 1, 2) bands of Fulcher α-system $H_2(d^3\Pi_u \rightarrow a^3\Sigma_g)$. The rotational temperature of the ground state of H_2 is calculated taking into account the ratio of the rotational constants of the ground and excited states. The temperature had a flat radial distribution and does not exceed 1000 K at pressures below 10 Torr [13].

To determine the rotational temperature of the nitrogen plasma the analysis of non-resolved spectra of the 1st negative and 2nd positive system of the nitrogen emission has been used. In our conditions this temperature is assumed to be equal to the gas temperature. Gas temperature was less than 1000 K and the radial temperature profile was flat [21,26].

Since the discharge is a non-uniform gas dynamic system, there is a possibility of generating vortex flows that intensively mix the gas. Non-uniform gas dynamics can be caused both by gas feeding system and by the non-uniform heating of the electrode tip. This results in a non-uniform distribution of the gas temperature in the entire volume of the spherical plasma.

3.5. The EMD as "self-sustained-non-self-sustained" system

The analysis of the publications on the experimental results of microwave discharges and the modeling results of the EMD allows us to formulate the conception of "self-sustained-non-self-sustained" mechanism of the discharge. The discharge consists of the near-electrode region of self-sustained discharge, with an overcritical plasma density, surrounded by a ball-shaped region of non-self-sustained discharge, with an under-critical plasma density [19,21,23,26]. A layer of charge separation (similar to the double layer) exists on the outer plasma ball boundary and can be responsible for the abrupt decrease of the plasma density there. This layer might be a consequence of the self-consistency of plasma properties.

3.6. After-effects of EMD non-uniformity

The EMD is strongly non-uniform and this is a consequence of the electric field heterogeneity: the highest field exists near the electrode [11,15,16,29,23]. The strong heterogeneity of the electric field in the EMD leads to different physical processes in different parts of it [20,21,23,26]. Electron impact is the main channel for heavy particles excitation and ionization near the electrode, while secondary processes with the participation of excited particles determine the processes in the plasma ball region. In the nitrogen plasma this leads to the fact that the 2^+ nitrogen band emissions prevail over others in the near-electrode region with main power consumption while the 1^+ nitrogen band emissions give the principal part of the emission in the spherical part of the EMD. The behaviour of the ion profiles is an illustration for this effect: the N_2^+ ion is the dominant one near the electrode, while N_4^+ prevails over others in the ball region. The near-electrode region is the source of excited particles for spherical part of plasma. The non-uniformity also leads to changes of the channels of charged particles losses along the radius (figure 5).

3.7. Generation of layers of charge separation

Recent simulations showed that charged layers can be created in the plasma with a sudden increase of the electric field and of the ionization rate. A one-dimensional, quasi-static model for a stationary microwave nitrogen discharge inside spherically symmetric system of electrodes has been used [20]. The model consists of the microwave field equation in quasi-static approximation, the Poisson equation, the balance equations which describe the kinetics of the charged (e, N_2^+, N_4^+) and neutral ($N_2(A^3\Sigma^+_u)$, $N_2(B^3\Pi_g)$, $N_2(C^3\Pi_u)$, $N_2(a'^1\Sigma^-_u)$) plasma components, and the time-independent homogeneous Boltzmann equation for electrons obtained in the two-term approximation. The kinetic scheme involves direct electron impact ionization, step and associative ionization, volume and wall recombination of charged and neutral particles, ion-molecular reactions, and excitation and de-excitation processes for molecules. Processes with vibrationally excited molecules are accounted for by using the well-known analytical expression for the vibrational distribution function (VDF) obtained in the diffusion approximation [27]. This condition is satisfied in strongly non-uniform microwave discharges in the plasma resonance region (figure 6). The charge separation ($\Delta n = (\Sigma N_i^+ - n_e)$) induces the static electric field and this can be considered as one of the possible reasons for the sharp external boundary of the EMD.

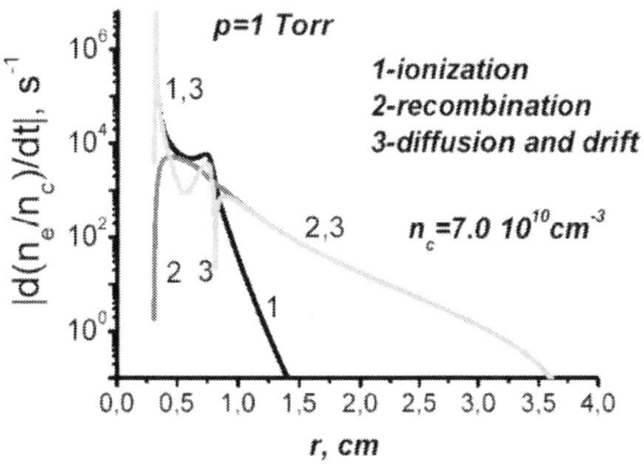

Figure 5. Calculated rates of the charged particle generation and losses in EMD with the spherical symmetry (n_c is the critical electron density).

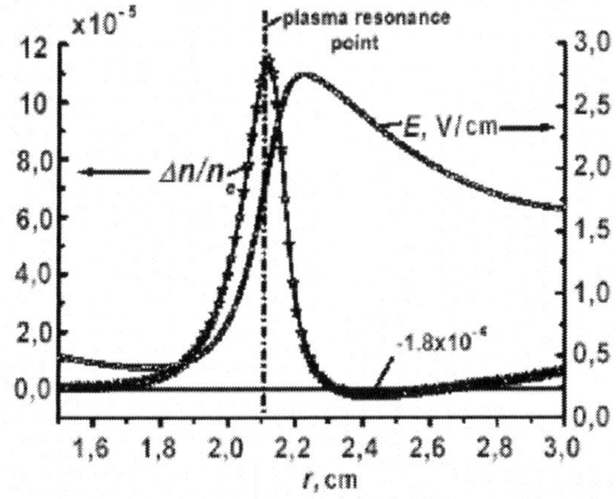

Figure 6. Calculated layer of charge separation and the static field E in the region of plasma resonance.

4. EMD applications

The EMD gives the possibility: to study and to produce plasma chemical processes, to intensify physical-chemical processes in gas phase, including the burning process, to study the interaction of plasma with solids, to study the behavior of solids surrounded by a plasma with gas flows, and to study the interaction of plasma formations with gas flows.

Notwithstanding the fact that our knowledge on EMD physics is not complete, we have some examples of successful and promising EMD applications. It is necessary to mention some obvious disadvantages of the EMD, which can be important in research perspectives of this type of discharge. The considerable non-uniformity of the plasma leads to low spatial homogeneity of plasma-surface processing. We also indicate a low level of plasma absorbed power.

4.1. Diamond growth

The EMD have been used for diamond growth and the deposition of diamond-like films [1,3,28]. The process is held at a 15 Torr pressure in the mixture H_2 + (1–10%) CH_4 with a stainless steel electrode [1,3] and at a 20 Torr pressure in the mixture H_2 + 0.4% CH_4 with a graphite electrode [28]. Several results are shown in figures 7, 8 [3]. Deposition is made on silicon wafers at temperature of 1000 °C. It is known that the quality of the deposit is increased by adding oxygen and this is illustrated by curves 1 and 2 in figure 8. Deposition rate was of 1 μm/h. It is shown that the risk of contamination of the plasma vessel and of the treated surfaces in EMD by products from electrode erosion is small [1,3]. The method of electron spectroscopy for chemical analysis showed no traces of products from electrode erosion in plasma-treated substrates in experiments on the diamond growth when the electrode was heated up to the red-heat temperatures and the temperature of the substrate, located at the distance of 1 cm from it, was 1000 °C. The problem of electrode erosion is very important for RF and DC discharges. As the field frequency increases, the thickness of the electrode sheath decreases (the electron density is high enough to exceed the critical density) and the energy of the ions bombarding the electrode decreases as well. This fact is also a distinctive feature of the EMD plasma. The role of the electrode is only in the formation of the microwave field structure.

The EMD is a strongly non-uniform system and the deposition rate depends both on the distance of substrate from the electrode (d_0) and the distance from the center of the substrate. Experimental deposition rates can be approximated as $K(r) \approx \alpha(d_0^2 + x^2)^{-1}$ where r is the distance from the point on the substrate from the center of plasma ball [28]. It is important that deposits on the substrate were observed also outside the luminous plasma ball. This means that the dark discharge region is active and the above formula reflects the distribution of active particles around the electrode.

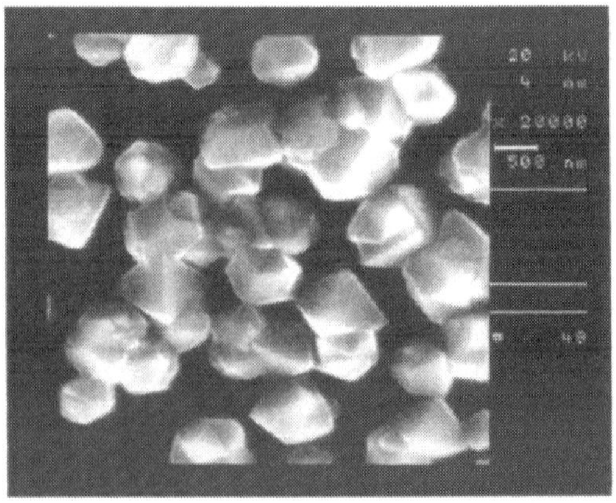

Figure 7. Microphotograph of the substrate after deposition [3].

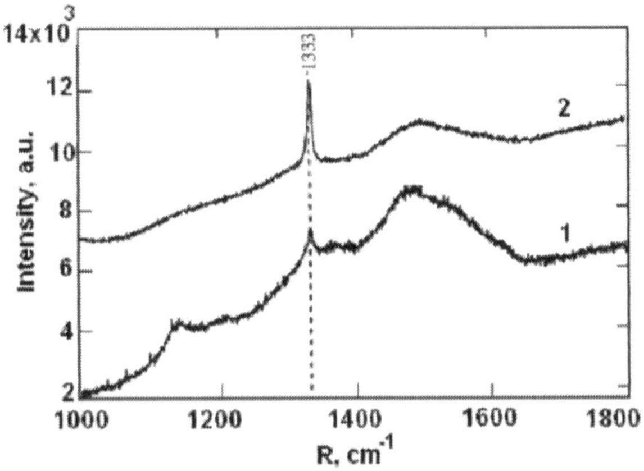

Figure 8. Raman spectra of deposits. Curve 1 – 400 sccm H_2 + 16 sccm CH_4 , curve 2 – 400 sccm H_2 + 16 sccm CH_4 + 1 sccm O_2 [3].

There are three methods to increase the size of homogeneously treated surfaces. The first is obvious: one has to decrease the microwave frequency. The second one relies on using a discharge system with several electrodes [28]. The third relies on using a dilative, conical electrode. Figure 9 shows that the radial distribution of the ion probe saturation current for a conical electrode is flatter than that for cylindrical one (the plane of measurements is placed at the distance of 5 mm below the tip of the electrode and perpendicular to its axis) [5].

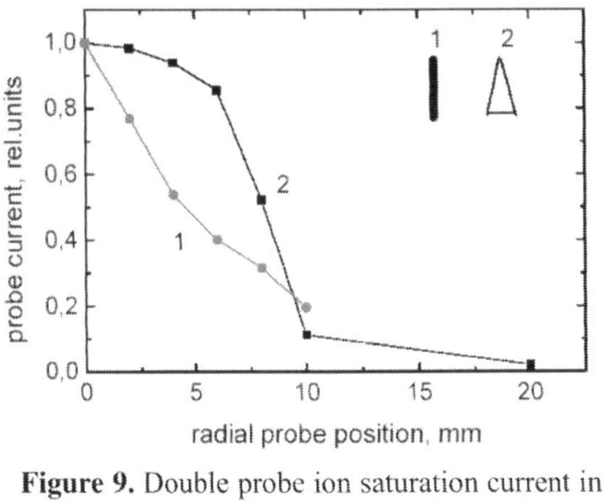

Figure 9. Double probe ion saturation current in the hydrogen EMD for a cylindrical and for a conical electrode [5].

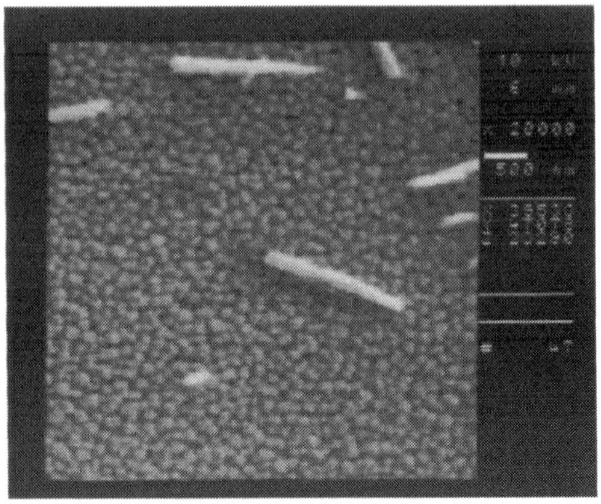

Figure 10. Microphotograph of a silicon substrate after CN–film deposition [30].

4.2. Coating deposition

Mixtures of different plasma gases with hydrocarbons are used for deposition of amorphous and polycrystalline coatings. The EMD with a graphite electrode in the mixture 0.9 Torr H_2 + 0.1 Torr CH_4 has been used for the deposition of nanotubes on a heated Ni substrate at an incident power of 250 W [29]. The density of the deposit was inhomogeneous over the substrate surface.

The possibility of deposition of CN_x-films with a mixture of N_2 and C_2H_2 at a pressure of 1 Torr (power consumption 2–40 W) on silicon, glass, Al, paper and polymer films have been demonstrated in [30]. The temperature of substrates is varied between 30 and 700 ^{0}C. The deposit is non-uniform and the highest thickness is observed under the electrode. Coatings on silicon wafers have a columnar structure and columns are oriented perpendicular to the surface. The tops of the vertical columns are seen in figure 10 along with a few columns deliberately placed on the surface for illustration. The hardness of the coating decreases with the content of C_2H_2 in the mixture. The nitrogen content in the deposit does not exceed 10 atomic percent. The nitrogen content in the polymer-like films deposited on thermal-sensitive substrates does not exceed 5 atomic percent.

Melting Cu and Al electrodes at high microwave powers have been used for the deposition of Cu and Al films on Pyrex and stainless steel substrates in hydrogen plasmas. Deposits include droplets of the metals [29]. Mixtures of hydrogen with nitrogen or oxygen have been used for the deposition of TiN and TiO_2 films on stainless steel substrates [29].

4.3. The EMD as a source of $O_2(a^1\Delta)$

The study of the conditions for the generation of the oxygen $O_2(a^1\Delta)$ singlet is of interest for researchers in the frame of numerous tasks in organic chemistry, plasma chemistry, and laser physics [31]. Gas discharges are one of the sources of $O_2(a^1\Delta)$. The process of $O_2(a^1\Delta)$ generation has been tested for the EMD on the base of a 1D quasi-static model for oxygen discharges at different pressures [25]. Several results are presented in figure 11. At a pressure of 1 Torr the calculated concentration of $O_2(a^1\Delta)$ is close to the theoretical limit (25%) and the radial distribution concentration is close to homogeneous. A pressure increase leads to a reduction of the concentrations and its homogeneity becomes worse. This model gives a good description of the EMD, so we can expect that an EMD at low pressures would be an effective generator of $O_2(a^1\Delta)$ for different applications.

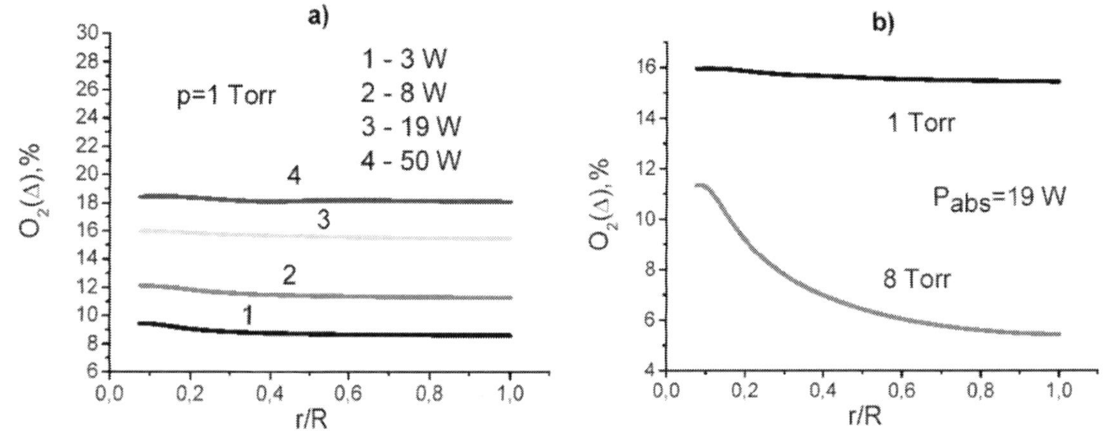

Figure 11. Axial distributions of $O_2(a^1\Delta)$ concentrations in an EMD at different powers (a) and pressures (b) [25].

4.4. The EMD as a plasma cathode

The installation shown in figure 1 allows the operation in the absence of galvanic coupling between the antenna and the chamber and thus to study the EMD properties in the external, static field. This field can be produced by an external DC power source. The DC voltage-current characteristics of the EMD have been studied in [22] (see figure 12). It was shown that the dark space of the discharge is a conductive medium. The DC current grows with the microwave power at constant pressure and falls with pressure at constant power. In the range of voltages shown in figure 13 the EMD does not change its structure. At higher positive DC voltages with respect to the chamber the DC current increases by two orders of magnitude and a secondary surface discharge (non-self-sustained) appears near the walls of the chamber. However, the magnitudes of these voltages are still not enough for the generation of an ordinary DC discharge. At high negative electrode voltage the EMD can become extinguish.

The peculiarity of this DC circuit is that the conductivity of the dark part of EMD is much less than that for the active part of the EMD with effective microwave ionization. This is why the external DC voltage is applied to the outer dark part of the discharge but not to the active part. The voltage-current characteristics in figures 12 are determined by the dark space and the sheath near the discharge walls. This fact is confirmed by simulations (figure 13) [22].

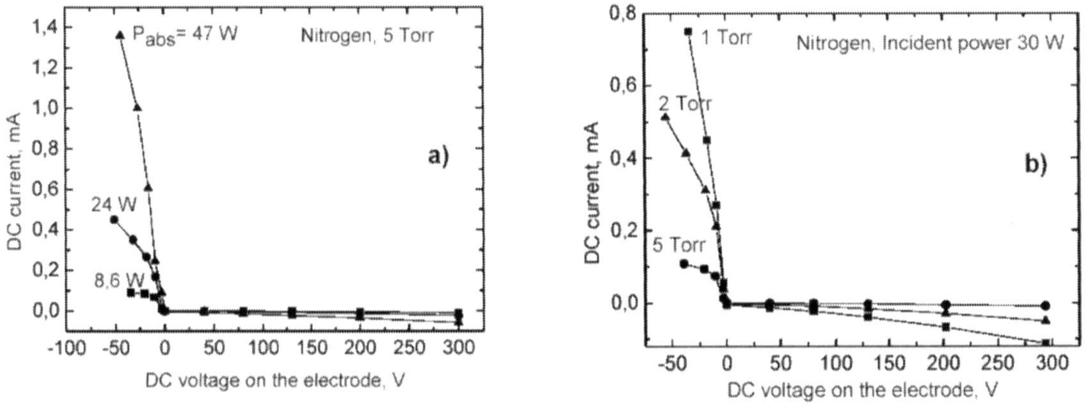

Figure 12. Experimental DC voltage-current characteristics of the nitrogen EMD at different powers (a) and pressures (b) [22].

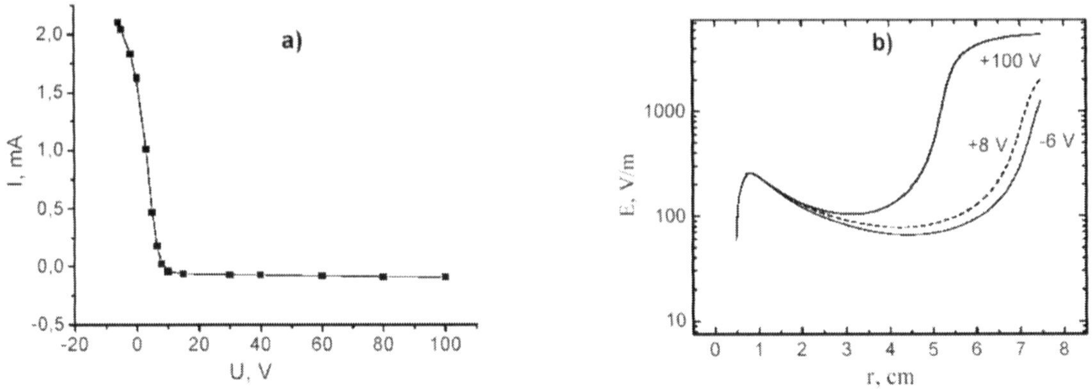

Figure 13. Calculated DC voltage-current characteristic of the EMD (a) and radial distribution of DC field E in nitrogen EMD at different applied voltages (b) [22].

The EMD plays the role of a plasma cathode. Thus the low power microwave source and the external DC voltage can activate the medium in a large volume and gives a possibility to process the internal surface of details with large dimensions. The danger of wall atomization is small because the secondary surface discharge is a low voltage one and is non-self-sustained.

5. Conclusion

Non-equilibrium electrode microwave discharges attract the interest because of its unique physical peculiarities and of its promising applications. This type of discharge is representative of strongly non-uniform discharges and thus combines the properties of quite different discharges. Physical processes in the EMD are different in various regions of the discharge. The plasma content (excited and charged particles) is also different. This discharge can be attributed to the class of "self-sustained-non-self-sustained" ones. Although experiments and modeling already gave answers to plenty of questions, many problems are still unsolved and a detailed study of the nature of the abrupt external discharge boundary, of the behavior of the EMD in external fields and in gas flows, etc., is needed. The EMD can be effectively used in plasma chemistry for the treatment of surfaces, for coating deposition or as a plasma cathode. The search for new applications of the EMD is also a subject for further studying.

Acknowledgments
This study was partly supported by the Program # 9 of Fundamental Researches of the Presidium of the Russian Academy of Sciences and RFBR Grant 07-08-00020.

References

[1] Bardos L, Barankova H, Lebedev Yu A, Berg S and Nyberg T 1996 Proc. 7th European Conf. on Diamond, Diamond-like and Related Materials, Tours, France, 4.1
[2] Brandt A A and Tikhomirov Yu V 1974 *Plasma multiplicators of frequency* (Moscow: Nauka, in Russian)
[3] Bardos L, Barankova H, Lebedev Yu A, Nyberg T and Berg S 1997 *Diam. Relat. Mat.* **6** 224
[4] Bardos L and Lebedev Yu A 1998 *Tech. Phys.* **43** 1428
[5] Bardos L and Lebedev Yu A 1998 *Plasma Phys. Rep.* **24** 956
[6] Lebedev Yu A, Mokeev M V and Tatarinov A V 2000 *Plasma Phys. Rep.* **26** 272
[7] Lebedev Yu A and Mokeev M V 2000 *High Temp.* **38** 338
[8] Bardos L and Lebedev Yu A 2000 *High Temp.* **38** 528
[9] Bardos L and Lebedev Yu A 2001 *Plasma Phys. Rep.* **27** 418
[10] Lebedev Yu A and Mokeev M V 2002 *Tech. Phys.* **47** 135
[11] Lebedev Yu A, Tatarinov A V and Epstein I L 2002 *Plasma Sources Sci. Technol.* **11** 146
[12] Lebedev Yu A and Mokeev M V 2003 *Plasma Phys. Rep.* **29** 251
[13] Lebedev Yu A and Mokeev M 2003 *Plasma Phys. Rep.* **29** 1059
[14] Lebedev Yu A and Mokeev M 2003 *High Temp.* **41** 725
[15] Lebedev Yu A and Tatarinov A V 2004 *Plasma Sources Sci. Technol.* **13** 1
[16] Lebedev Yu A, Mokeev M V, Tatarinov A V and Epstein I L 2004 *Plasma Phys. Rep.* **30** 91
[17] Lebedev Yu A and Shakhatov V A 2006 *Plasma Phys. Rep.* **32** 58
[18] Lebedev Yu A, Tatarinov A V and Epstein I L 2006 *High Temp.* **44** 317
[19] Lebedev Yu A, Epstein I L, Tatarinov AV and Shakhatov V A 2006 *J. Phys.: Conf. Ser.* **44** 30
[20] Lebedev Yu A and Epstein I L 2007 *Plasma Phys. Rep.* **33** 63
[21] Lebedev Yu A, Solomakhin P V and Shakhatov V A 2007 *Plasma Phys. Rep.* **33** 157
[22] Lebedev Yu A, Tatarinov A V and Epstein I L 2007 *High Temp.* **45** 283
[23] Lebedev Yu A, Tatarinov A V and Epstein I L 2007 *Plasma Sources Sci. Technol.* **16** 726
[24] Shakhatov V A and Lebedev Yu A 2008 *High Energ. Chem.* **42** 207
[25] Lebedev Yu A, Tatarinov A V and Epstein I L 2008 *High Temp.* **46** 645
[26] Lebedev Yu A, Solomakhin P V and Shakhatov V A 2008 *Plasma Phys. Rep.* **34** 614
[27] Gordietz B, Osipov A and Shelepin L 1980 *Kinetic Processes in Gases and Molecular Lasers* (Moscow: Nauka)
[28] Taniyama N, Kudo M, Matsumoto O and Kawarada H 2001 *Jpn. J. Appl. Phys.* **40** L698
[29] Eto A, Kimura S and Kando M 2003 *Microwave Discharges: Fundamentals and Application* ed Ohl A (Greifswald, INP) 65
[30] Bardos L, Barankova H and Lebedev Yu A 1999 42nd Ann. Conf. of Soc. of Vac. Coaters, Chicago, IL, Proc. SVC TC paper E-7
[31] Ionin A A, Kochetov I V, Napartovich A P and Yuryshev N N 2007 *J.Phys.D: Appl.Phys.* **40** R25

Data processing of electric probe data acquired under locally close-to-coherent noise conditions

F M Dias

Instituto de Plasmas e Fusão Nuclear – Instituto Superior Técnico, 1049-001 Lisboa, Portugal

E-mail: francisco.dias@ist.utl.pt

Abstract. A numerical procedure able to effectively eliminate close-to-coherent noise from measured data is presented. Although the data processing of probe measurements in Tokamak machines was the initial motivation, the current procedure is quite general concerning the kind and the source of data it can be applied to. As a result, the procedure can be useful for decreasing drastically the acquisition-time that otherwise is required to achieve acceptable signal-to-noise ratios in a very wide range of measurements.

1. Introduction

Removing high-level noise from probe characteristics is always a complex task due to the non-linearity of the current-voltage (*I*, *V*) function. Indeed, though in some cases noise removal may be effective enough so that "look nice" characteristics are achieved, distortions are unavoidable whenever fluctuations, whatever the physical processes involved, are detected as voltage-noise or as electron energy distribution function (EEDF) modulations. As a result, noise itself becomes "information" in the sense that its evaluation is as required as that of what is intended to be measured because actual fluctuations must be known in order to correct their effects in non-linear systems.

The simplest filtering procedures can be applied when the noise and the signal have spectral contents in completely different frequency regions, *e.g.* a low-frequency signal having a high-frequency noise. Conversely, filtering is more and more challenging as the above spectral regions merge. As a result, the instrumentation must be a guardian against noise by imposing a bandwidth not wider than that required for the signal being measured. In case of data acquisition systems, the above means a wise choice of the sampling rate and of the matching anti-aliasing filter, and eventually the need for external triggering in low signal-to-noise ratio situations.

Considering that the spectra of digitised data are commonly evaluated using the standard Fourier series, *i.e.*, based on n^{th}-order harmonics where n is an integer, filtering based on the above approach becomes extremely ineffective when the total acquisition-time is not an integer-multiple of the fundamental period of the noise. Then, a purely sinusoidal function and the $\sin(n\omega t + \varphi)$ terms of the Fourier series are not, in general, orthogonal functions, which means that the above series converges very slowly, and that spectra becomes too broaden. As a result, common filtering techniques may remove also a non-negligible part of the signal spectrum.

Unfortunately, spectral broadening is quite common in many kinds of measurements, either due to mains' induced fluctuations in case of non-triggered data-acquisitions based on modern, non-real-time operating systems, or in case of measurements in magnetised plasmas. In the former case, it cannot be exactly predicted when data is actually sampled (there are exceptions but they are expensive and not too common). In the latter case, fluctuations do not feature a fixed, fundamental frequency during the whole acquisition-time. In all cases, and as a final result, the acquired data seems to have random- or variable-frequency superimposed disturbances. To our knowledge, the previous, simple solution for handling non-coherent disturbances was based on an increased total acquisition-time so that disturbances could be considered as random variables, which obviously implied a waste of precious time.

Although the above is already a challenging noise-problem, noise can be too complex to be described accurately as a simple, variable-frequency disturbance because its amplitude is usually also time-modulated. Yet, in most cases it is acceptable to assume that amplitudes and frequencies are slow-varying functions of time. This is what we hereby define as "locally close-to-coherent noise" and the aim of the current work is precisely to provide a way on how such type of noise can be eliminated from measured data without an acquisition-time waste.

As a logical continuation of our previous works on the same subject [1,2], here we present an improved, adaptive, self-tuning, numerical procedure able to identify, to evaluate, and to eliminate locally close-to-coherent noise without the need for a useless waste of the acquisition-time. In addition, the pertinent information concerning noise is not lost, which means that *a posteriori* deconvolution and reconstruction procedures can be applied.

2. A glance into the real world

In order to avoid possible misunderstandings, we consider as noise whatever is present in the measured data other than what was not initially intended to be measured.

Unfortunately, the idea that noise is just a minor, random, high-frequency disturbance is too widely spread. As a result, too many people still believe that noise can always be readily eliminated using some simple, low-pass filter. Also unfortunately, but now under an educational point of view, manufacturers of lock-in amplifiers, boxcars, etc., use to "sell" the idea that such devices readily eliminate whatever noise problems users may have. The real problem is that new users "buy" such idea without paying attention for what kind of noise those equipments are actually effective as filters.

Here we show the difficulties that automated procedures based on the standard Fourier series have while trying to identify a purely sinusoidal function, F, which is the simplest kind of disturbance. Let us consider the case when the frequency, f, of the sinusoidal function is an odd-multiple integer of one-half of the total acquisition-time, T, *i.e.*, $f = (2n + 1)/(2T)$, and let us calculate how many terms of the series are required to recover F within a given error. Such results are shown in figure 1 for five values of n.

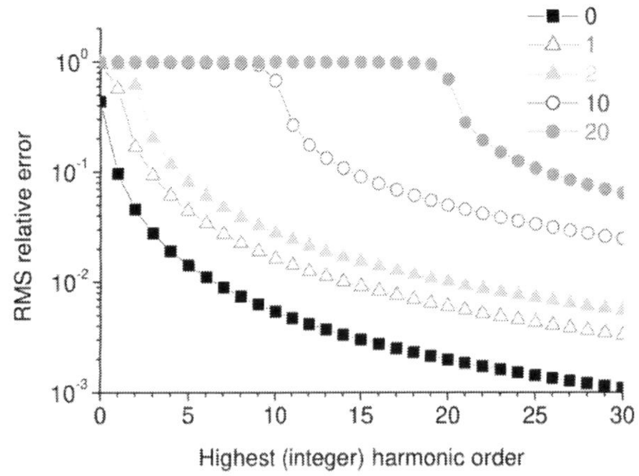

Figure 1. RMS relative error as a function of the highest order of the Fourier series terms taken into consideration, for five values of the n parameter (see text).

As can be seen in figure 1, actual frequency values are somewhat difficult to be identified for small values of n. Conversely, a huge number of the Fourier series terms need to be taken into consideration for high values of n. Yet, an inexperienced observer can readily and accurately identify amplitude and frequency values in a plotted sequence of half-periods, which means that a standard Fourier series is not the general, best approach. Conversely, one single term of a modified Fourier series (based on a non-integer order of the harmonics) is all what is needed to evaluate the amplitude if the frequency is known *a priori*. Although such means an *eigenvalue* approach, the overall numerical procedure may run much faster.

Unfortunately, single-frequency functions do not belong to the real world of instrumentation. So, let us be realistic: what if a non-sinusoidal disturbance, whose amplitudes or/and frequencies are time-modulated, is superimposed over a signal? Well, then the fluctuation spectrum is broader and filtering operations are problematic because they will noticeably deplete information from the signal in the regions where the signal and the fluctuation spectra merge.

The very short analysis in this Section was solely intended to aware the reader that, within the limitations imposed by particular constraints, the spectral contents of a disturbance (or its time-variation) should be known as accurately as possible in order to choose the best filtering approach and to minimise the inherent errors.

3. Filtering procedures

It should be kept in mind that filtering and fitting are essentially equivalent procedures, which means that what are the sound mathematical guidelines for a fitting [3] should also be followed for a filtering operation. Namely, the user must have an educated guess on what actual signal and noise functions are before attempting to separate them. Otherwise, filtering is not too different from "witchcraft".

A common educated guessing is when measurements are intended to validate a theory other than the one that supports the measuring method. For instance, if, and only if, probe measurements are intended to prove that EEDFs are Maxwellian, then fittings are the best filtering procedures. Yet, and by definition, then the experimentalist should not be able to achieve anything else than Maxwellian EEDFs. Fortunately, the number of degrees of freedom is often large enough for the experimentalist to have good grounds to suggest that EEDFs are something else, *e.g.* bi-Maxwellians [2,4–5].

Another educated guessing is the so called optimal (Wiener) filtering [6]. Although it does not rely on a completely objective approach, its name clearly suggests that even applied mathematicians consider it as "the best one may get". The idea is simple: considering that each kind of signal and noise have spectral *signatures* of their own, which experienced users are expected to be readily able to identify, whatever unexpectedly deviates from a signal-trend is noise, and vice-versa; optimal filtering is the procedure that corrects the above deviations so that the expected trends are kept. This approach is well-suited for *white* noise and for coherent noise because they feature precise spectral *signatures*.

As the reader may have already understood, filtering becomes as problematic as the *a priori*, required information about the signal and/or the noise functions are missing. Concerning probe measurements in Tokamak machines, though it has been assumed that EEDFs could be considered as Maxwellian, it was shown later that such was not the case (see references above). In addition, this author prefers to evaluate EEDFs without any *a priori* assumptions concerning the EEDF. Fortunately, an analysis of several probe characteristics has shown that disturbances and signals were not correlated (note: there is an apparent correlation between the signal and the amplitude of the noise, which we interpret as inherent to how the plasma fluctuations are detected by a probe, and which the reader should not misunderstand as an actual signal-noise time-correlation). As a result, the soundest educated guess is the peculiar locally close-to-coherent noise feature, which makes our approach somewhat non-orthodox. Indeed, instead of identifying the signal and then the noise is "what else", which is the most trivial approach, we start by identifying the noise and the signal is what remains after noise is removed. Hence, it is not strange that our procedure becomes more accurate for higher noise amplitudes (see below) because then it is easier to identify what actually is the noise component.

We wish to emphasise a few more filtering basics before starting a more comprehensive description of our current procedure.

1) Filtering operations carried out in the time-domain typically use finite impulse response filters [7] due to practical reasons, *i.e.*, finite sets of data.

2) Coherent (or close-to-coherent) noise typically asks for filter functions whose duration [7] must keep a close relation with the period (or the characteristic period).

3) The larger the filter function duration the lower the frequency of the signal components that will be removed by the filtering operation, hence the higher the distortion in those particular cases where the spectral density of the signal is a decreasing function of the frequency, which is typically the case for probe measurements.

In simpler words, the above means that an effective filtering can only be achieved if the filter-size and the characteristic period values are similar in value and that the signal will be unavoidably distorted for large values of the characteristic period unless corrective measures are applied (see [8] for possible solutions).

Readers more interested in the theoretical grounds of less-common, time-frequency transforms are hereby suggested to look for *wavelet*, in general, and for Short-Time Fourier Transform (STFT), in particular, in general mathematical text books. Due to the possible similarity between our procedure and a STFT, we wish to aware the readers that we are not calculating "local" spectra. Instead, what we do is to locally fit the noise by a function that satisfies our locally close-to-coherent noise definition.

3.1. The locally close-to-coherent noise filtering procedure

Here we describe how our numerical procedure handles locally close-to-coherent noise while processing the 1st derivative of probe data, which is the starting point for EEDF evaluation under collisional conditions, *e.g.* in magnetised plasmas. Nevertheless, and considering that the actual noise processing only starts at step 3) below, readers are free to "input" whatever data they may like to be filtered at that step. There are other reasons why we use the 1st derivative instead of the original data: derivatives enhance the amplitudes of high-frequency components, our procedure works better for higher noise levels, and we expect that noise will have a high-frequency content higher than that of the signal.

Our procedure works in the following way:

1) The (I, V) measured data is sorted, and interpolated in order to provide 1024 uniformly V-spaced bins as a way to speedup the whole processing (explanation: considering that sampled voltages are usually less noisy than those of the currents, this step avoids the need for a more complex V-filtering; we currently follow a linear interpolation based on the 3, closest measured V-points but the reader is free to use any other sorting procedure).

2) A raw, first derivative, dI/dV, is numerically calculated from the convolution of the above data with the first derivative of an instrument function (IF), whose full with at half maximum (FWHM) is adaptively chosen aiming for locally equal noise and error values. Noise is deduced from 2nd derivatives using 3 adjacent data points [8], and error from a one-step algebraic reconstruction technique (ART) based on a deconvoluted form of Hayden procedure [9] as mentioned in [10]. As in previous works (see e.g. [1]), our basic IF, $g(j)$, is a Hann-like window

$$g(j) = \frac{1 + \cos(\pi j / n)}{2n}, \tag{1}$$

where n is the initial FWHM value. Anyway, each convolution is followed by an one-step ART, hence the actual IF is $2g - g * g$, where "$*$" stands for convolution.

3) The characteristic period of the disturbance is evaluated as the lowest value of τ $(0 < \tau < T/2)$ that maximizes the function:

$$\left| \int_0^{T/2} I(t) I(t + \tau) dt \right| \Big/ \sqrt{\left| \int_0^{T/2} I^2(t) dt \int_0^{T/2} I^2(t + \tau) dt \right|} . \tag{2}$$

We found that the above autocorrelation-like procedure is more reliable than all the other methods that we have tested. The so calculated characteristic period is currently only used as a seed for the first point of the iterative process carried out as described in steps 5) – 9) below.

4) An approximated value of the local period, which is V-dependent, is deduced at each point, *i.e.*, for each window, as being the window size that maximizes the amplitude of the fundamental under a Fourier series approach and approximated values of its COS and SIN components are calculated.

5) The first derivatives of the disturbing signal amplitude, phase, and frequency are calculated (note: in order to avoid an increased processing-time, which seems to be useless for Tokamak data, we are not yet accounting for the 1st derivative of the frequency variation).

6) The period of a dummy, amplitude- and frequency-modulated sinusoidal function matching the apparent period found in step 4) is calculated accounting for the above first derivatives, and a more accurate amplitude value is calculated accounting for the 2nd derivative, d^2I/dV^2, which is evaluated at step 8) during the previous loop (explanation: we wish to emphasize our non-orthodox approach as mentioned in section 3; indeed, what we are actually evaluating is noise, and higher derivatives of the signal are being used to achieve a higher level of accuracy concerning noise evaluation). Note that it is in this step that the actual close-to-coherent fitting takes place.

7) dI/dV is filtered using the same FWHM that leads to a notch-filter for the dummy function (explanation: the noise will be absolutely made equal to zero if it satisfies our initial assumptions).

8) d^2I/dV^2 is calculated from the filtered dI/dV achieved in the previous step.

9) The procedure loops back to step 5) until convergence is reached.

10) An additional ART is carried out to guess the error of the whole procedure.

In a former version of our numerical procedure, and as shown in [1], the error values evaluated at the above step 10) were in good agreement with the actual error under model test conditions. Although in the past we were pleased by such result because it provided a means to have a likelihood parameter, latter we realised that values achieved at step 10) and those of the initial disturbance were too closely correlated to be considered as an error, hence they should be considered as information instead. As a result, in the current work we decided to use the results achieved at step 10) as a further improvement to the filtering ability of our procedure. The drawback from such decision as well as the directions for further improvements will be discussed below at the appropriated places.

4. Model tests

To our best knowledge, the current numerical procedure is the only that effectively filters data measured under locally close-to-coherent noise conditions, *e.g.* that achieved in Tokamak machines, without imposing unnecessary, restrictive fittings. Although our results have been confirmed by those obtained using less-detailed approaches [2,4–5], we think that a model test is the best way to prove the correctness of our procedure. Considering that the procedure was initially intended to process probe data measured in collisional conditions, we have chosen a test probe characteristic whose 1st derivative of the signal component is (see *e.g.* [5] and references therein):

$$\left(\frac{dI}{dV}\right)_{\text{signal}} = \begin{cases} a\left(V_{\text{p}} - V\right)e^{\frac{V-V\text{p}}{b}} & V \le V_{\text{p}} \\ 0 & V > V_{\text{p}} \end{cases}. \tag{3}$$

The value of the plasma potential, V_{p}, was set to 70 V, and the parameter b, which provides information on the electron temperature, was set to 20 V. The 1st derivative of the noise component (see solid curves on the left side of the appropriate figures 2–4 below) is given by:

$$\left(\frac{dI}{dV}\right)_{\text{noise}} = f(j)\sin\left(\frac{2\pi j}{c\left(1 - 5\times10^{-7} j^2\right)}\right), \tag{4}$$

where j is the bin order (0 – 1023), c is a scaling parameter for the local period, and $f(j)$ provides an amplitude variation similar to that found in Tokamak data and scales amplitudes. Scaling is hereby used to analyse how much the characteristic period and amplitude of the noise have an influence on the accuracy of our procedure. As reference conditions we shall consider $c = 120$ and a maximum amplitude value equal to 5.93 a.u..

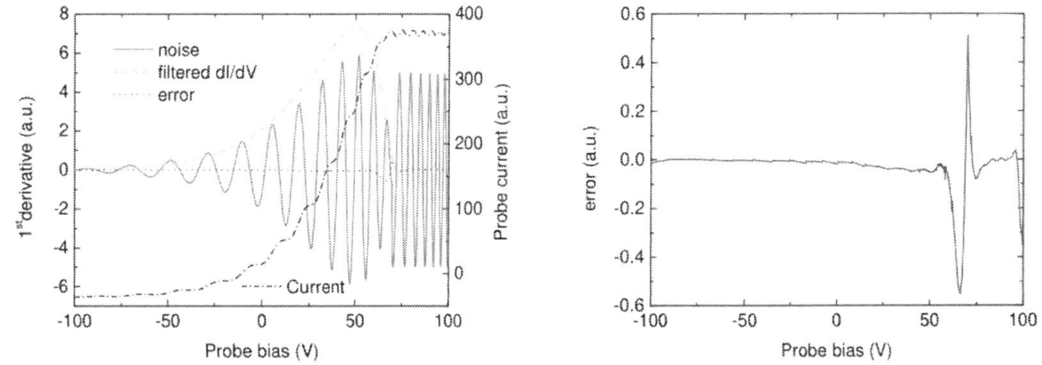

Figure 2. Model test results for the reference conditions (see text). Left: 1st derivative of the noise component (solid), filtered 1st derivative of the signal (dash), error (dot), and the associated probe characteristic (dash dot). Right: error (solid) plotted in an expanded scale.

The model test results for the reference conditions are depicted in figure 2, while figures 3 and 4 show those achieved, respectively, when the amplitude and the local period was made twice as large.

The error, which we consider the most pertinent result, can be seen in an enlarged scale on the right side of figures 2–4, and its analysis shows that: *i*) the initial noise is virtually absent in the filtered data; *ii*) error values are rather small (< 1%) except in the vicinity of the plasma potential and close to the boundaries of the data-interval; *iii*) the filtering efficiency increases as noise amplitude increases (note that the span of the error scale is the same in figures 2 and 3) but, conversely, error values in the vicinity of the plasma potential are very sensitive to that of the noise period.

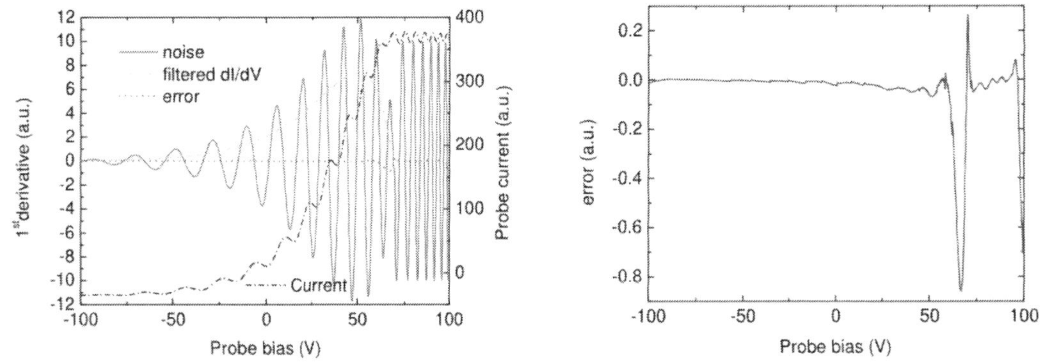

Figure 3. Model test results when noise amplitude is doubled. Left: 1st derivative of the noise component (solid), filtered 1st derivative of the signal (dash), error (dot), and the associated probe characteristic (dash dot). Right: error (solid) plotted in an expanded scale.

A more precise analysis can be achieved checking the accuracy as the current procedure identifies the value of the plasma potential, which is a key parameter in probe diagnostics. The resulting errors are 2.5, 1.4, and 7 V, respectively, in the reference, the doubled-amplitude, and the doubled-period case. Under the same conditions, the FWHM of the Hann-like filter are, respectively, 7, 7, and 13 V.

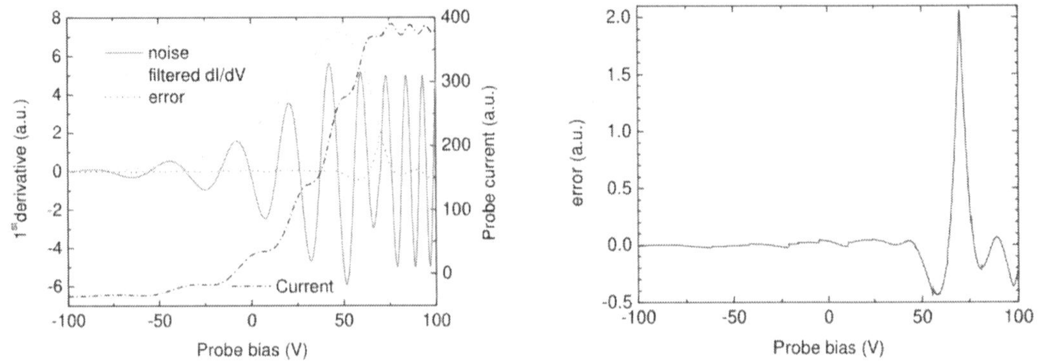

Figure 4. Model test results when the noise period is doubled. Left: 1st derivative of the noise component (solid), filtered 1st derivative of the signal (dash), error (dot), and the associated probe characteristic (dash dot). Right: error (solid) plotted in an expanded scale.

Concerning the doubled-amplitude case, and taking into account that the error value is only 1/5 of that of the FWHM, we consider the above result as very good. Conversely, the similar, relative result for the reference case is 1.75 times worst, which suggests that the current numerical procedure still has some limitations for lower-level amplitudes, whose explanation we attribute to the fact that the 1st derivatives of the period were not taken into account. Concerning the doubled-period case, the error was expected to increase due to the unavoidable increase of the FWHM values, which the current results confirmed. Nevertheless, we wish to emphasise that (3) leads to a discontinuity of $(d^2I/dV^2)_{signal}$ at V_p, which suggests that the expected error values should be smaller under a real condition situation. The reader should keep in mind that the electron temperature is accounted in (3) through the b parameter, whose value was set to 20 V, which means that all the above errors concerning the plasma potential evaluation would be fully acceptable as a first, uncorrected probe diagnostic.

Cross-correlations between the actual noise amplitudes and those evaluated by our numerical procedure lead to an analysis similar to that achieved above concerning the plasma potential. Indeed, cross-correlation values are 0.999245, 0.999634, and 0.991431, respectively, in the reference, the doubled-amplitude, and the doubled-period case (note: in the above calculations the data-interval was shortened by "1" FWHM at each boundary to minimise end-effects; see explanation below). Under the same model test conditions, local periods are evaluated with 2.02, 1.42, and 6.58 % RMS averaged errors.

The increased error close to the boundaries of the data-interval is trivial as a result and always arises in case of a finite set of data, which means that the size of the input data-interval should be always larger than that of the data to be actually evaluated. As a guideline, each extended region should be larger than the respective FWHM value at each boundary.

5. Discussion and Conclusions

We have improved a previous version of our numerical procedure intended to filter locally close-to-coherent noise, *e.g.* that found in Tokamak probe data, and we have analysed how much the characteristic values of the noise amplitude and of the period have an influence on the filtering efficiency. The improvement was made possible by considering as "information" what previously was assumed as just a measure of the error. This approach is justified by the fact that we found that the above evaluated "error" and the noise were closely correlated. Although this new approach made us lose our previous likelihood parameter, we think that such is just an intermediate phase that, in a near future, will lead to further improvements and to a more accurate error evaluation because an increased filtering efficiency is the basic condition so that an increased number of deconvolution iterations are possible without retrieving back a noticeable amount of noise.

Considering that filtering in the time-domain, *i.e.*, a convolution, is a reversible operation, readers may wonder about the validity of the last statement in the above paragraph (note: we shall skip here the practical limitations when discrete convolution operations are applied to finite sets of data). Yet, when deconvolutions are carried out using iterative procedures, which is the scheme we use, the recovering speed is spectrum-dependent, which means that it is not the same for the signal and for the noise components. The most striking example is when a DC signal having a superimposed, constant amplitude and period, sinusoidal noise is filtered by a rectangular window whose width is precisely equal to the period. Then, the result of deconvolutions intended to sharpen the IF is errorless and noiseless independently of the number of iterations carried out. Our wishful approach is precisely to build a filtering procedure able to eliminate completely the noise superimposed on data.

Following the same line of thought, we expect that further improvements will be possible when the 1st derivative of the period is taken into account because then the filtering efficiency should be even greater hence also the possibility to increase the number of deconvolution iterations without a noticeable increase of the noise level.

Concerning the influence of the noise amplitude on the filtering efficiency, we have shown that the latter increases as amplitudes increase, which was explained as a direct result from the so increased ability of the numerical procedure to identify what noise is. As a corollary, errors decrease under the same conditions. Conversely, an increased noise period leads to a similar error variation, which was the expected result due to the inherent FWHM value increase of the IFs and due to the merging of the signal and the noise spectra.

The current numerical procedure keeps, as a heritage from its previous versions, features that are not the most appropriated for high-accuracy, numerical processing. For instance, some of the variable arrays are of the integer-type. Yet, the wishful idea of the author was (and still is) that the filtering program may be included in the firmware of a front-end, digital data processor, in which case speed constraints need to be considered very seriously whenever data-processing operations are to be carried out during the time between consecutive data-acquisitions. Conversely, we have good reasons to expect that the filtering efficiency will be further increased as soon as all numerical variables are handled as floating-point entities.

As a final remark, we wish to emphasise that our filtering procedure is able to accurately retrieve noise amplitudes and frequencies, which are basic requirements for EEDF reconstruction, and which is the starting point to achieve accurate values of the electron density and of the plasma potential. We have the knowledge on how to reconstruct EEDFs from averaged probe data provided that additional information concerning EEDF modulation is made available, which is not possible from one-probe measurements. As a result, the filtered results that we can achieve are bound to be close-to-useless concerning EEDF evaluation unless information achieved from other simultaneous diagnostics, or any sound theoretical guidelines, are provided so that the appropriated EEDF reconstruction can be applied. Otherwise, many thousands of probe measurements, which have never been processed for "having some superimposed weird disturbance", will be forever lost.

Acknowledgments

This work was partially supported by FCT/FEDER in the framework of the project "Ecological Engineering Plasma Laboratory" POCI/FIS/61679/2004. The author wishes to thank all those who have provided noisy probe characteristics so that the reliability of the numerical procedure could be tested in real situations. The author also wishes to thank Dr. Tsv. Popov for the fruitful discussions concerning EEDF evaluation of probe measurements in magnetised plasmas.

References

[1] Dias F M and Popov Tsv K 2007 *Proc. 34th EPS Conf. on Plasma Physics (Warsaw)* ECA **31F** P-5.080

[2] Popov Tsv K, Ivanova P, Stöckel J, Dejarnac R and Dias F M 2007 *Proc. 34th EPS Conf. on Plasma Physics (Warsaw)* ECA **31F** P-5.104

[3] Press W H, Flannery B P, Teukolsky S A and Vetterling W T 1992 *Numerical Recipes in C: The Art of Scientific Computing* (Cambridge: Cambridge University Press) chapter 15.0 pp 656–7

[4] Popov Tsv K, Ivanova P, Benova E, Bogdanov T, Dias F M, Stöckel J and Dejarnac R 2008 *Proc. 35th EPS Conf. on Plasma Physics (Hersonissos)* ECA **32** P-4.081

[5] Popov Tsv K, Ivanova P, Dimitrova M, Stöckel J and Dejarnac R 2008 *J. Phys.: Conf. Ser.* **113** 012004

[6] Press W H, Flannery B P, Teukolsky S A and Vetterling W T 1992 *Numerical Recipes in C: The Art of Scientific Computing* (Cambridge: Cambridge University Press) chapter 13.3 pp 547–9

[7] Press W H, Flannery B P, Teukolsky S A and Vetterling W T 1992 *Numerical Recipes in C: The Art of Scientific Computing* (Cambridge: Cambridge University Press) chapter 13.1 p 538

[8] Dias F M and Popov Tsv K 2007 *J. Phys.: Conf. Ser.* **63** 012005

[9] Hayden H C 1987 *Comput. Phys.* **1** 74–5

[10] Dias F M and Popov Tsv K 2002 *Vacuum* **69** 159–63

Resonance phenomena in microwave nonmagnetized plasma source

I.A.Kossyi

A.M.Prokhorov General Physics Institute of RAS, Moscow, Russia

Email: kossyi@fpl.gpi.ru

Abstract. Scheme of coaxial microwave gas-discharge plasma source based on a "plasma resonance" phenomenon is presented. Possibility of conditions realization at which energy of accelerated in the "resonance" electrons goes into ionization processes (in volume of chamber outside of "resonance" region) is discussed. Results of experimental investigation of a coaxial microwave plasmatron in which "resonance" mechanisms have been manifested are presented.

1. Introduction

In the present paper we report the results of the development and investigation of microwave gas-discharge plasma sources based on the phenomenon which is known as the "plasma resonance". This name comes from the fact that the plasma resonance involves nonlinear processes occurring in a spatially inhomogeneous nonmagnetized plasma in which the local value of the plasma density approaches the critical value $n_{ec} = \frac{m\left(\omega_0^2 + \nu_{eff}^2\right)}{4\pi e^2}$ (ω_0 is the angular frequency of the incident microwave radiation, ν_{eff} is effective electron–neutral collision frequency).

In a fairly narrow layer adjacent to the resonance surface, the microwave energy can be absorbed with high efficiency and converted into the energy of plasma electrons. The generation of fluxes of accelerated electrons in a plasma under the action of a pump electromagnetic field is a fundamental phenomenon which has been much investigated both experimentally and theoretically (see, e.g., [2–4]). Here, it is pertinent to note that these studies mostly concerned an inhomogeneous plasma in a highly rarified gas, in which the processes of interaction between electrons and gas molecules (atoms) can be ignored in describing fast electrons emission into the surrounding medium.

The problem of generation of fast electrons in the region of local resonance in inhomogeneous plasma was studied in a number of theoretical works published in the 1970s. In these works, the appearance of electrons with energies far above the thermal energy is associated with the presence of singularity of electromagnetic field, in which case the electric field substantially increases in the plasma resonance region.

Electrons interacting with this localized microwave field may gain energy through various mechanisms, such as collisions [5], the Cherenkov resonance acceleration mechanism [6], and the wave break at the nonlinear stage of plasma resonance developing in a strong pumping field [7,8].

In the 1980s, there appeared publications of experimental works devoted to investigation of the phenomenon of resonance acceleration and heating of electrons in inhomogeneous "overcritical" plasma by microwaves over a wide (from mm to dm) wavelength range (the review of these works can be found in [4]). Experiments were mostly carried out with high-power microwave beams interacting with a plasma produced immediately by these beams or with a plasma produced in other way. The electrons accelerated under conditions of resonance collisionless absorption of microwave beam energy were detected by using multigrid probes and electrostatic differential analyzers of electron energy distribution. In some experiments, when measuring the electron energy distribution function, provision was made for compensation of an adverse effect of the flow of electrons leaving the plasma. For this purpose, an electrode emitting slow electrons was introduced in the "overcritical" region to neutralize the positive plasma potential impeding the escape of electrons toward the walls.

The experiments demonstrated not only generation of beam of fast electrons but also abnormally strong heating of the bulk of electrons in the plasma resonance region. In case the emitter is absent and the plasma potential increases substantially, the acceleration and heating of electrons in the resonance layer is accompanied by acceleration of ions to energies close to the average energy of the beam of hot electrons. A detail description of the experiment and analysis of the results can be found, for example, in [9,10].

Theory and experiment in the 1970s and in the1980s [2–10] allow us to formulate in the general form the conditions for resonance acceleration and heating of electrons in inhomogeneous plasma irradiated with microwaves (see section 3 of the present paper).

In the 1990s and to date, the development of a theory for plasma resonance and related experiment proceeds along the path very interesting and thorough analysis of so-called coaxial and "plane" microwave plasma sources which found wide applications, including in physical laboratories. A theory for the coaxial version was developed in [11–13]. The "plane" version (with a slit antenna) was analyzed in [14,15]. Conclusions following from recent theoretical studies agree with results obtained in the 1970s and in the 1980s.

Experiments carried out with the above microwave sources (see, e.g., [16–20]) demonstrated the presence of hot suprathermal electrons predicted theoretically.

In the present work, we attempted to realize the conditions such that the energy of electrons accelerated at the resonance is expended predominantly for ionization processes, resulting in the formation of a fresh plasma and glow surrounding the plasma-resonance region. A control parameter in this case is the gas pressure. It is essential that the pressure be low enough, lest the resonance be suppressed, and at the same time, it must be high enough to bring into action ionization processes in the gas surrounding the resonance region.

As an experimental device suited for realization of the resonance mechanism of the plasma production, we have chosen a coaxial microwave plasmatron close to the "Duo-Plasmaline" developed and investigated in [21]. As a "Mono-Plasmaline" this system with tailor made coaxial waveguide end has been realized and investigated at the General Physics Institute [22,23]. Gas-discharge sources described in [21–23] are operable over a wide range of the working gas pressures and are capable of producing a cylindrical plasma region where the plasma density exceeds the critical density.

Results of experimental investigation of a coaxial microwave plasmatron in which "resonance" mechanisms have been manifested are presented. A plasmachemical reactor based on a "resonance" plasmatron is described.

It is reasonable to note that resonance phenomenon has long been known and applied in a magnetized plasmas (electron-cyclotron (ECR) resonance at which cyclic frequency of microwave radiation equals to electron cyclotron frequency: $\omega_0 = \omega_c$. However field of ECR-discharges application is limited due to the necessity to impose of a strong magnetic field on a discharge volume. It adds complexity and price to a plasmatron construction.

2. "Resonance" coaxial microwave gas-discharge plasma source

2.1. *Description of the Coaxial Microwave Plasmatron. Experimental Layout.*
The coaxial microwave plasma source is shown schematically in figure 1. Radiation of a magnetron (1) is fed to a coaxial waveguide consisting of an outer cylindrical electrode (2) and an inner rod-electrode (5). The inner electrode is inserted into a quartz tube (3) which is welded up on the vacuum-chamber side. The outer electrode is shorter than the inner electrode; it tapers until it reaches the dimension of the quartz tube. Length of a quartz tube came out from the "truncated" external electrode is $L_{qt} \approx 10 - 15$ cm. Quartz tube diameter is $\theta_{qt} \approx 1.2 - 1.5$ cm. Diameter of coaxial waveguide external electrode (up to its conversion into a conical constriction) is $\theta_{ee} \approx 1.2 - 1.5$ cm.

The chamber is filled with a working gas. When the magnetron is switched on, a gas-discharge plasma (4) is produced near the surface of the quartz tube, on the length corresponding to the projection of the rod-electrode beyond the coaxial waveguide.

As a microwave source, we use a production magnetron like those commonly used in household appliances (in microwave ovens). The microwave wavelength is $\lambda_f \cong 12.5$ cm, and the mean power is $\overline{P} \le 1$ kW. The magnetron generates a sequence of microwave pulses with a duration $\tau_f \cong 8$ ms and a time interval between pulses $\Delta\tau \cong 12$ ms.

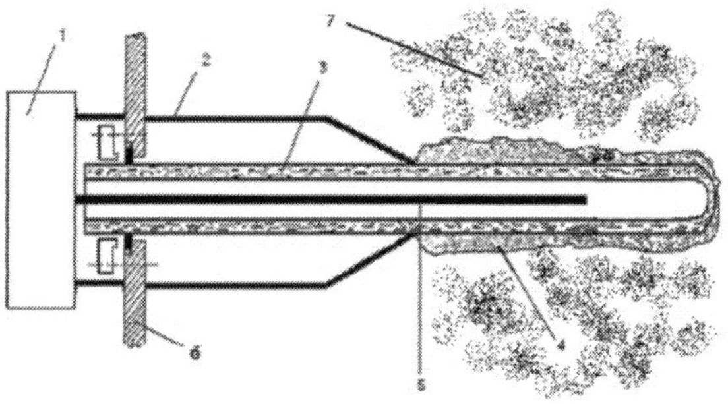

Figure 1. Schematic of the coaxial microwave plasmatron: (1) magnetron, (2) outer electrode, (3) quartz tube, (4) near-the-tube overdense plasma, (5) inner rod-electrode, (6) wall of the reactor chamber, (7) volumetric "aureole".

Figure 2 shows the experimental layout for studying the plasma produced by the coaxial microwave source. A plasmatron (1), which is similar to that exhibited in figure 1, is introduced into a cylindrical metal chamber of diameter about 350 mm (5). The preliminarily evacuated chamber is filled with the working gas (argon) at pressure $0.02 \le p \le 5$ Torr.

The chamber has two opposite quartz windows (8) for input and output of diagnostic radiation. Horn-lens antennas (2, 3) are used to transmit and receive the diagnostic beam of a microwave ($\lambda_d \cong 8$ mm) interferometer measuring the electron density in the plasma produced by the plasmatron.

Radial distribution of electron concentration is determined with help of single and double Langmuir probes (7). The single probe measured the floating potential.

Collimated photomultipliers (4) measured the radial profiles of the plasma radiation.

Spectral characteristics of microwave gas-discharge plasma have been determined with the help of AvaSpec 2048 FT (Avantis) spectrographs (6).

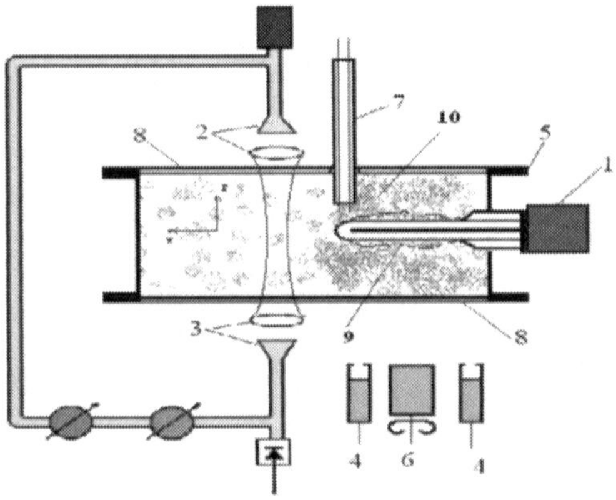

Figure 2.Experimental layout: (1) plasmatron, (2, 3) horn-lens antenna of the microwave interferometer, (4) photomultiplier, (5) vacuum chamber of the reactor, (6) spectrograph, (7) Langmuir probe, (8) quartz windows, (9) overdens microwave gas-discharge plasma, (10) volumetric plasma ("aureole").

2.2. *Experimental results*

The most reliable technique in this experiment is microwave interferometry which does not perturb the plasma. The plasma density in the chamber was measured using a diagnostic microwave beam and the experimental arrangement shown in figure 2. The results of measurements of the average density are presented in figure 3. The diagnostic beam crosses the plasma at a distance of 2 cm from the end of the quartz tube of the microwave plasmatron.

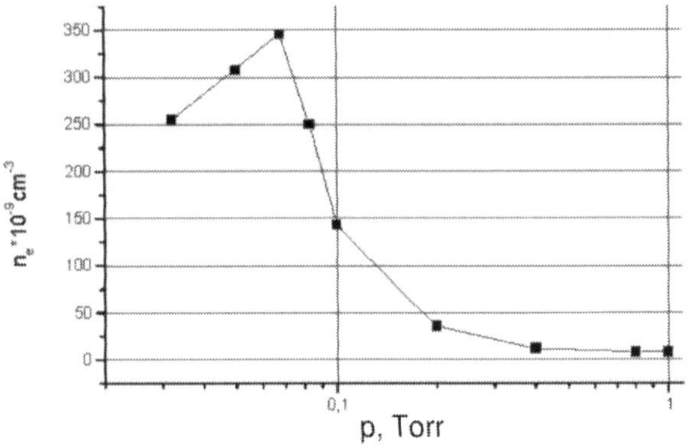

Figure 3.Pressure dependence of the electron density averaged over the path length of the diagnostic beam in the plasma. The working gas is argon.

Figure 4 shows the dependence $n_e(p)$ and floating probe potential $\varphi_f(p)$ measured by probes at a distance of $r \cong 6.0$ cm from the axis. The radial profiles of the density measured by the probes are demonstrated in figure 5 (for $p = 0.1$ Torr).

In [4,9,10], where the plasma resonance in collisionless inhomogeneous plasma was studied, the use of special constructions of multielectrode probes and electrostatic differential analyzers placed near the walls of the vacuum chamber made it possible to construct the electron energy distribution function over a wide range of energies and not only to determine the bulk electron temperature but also to detect a group of suprathermal fast electrons in the tail of the distribution.

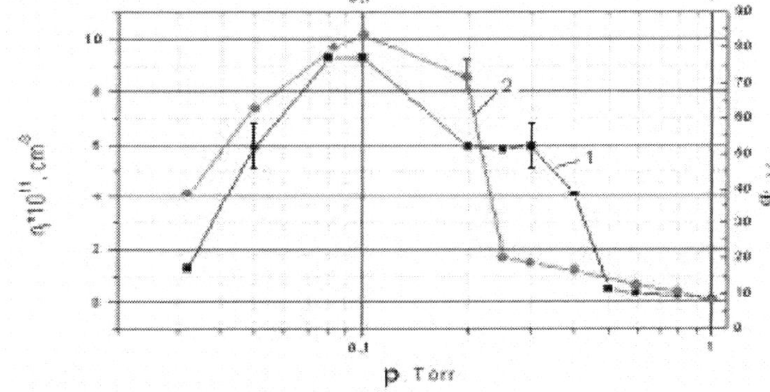

Figure 4. Electron density (1) and probe floating potential (2) as functions of the argon pressure in the chamber. $t = 5$ ms.

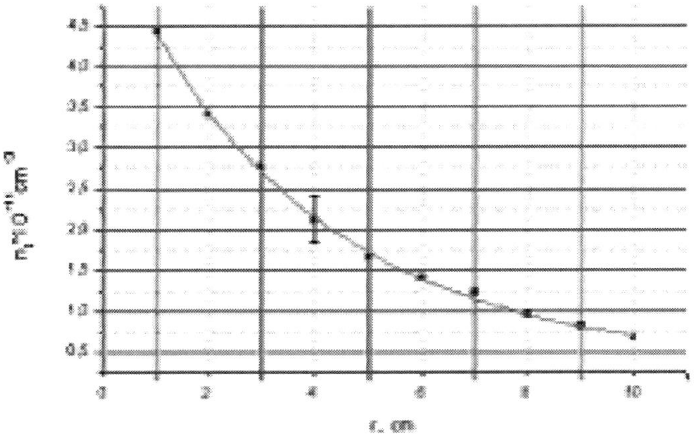

Figure 5. Radial profile of the electron density at $p = 0.05$ Torr, $\Delta Z = 0.3$ cm, $t = 5$ ms. The experimental dependence is approximated by the function $\frac{n_e}{10^{11}} = 0.35 + 4.06 \times \exp\left(-\frac{r}{3.68}\right)$ (solid line).

In the present experiment, however, the electron mean path lengths are comparable with, or less than, sizes of the chamber, the use of energy analyzers similar to those employed in [4,9,10] is ruled out. The form of the *I-V* characteristic of the single Langmuir probe used in the experiment described in the paper in the electron brunch is far from being ideal (this refers to both cylindrical and plane probes). In particular, as follows from the present experiments and also from experiments described in [16,18], the characteristic does not show a distinct saturation of the electron current. This makes the interpretation of this characteristic difficult, because the probe taking high electron currents from the plasma introduced strong perturbations into the plasma medium.

The probes in our experiments were usually in the regime of measurement of floating potential (the load resistance is high – at a level of hundreds of kΩ to 1 MΩ), in which case the perturbing effect of the probe is minimal. At the same time, the presence of a distinct saturation in the ion branch allows us to use this brunch of the characteristic for estimating the ion density.

A characteristic thickness of the plasma shell can be inferred from the point-by-point measurements with collimated photomultipliers displaced in a direction orthogonal to the source axis.

3. Discussion

The experimental results allow us to conclude that our microwave gas-discharge source makes it possible to realize the situation that, at relatively low pressures of the working gas, the nonlinear processes occurring in the plasma-resonance region play a decisive role in the plasma formation. This view on the plasma formation mechanism is consistent, in particular, with very large values of the electron density and plasma potential in the chamber volume that fall outside the range predicted by conventional models of a microwave discharge (see, e.g., [24]):

$$n_e \geq 3\times10^{11}\ \text{cm}^{-3} >> n_{ec} \approx 6.4\times10^{10} \quad \text{and} \quad \varphi_p \geq 70\ \text{eV}.$$

These values can be consistently explained at the assumption that immediately near the plasmatron there is a resonance layer in which the microwave energy is efficiently converted into the electron energy. A narrow plasma layer adjacent the resonance layer (with the critical density) is a region where the energy of the electromagnetic wave is efficiently absorbed, resulting in the acceleration of the plasma electrons. The generation of fast electrons whose energy far exceeds the thermal energy was observed in a number of experiments [9,10,16–20]. Theory explains this as being due to the presence of singularity of an electric field in an inhomogeneous plasma, i.e. due to a considerable enhancement of the electric field in the plasma-resonance region. For electrons interacting with this localized microwave field, a change in their energy may be caused by different mechanisms: collisions, the Cherenkov resonance acceleration mechanism or the wave break in the nonlinear stage of the developing plasma resonance in a strong field of the pump wave. If the ion density has a linear profile, the electrons in the plasma-resonance region are accelerated primarily toward a lower plasma density.

A prerequisite to the definitely nonlinear interaction of an electromagnetic wave with the inhomogeneous plasma whose density is over the critical level is the presence of the electric field component directed parallel to the plasma density gradient.

The main characteristics of the plasma-resonance region can be calculated within the framework of available theories (see, e.g., [4]).

Thus, the width of this region is defined by the relation:

$$\Delta R_{res} \approx R_p S. \tag{1}$$

The strength of the enhanced electric field of the electromagnetic wave in this region is

$$\widetilde{E}_{res} \approx \frac{\widetilde{E}_{0r}}{S}. \tag{2}$$

The characteristic energy of suprathermal electrons generated in a collisionless plasma in the resonance region can be estimated as

$$\varepsilon_{eh} \approx e\widetilde{E}_{0r} R_p. \tag{3}$$

In equations (1–3), the variable S is greatest of three dimensionless quantities:

$$S = \max\left\{\left(\frac{\nu_{eff}}{\omega_0}\right); \left(\frac{r_{De}}{R_p}\right)^{2/3}; \left(\frac{r_{osc}}{R_p}\right)^{1/2}\right\} \tag{4}$$

Here R_p is the characteristic dimension of the plasma inhomogeneity, λ is the microwave wavelength, r_{De} is the Debye radius, r_{ocs} is the electron oscillation amplitude in the pump field, and $\widetilde{E}_{0r}$ is the amplitude of the radial component of the vacuum electric field of the pump wave.

Turning to the experiment described in this paper, we note that almost all of the necessary conditions for the plasma resonance are satisfied at pressures $p \ll 1$ Torr. The electron density near the plasmatron is well over the critical level and at a certain distance r_c from the source takes the value n_{ec}.Thus, there exists a resonance region in the spatially inhomogeneous plasma, and the density gradient of this plasma is orthogonal to the plasmatron axis, i.e., it is in the radial direction. The electromagnetic wave emitted from the plasmatron also possesses the electric field component $\tilde{E}_{0r}$ in the radial direction.

It should be pointed out that existence of a considerable radial component of microwave electric field $\tilde{E}_{0r}$ in the vicinity of dielectric tube in investigated device (as well as in "Duo-Plasmaline" system [21]) arises from constructive features of microwave energy outlet from coaxial waveguide. Standard configurations where plasma is produced within a dielectric tube (inner surface wave configuration) does not provide means for considerable $\tilde{E}_{0r}$ component appearance and can not satisfy the conditions for resonance acceleration (heating) of plasma electron component.

Processes of electron acceleration developing in a resonance region are adequately described by equations (1 – 4) in a broad area of microwave power including our experimental value as well as values characteristic for [2–6,9,10].

Argon pressures used in our experiment satisfy the inequality $\nu_{eff} \ll \omega_0$ which is the necessary condition for resonance processes. (Using the approximate relationship for argon [24] $\nu_{eff} \cong 7\times10^9 p$, where p expressed in Torr is the pressure which might be expected to bring about strong effects at the resonance, we find $p \ll 2$ Torr.)

Under our experimental conditions, the parameter S which characterizes the intensity of nonlinear processes at the resonance is defined by the ratio $\frac{\nu_{eff}}{\omega_0}$ and at $p = 0.1$ Torr amounts to

$$S \approx 5\times10^{-2}.$$

In the experiment we have $R_p \cong (1-2)$ cm, so that the width of the discharge region (see equation (1)) is

$$\Delta R_{res} \approx (0.05-0.1) \text{ cm}.$$

The radial component of the vacuum microwave electric field does not exceed $\tilde{E}_{0r} \approx 100$ V/cm, and from equation (2) it follows that the electric field is enhanced at the resonance to

$$\tilde{E}_{res} \approx 2\times10^3 \text{ V/cm}.$$

The characteristic energy of the suprathermal electrons generated at the resonance is estimated by equation (3) as

$$\varepsilon_{eh} \approx 80 \text{ eV}.$$

At p =1.0 Torr we have $S \approx 0.5$. This means $\Delta R_{res} \approx 1$ cm and $\tilde{E}_{res} \approx 200$ V/cm. The electron mean free path at this pressure $l_e \approx \frac{1}{n_m \sigma} \approx 3\times10^{-2}$ cm (σ is the electron elastic scattering cross-section) is much smaller than the dimension of the resonance region, and the enhanced field $\tilde{E}_{res}$ differs only slightly from the vacuum field. Under these conditions, the average energy (or temperature) acquired by the electrons at the resonance can be estimated in the conventional models [24] by the formula

$$T_{e\,res} \approx \frac{\tilde{E}_{res}^2 e^2}{3m_e \eta_e \left(\omega_0^2 + \nu_{eff}^2\right)}, \tag{5}$$

where η_e is the fraction of the electron energy lost at collisions with neutrals. For typical values of a reduced electric field $\frac{\tilde{E}_{res}}{n_m}$ in our experiment, we have $\eta_e \approx 0.1-1$ [24]. Using these values of η_e we find that the average energy acquired by the electrons at the resonance would be on the order of $T_{er} \approx (0.1-1)\text{eV}$ – much smaller than the calculated electron energy for the case of low argon pressures ($\varepsilon_{eh} \approx 80$ eV for $p = 0.1$ Torr).

The suprathermal electrons that are generated at the resonance and then are accelerated in the direction along the plasma density gradient can lose their energy at collisions with neutrals, thereby producing the ionization in the chamber volume.

Presumably, another source of the volume ionization is the flux of UV photons originated on the surface of the quartz tube bombarded by electrons that are reflected from the potential barrier formed by quasi-steady fields of separated charges in the plasma. According to [25], the most likely mechanism of photoionization under conditions of our experiments is the stepwise photoionization.

Thus, it is quite possible that, under conditions realized in our experiment, the electrons, oscillating in the pump field in a collisionless plasma within a narrow resonance region, gain energies on the order of 100 eV and then expend it efficiently for ionizing the gas in the chamber volume outside the resonance region.

As demonstrated in the experiments (see, e.g., [4,9,10]), the energy distribution function (EDF) for electrons that passed through the resonance region contains a group (or groups) of high-energy electrons (beams) whose energy is determined by expression (3). In this case, the EDF is characterized by a high temperature of the bulk electrons. (The high temperature is a result of high effective collision frequencies which are due to developing plasma instabilities at the resonance and are substantially higher than the electron-ion and electron-neutral collisions frequency).

Depending on the specific form of the EDF, the plasma potential can be determined by the temperature of the bulk (Maxwellian) of the EDF [24]:

$$\varphi_p \approx \frac{kT_{e\,res}}{e}\ln\sqrt{\frac{M_i}{m_e}} \tag{6}$$

or the energy of the suprathermal electron beam:

$$\varphi_p \approx \frac{\varepsilon_{eh}}{e} \tag{7}$$

(according to which part of EDF makes the main contribution in an electron flax on a chamber walls).

One more demonstrative result is the time evolution of the electron density n_e and the probe floating potential φ_f early in the discharge pulse. Characteristic of the time behavior of n_e and φ_f is the presence of a delay of $\Delta t \cong (1-2)$ ms relative to the beginning of the microwave pulse (and relative to the appearance of radiation, to judge from the multiplier signal). After this delay time, the plasma density and the plasma potential rapidly grow. The analysis of photomultiplier signals and high-speed photographs of the discharge leads us to the conclusion that a plasma shell around the quartz tube of the plasmatron forms in a time shorter than $\Delta t \cong (1-2)$ ms. The observed delays may be associated with the time required for the electron density in the formed plasma shell to reach and exceed the critical level, after which time there appears a plasma-resonance region with singularity in the electric field of the electromagnetic wave.

For an electromagnetic wave travelling in the radial direction from the maximum-density side toward lower densities, the depth of penetration into plasma is on the order of

$$\delta \approx \frac{c}{\omega_p}$$

that under our experimental conditions amounts to several centimeters and is comparable with a characteristic radial sizes of the plasma layer. This means that the nonlinear processes of interaction of the electromagnetic wave with the plasma can develop on the inner slope of the electron-density profile toward the quartz tube as well as on the outer slope toward the chamber wall.

4. Conclusion

The results of our investigations lead to the conclusion that, under conditions realized in the experiments, the parameters of the plasma created in the volume of the reactor chamber by the microwave plasmatron are governed by nonlinear processes occurring in the plasma-resonance region. The high electron density n_e and the high plasma potential φ_p which bear witness to the definitely nonlinear interaction of the electromagnetic wave with the plasma created by this wave, are observed in the range of the working gas (argon) pressures $0.03 \le p \le 0.5$ Torr. In most of this interval, the plasma density at a distance of several centimeters from the plasmatron is substantially (three to five times) higher than the critical density for incident microwaves

This "resonance" microwave source of plasma can find application in various fields, including plasma chemistry (in particular, for chlorine and fluorine containing gases utilization), gas-discharge sources of visible light and ultraviolet radiation, etc. There is a further application for which excitation of plasma along the dielectric by means of surface electromagnetic wave under the "plasma resonance" conditions is of immediate interest. The case in point is problem of supersonic aircraft flow past regime control. Discharge supported by surface electromagnetic wave has been considered as a possible way of this problem handling in number publications (see for example [26]). Overdense plasma production under the resonance conditions offers a possibility to enhance the action on a supersonic flow structure due to the high efficiency of energy input in a near-the-surface plasma layer.

Acknowledgements
This work was supported in part by Samsung Electronics Co., LTD, the Netherlands Organization for Scientific Research (project NWO 047.016.019), grant of Russian Foundation for Fundamental Research No 07-02-00011-a and by grant of Russian Federation President NSH-5382.2006.2. Authors are grateful to Prof. Batanov G.M. and Dr. Silakov V.P. for fruitful discussions.

References

[1] Ginzburg V L 1960 *Propagation of Electromagnetic Waves in Plasma.* (Moscow: Fizmatgiz) 271 p. (in Russian)
[2] *Problems of Plasma Physics and Electronics. Transact. of Lebedev Physical Institute*, ed. Kovrizhnykh L M 1985 vol. 160 (Moscow: Nauka) 221 p. (in Russian)
[3] *Dissipation of Electromagnetic Waves in Plasma. Transact. of Lebedev Physical Institute*, ed. Basov N G 1977 vol 92 (M: Nauka) 131 p. (in Russian)
[4] *Generation of Nonlinear Waves and Quasi-steady Currents in Plasma. Transact. of Genetal Physics Institute* , ed. Kovrizhnykh L M 1988 vol 16 (Moscow: Nauka) 168 p. (in Russian)
[5] Bulanov S V and Kovrizhnykh L M 1975 About acceleration of particles in a region of plasma resonance *Fizika Plazmy* **1** 1016-18, (in Russian)
[6] Kovrizhnykh L M and Sakharov A S 1979 Acceleration of electrons in a plasma resonance region *Fizika Plazmy* **5** 840-48, (in Russian)
[7] Albriton J and Koch P 1975 Cold plasma wavebreaking: production of energetic electrons *Phys. Fluids* **18** 1136-39
[8] Bulanov S V, Kovrizhnykh L M and Sakharov A S 1977 Langmuir oscillations of a finite amplitude in plasma resonance region *Zhurnal Experimental'noi i Teoreticheskoi Fiziki* **72** 1810-24 (in Russian)
[9] Batanov G M, Berezhetskaya N K, Dorofeyuk A A, Kossyi I A and Nikuradze P O 1983 Electron spectrum in a plasma device for driving a direct current with microwaves *Sov. J. Plasma Physics* **9** (3) 352-57
[10] Batanov G M, V.A.Ivanov V A, Kossyi I A and Sergeichev K F 1986 Large-amplitude plasma

waves and particle acceleration in the plasma corona of a microwave discharge *Sov. J. Plasma Phys.* **12** (5) 317-25.

[11] Aliev Yu M, Bychenkov V Yu, Maximov A V and Shluter H 1992 High energy electron generation in surface-wave-produced plasmas *Plasma Sources Sci. Technol.* **1** 126-31.

[12] Aliev Yu M, Maximov A V, Kortshagen U, Schluter H and Shivarova V A. 1995 Modeling of microwave discharges in the presence of plasma resonances *Physical Review E* **51** 6091-103

[13] Alves L L, Letout S, Boisse-Laporte C and Leprince P 2006 *Microwave Discharges. Fundamentals and Applications VI Int. Workshop, Proceedings* (Yanus-K, Moscow) pp 177-186

[14] Sugai H, Ghanashev I and Nagatsu N 1998 High-density flat plasma production based on a surface waves *Plasma Sources Sci. Technol.* **7** 192-205

[15] Ganachev I P and Sugai H 2002 Production and control of planar microwave plasmas for materials processing *Plasma Sources Sci. Technol.* **11** A178-A190

[16] Letout S, Alves L L, Boisse-Laporte C and Leprince P 2005 *Proceedings XXVIIth ICPIG* (Eindhoven, the Netherlands) p 18-22

[17] Letout S, Boisse-Laporte C, Alves L L 2006 Electron-drift detection using directional planar probes in a low-pressure coaxial surface-wave discharge *Appl. Phys. Letters* **89** 241502-1 – 241503-3

[18] Letout S, Alves L L, Boisse-Laporte C and Leprince P Characterization of a surface-wave sustained plasma discharge in a co-axial structure *Journal of Optoelectronics and Advanced Materials* **7** 2471-75

[19] Terebessy T, Kando M, Kudela J 2000 Detection of localized hot electrons in low-pressure large-area microwave discharges *Applied Physics Letters* **77** 2825-27

[20] Terebessy T, Siry M, Kando M, Kudela J and Koizec D 2003 Observation of localized electron beam in a large-area microwave discharge *Appl. Phys. Lett.* **82** 694-6

[21] Rauchle E. 1998 Duo-plasmaline, a surface wave sustained linearly extended discharge *J. Phys. IV France* **8** 99-108

[22] Gritsinin S I, Kossyi I A, Malykh N I, Misakyan M A, Temchin S M and Bark Y B 1999 *Plasma Coaxial Discharge as a new Type of the Microwave Surface Wave Discharge* Preprint of General Physica Institute no.1 24 p.

[23] Gritsinin S I, Kossyi I A, Malykh N I, Misakyan M A, Temchin S M and Bark Y B. 1999 *14th International Symposium on Plasma Chemistry, Proceedings* vol II (Praha) 675-80

[24] Raizer Yu P 1992 *Physics of Gas Discharge* (Moscow: Nauka) 536 p. (in Russian).

[25] Kossyi I A and Silakov V P 2005 Step Photoionization as a Mechanism Responsible for the Raether Paradox *Plasma Sources Sci. Technol.* **14** 594-98

[26] Shibkov V, Abramova A, Aleksandrov A, Chernikov V, Dvinin S and Ershov A. 2006 *44th AIAA Aerospace Sciences Meeting and Exhibi,* Report AIAA-2006-1455 (Reno Nevada)

Diagnostics of plasma plume produced by laser ablation using ICCD imaging and transient electrical probe technique

I Mihaila, C Ursu, A Gegiuc and G Popa

Department of Physics, Faculty of Physics, "Alexandru Ioan Cuza" University, 700506, Iasi, Romania

E-mail: ghpopa@uaic.ro

Abstract. The dynamics of transient plasmas generated by high-fluence nanosecond laser ablation has been investigated by means of fast ICCD imaging and electrical probes in transient regime. Measurements have been carried out on plasmas produced in vacuum (5×10^{-6} Torr residual pressure) by a pulsed Excimer laser (20 ns, XeCl, λ = 308 nm) irradiating stainless steel targets. Two plasma expansion velocities were estimated, one from probe measurements and another from recorded ICCD images. Electron plasma temperature (0.1 – 0.3 eV) and density ($1 - 5\times10^{11}$ cm^{-3}) were measured by electrical probes for a laser radiation power density of 6×10^{8} W/cm^{2}.

1. Introduction

Nanosecond laser ablation is a complex process, which implies interactions between laser radiation and target materials [1], between a laser beam and an ablation plasma plume [2,3], and, finally, between ablation cloud particles [4,5]. Experimental methods, theoretical models and numerical simulations were used to describe the laser-target interaction and the dynamic of the particles ejected and the plasma plume formed after laser irradiation on a solid target [6–8]. The early stage of the ablation process and the space and time dynamics of the involved species were investigated, so that useful information were obtained for improving the technology for thin layer deposition of materials under special conditions of purity.

In the current paper experimental results are presented and discussed on the dynamic properties of the plasma plume produced by so called laser ablation of a metallic target. The diagnostic techniques used were fast ICCD imaging and electrical probes in transient regime. The ICCD images can register the instantaneous spatial distribution of the light emission of the plasma plume at different moments of its temporal evolution. Plasma parameters as electron density and temperature were measured using electrical probes. The light intensity profiles recorded by the ICCD camera and the electrical probe measurements allow identifying the two stages of the plasma plume expansion with different velocities.

2. Experimental set-up and measuring circuits

The experimental set-up with the ICCD measuring circuit is schematically shown in figure 1. The laser radiation of a 20 ns excimer pulsed laser (XeCl, λ = 308 nm, 10 Hz frequency) beam was focused by an f = 18 cm lens into a stainless steel target placed in a cylindrical vacuum chamber (20 cm in diameter and 15 cm in height). A vacuum system (preliminary and turbomolecular pump) pumped down the chamber to a 5×10^{-6} Torr residual pressure. Due to the technical parameters of the laser design and of the optical system used, the laser beam was focused onto a line with length of 1.5 mm and thickness of 0.3 mm (an area of 0.45 mm^2). Before each experiment, the laser beam energy has been measured with a Joulemeter. The energy usually used was 50–55 mJ/pulse, which leads to a typical laser intensity of ~ $6\times10^8 W/cm^2$.

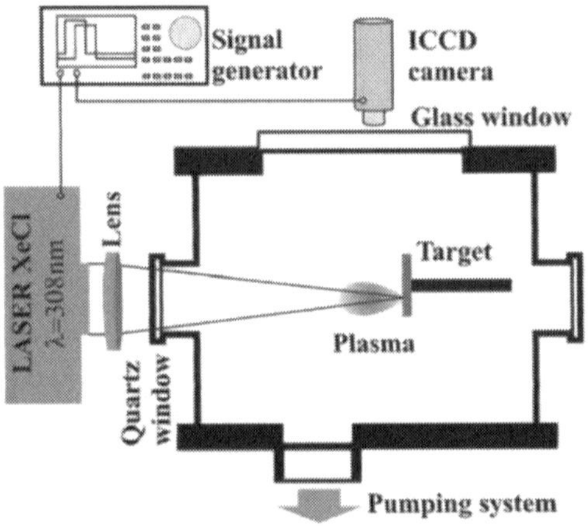

Figure 1. Experimental set-up with ICCD acquisition circuit.

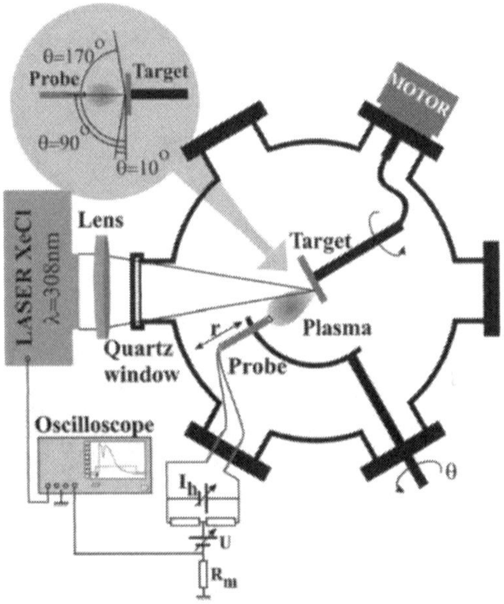

Figure 2. Experimental set-up with probe measurement circuit.

Both the formation and the dynamic of the plasma plume have been studied by means of a Hamamatsu digital camera (model C848-O5G) coupled with an image intensifier (model 9546-03) placed perpendicular to the plasma expansion direction. A signal generator was used as a master, which may control the delay between different measuring channels and to trigger the laser and the ICCD. Thus, the ICCD can acquire the total light emitted by plasma plume at different time instants during the temporal evolution of the plasma plume.

Figure 2 shows schematically the experimental set-up with the probe measuring circuit. The heated cylindrical probe was a tungsten wire of 0.23 mm in diameter and 3 mm in length. The probe was movable both radially (from 1.5 cm to 4 cm away from the target surface) and azimuthally (θ ranged from 10 degree to 170 degree).

The time evolution of the probe current for a constant bias of the probe was recorded by a Tektronix TD2500 oscilloscope and then transferred to a PC for further analysis. The bias voltage of the probe was changed in the range form −10V to +10V with respect to the target and the vacuum chamber. These probe signals recorded along time were used to reconstruct the probe characteristic at a certain moment of the time evolution of the plasma plume. The total time required for recording the signals needed for the reconstruction of a probe characteristic was about 30 minutes.

A double probe system was also use to detect and to measure the internal electric field within the plasma plume. Each probe of the system was made from tungsten wire of 0.23 mm in diameter and 3 mm in length. The time evolution of the potential for each probe was also recorded using the Tektronix TD2500 oscilloscope and then transferred to a PC for analysis.

Journal of Physics: Conference Series **207** (2010) 012005 doi:10.1088/1742-6596/207/1/012005

3. Results and discussion

3.1. ICCD imaging

The spatial distribution of the light emission produced by the plasma plume at different instants of time of the plasma plume evolution was found from the ICCD images. Each image was obtained for a separated laser shot. Because of that, special attention has been paid to keep similar conditions for the laser ablation process at each laser shot. In this view, an equal energy of 55 mJ was used for each laser pulse. Moreover, by a continuously target spinning around the axis each shot send the laser beam on a fresh target surface. Each picture was taken for a gate time duration of 200 ns. A set of such ICCD images taken at certain moments of the plasma plum is presented in figure 3. Because the images were recorded with different setting of ICCD gain, current, and exposure time, these images show only a qualitative evolution of the plasma plume.

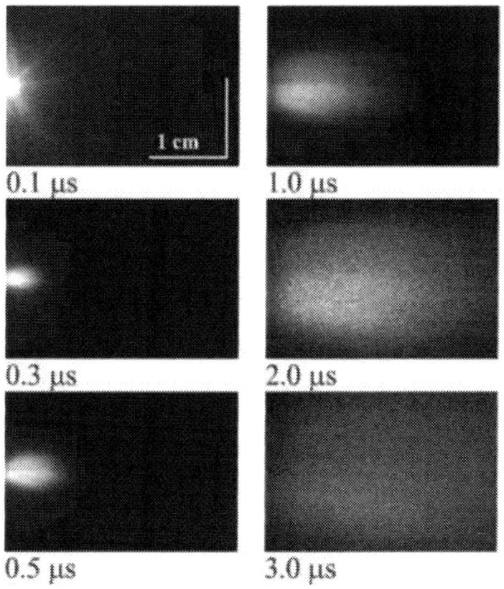

Figure 3. Temporal evolution of the plasma plume (ICCD gate time – 200 ns). Different images were recorded for successive laser pulses of equal energy (55 mJ/pulse).

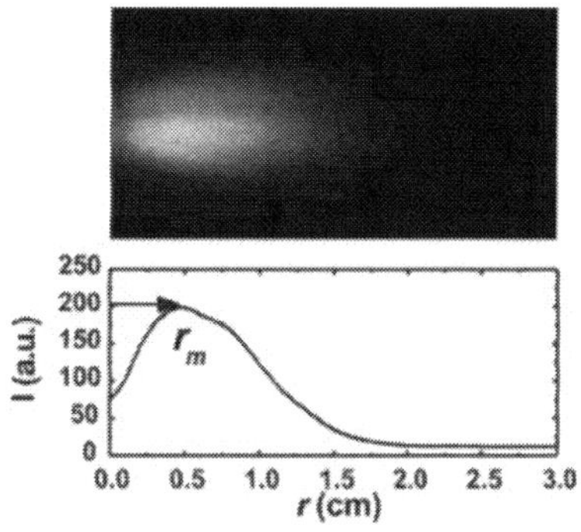

Figure 4. Spatial light intensity profile (1 μs after laser shot).

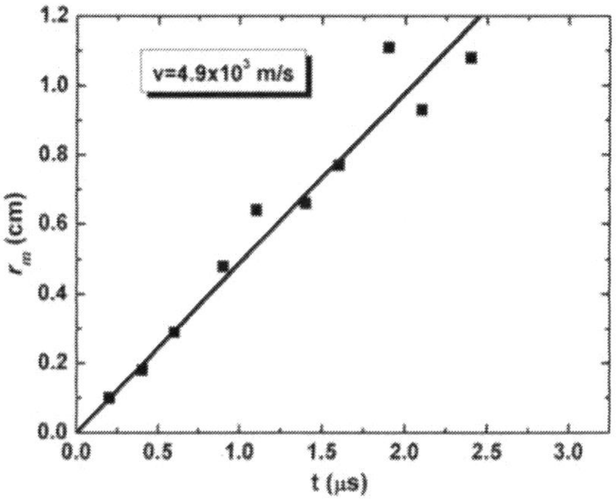

Figure 5. Dependence of the distance r_m given by the maximum value of light intensity versus time.

The pictures from figure 3 were used to find the spatial profile of the plasma plume light intensity. Thus, using free software (ImageJ), the spatial profile on the direction normal to the target surface corresponding to the ablation point was plotted. Such an example of a light intensity profile is presented in figure 4 for an image recorded at 1 μs after the laser shot. The distance $r = 0$ cm represents the surface of the target. The expansion velocity of the plasma plume was calculated as the slope of the line which depicts the distance from the target surface where the maximum value of light intensity (distance r_m in the figure 4) is obtained versus the time of the recorded image. This dependence is presented in figure 5. It was found that the expansion velocity of the plasma plume is about 4.9×10^3 m/s.

3.2. Probe measurements

The use of probe measurements in the diagnostic of the ablation plasma plume was previously reported [9–13]. First, experiments and testing of the probe method have been made taking into account the contamination process due to a possible thin layer deposition of the ablated metal onto the probe surface.

First, the probe was exposed for 30 minutes to the plasma plume. Then, the probe characteristic was reconstructed from the signals recorded (figure 6, dashed line). The dashed dot line in figure 6 represents the probe characteristic after a 90 minute total exposure of the probe to the plasma plume. It is clear the effect of the contamination due to a deposition of a thin layer of the target material on the probe surface.

Then, to avoid probe contamination, a heated probe was used. The probe characteristic for the heated probe is represented in figure 6 – full line. The probe was heated to a temperature at which deposition of the ablated material on the probe surface was not possible but taking care that the probe has not became a thermoemissive one. The latter fact can be proved by keeping the ionic saturation current at the same value as that of an uncontaminated cold probe.

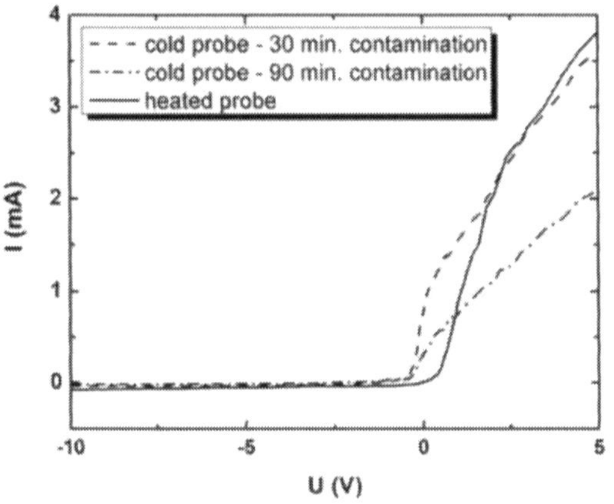

Figure 6. Cold probe versus heated probe characteristics.

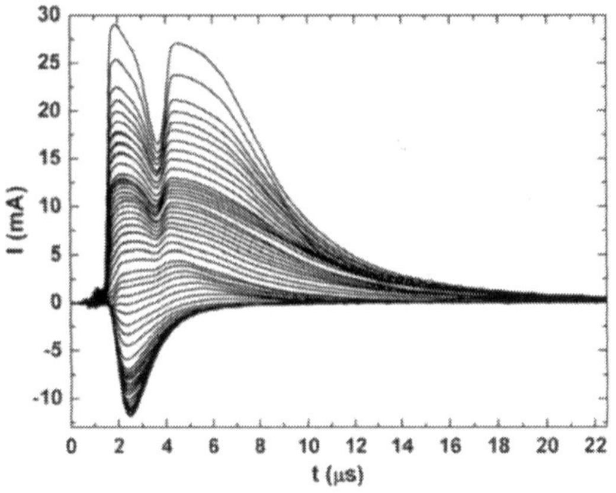

Figure 7. Time dependence of the probe signal for various biasing voltage (r = 3 cm and θ = 90°).

A large difference between the cold contaminated probe and the heated probe characteristics can be observed. It shows that measurements performed with contamination of the cold probe might lead to untrue plasma parameters, while the heated probe provides reliable data on plasma parameters.

A typical time-dependence of the probe signal for various biasing voltage (r = 3 cm and θ = 90°) is shown in figure 7. At certain values of time, the probe characteristics were reconstructed and plasma parameters were calculated. The electron temperature, T_e, and the electron density, n_e, were estimated using the standard procedure of processing the probe characteristic [14]. Time dependence of the electron temperature and of the electron density for various values of the θ angle are shown in figure 8 a) and b), respectively.

Moreover, the probe characteristics show the presence of two groups of electrons within the plasma plume. One group of electrons had the temperature in the range of 0.1 eV to 0.3 eV. The other group of electrons could not always be measured (due to the small number of points of the characteristic) and the temperature ranged from 1 eV up to about 4 eV. The density of the slow electrons ranged from 1×10^{11} cm^{-3} to 5×10^{11} cm^{-3} and decreases in time, while the second group is more than one order of magnitude lower than the density of the slow electrons.

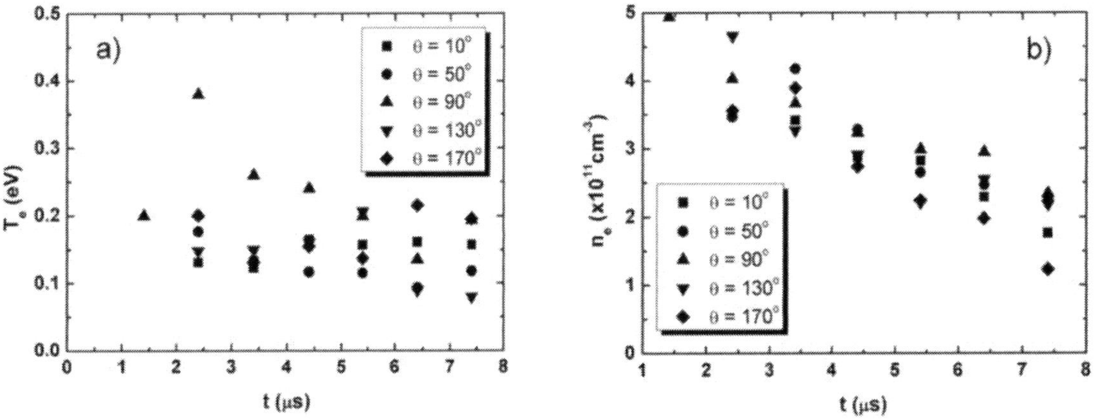

Figure 8. Time dependence of T_e (a) and n_e (b) for various θ (r = 3 cm and a laser intensity of 6×10^8 W/cm^2).

Another velocity estimation of the spatial expansion of the plasma plume was obtained from the time evolution of the ion saturation current of the probe registered at a different position, r, along the axis of the plasma plume. It can be seen that with increasing the r distance measured from the target (starting with 1.7 cm up to 4 cm) the maximum of the ionic probe signal is moving from early values of time toward later values of time (figure 9). Again a remarkable linear dependence is obtained when the distance r is plotted versus time t_m when the ionic probe signal has a maximum value. This dependence is shown in figure 10. The estimated value of the plasma plume expansion velocity is around 13×10^3 m/s, which is more than twice the expansion velocity estimated from the light intensities recorded with the ICCD.

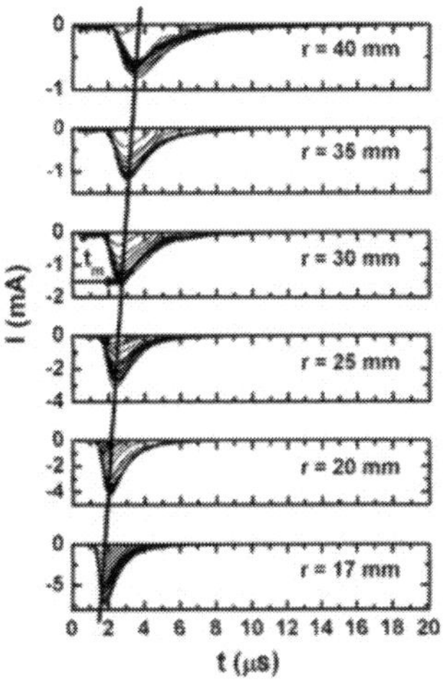

Figure 9. Ionic signals for various radial positions

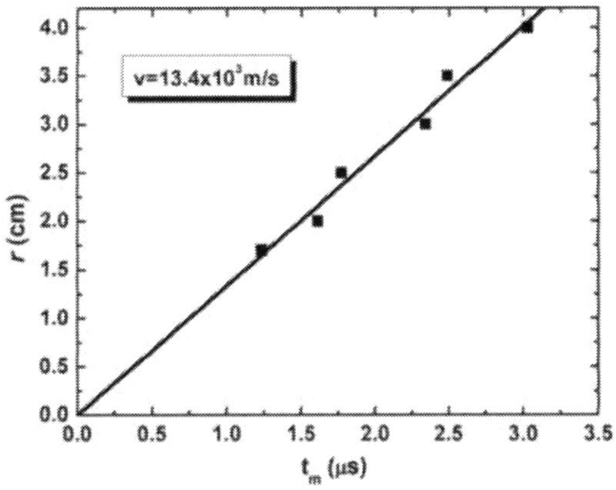

Figure 10. Estimation of plasma expansion velocity from ionic part of probe signals

Using the double probe instead, potential differences between the probe tips at various distances from target surface were recorded (figure 11). From the dependence of the distance at which the probe signal was recorded versus the time-value t_m that correspond to the maximum value of the potential difference, the plasma plume expansion velocity can be also estimated from the slope of the dependence. This dependence is shown in figure 12. The estimated value of the plasma expansion velocity is around 10×10^3 m/s which is in good agreement with the one estimated from the ionic probe signal and is twice the one estimated from ICCD images.

The presence of two different values of the plasma expansion velocity can be explained by the formation of two plasma structures. This process has already been observed in studies on targets as Cu [12,13], Al [15,16] or graphite [17]. The two structures exhibit distinct dynamics and velocities, usually in the range of 10^4 m/s for the first structure and 10^3 m/s for the second one. The first structure, with a high speed, is mainly formed by plasma charged particles, while the second one is mainly formed by neutrals [16]. Consequently, probe measurements allow measuring the velocity of the first structure of the expanding plume (with velocity in the range of 10^4 m/s), while using the light intensity recorded by ICCD images the expansion velocity of the second structure can be estimated (with the velocity around 10^3 m/s).

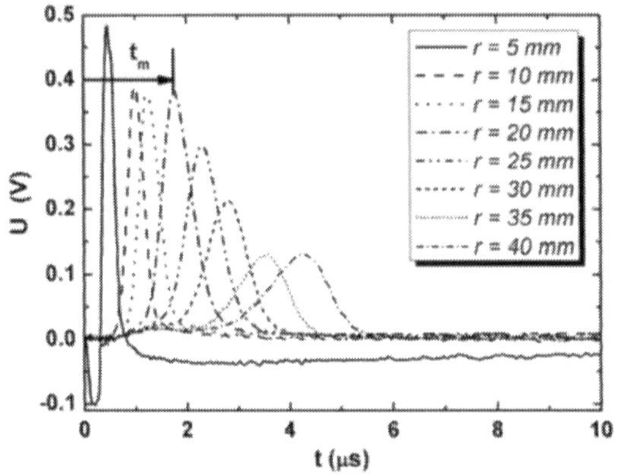

Figure 11. Double probe difference potential for various radial positions.

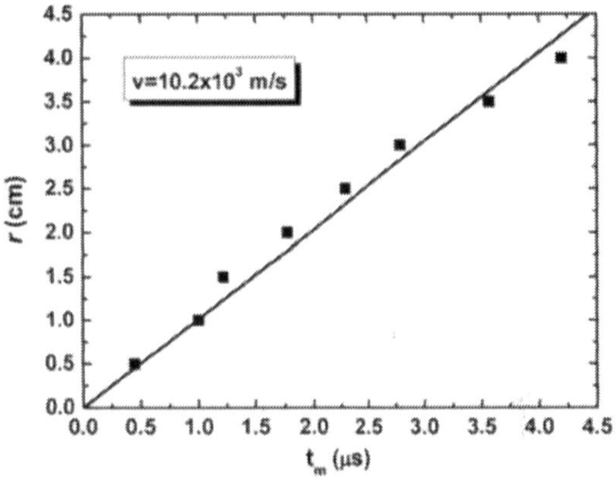

Figure 12. Plasma expansion velocity calculated from double probe measurements.

4. Conclusion

The transient plasmas generated by high-fluence nanosecond laser ablation of stainless steel targets have been investigated by means of fast ICCD imaging and by electrical probes in transient regime. The electron temperature and density calculated from the probe characteristics recorded by electric probes in transient regime were in the range of 0.1 eV to 0.3 eV and $1 - 5\times10^{11}$ cm^{-3} respectively.

The expansion velocity estimated from the total light intensity recorded by ICCD imaging is correlated with a second slower structure, mainly formed by neutral particles, which appears in the dynamics of plasma plume and has the value around to 5×10^3 m/s.

From the probe measurements a plasma expansion velocity was estimated that is correlated with a first faster structure, mainly formed by charged particles, and has a value around to 1×10^4 m/s.

Acknowledgements

This work is under financial support offered by Romanian Ministry of Education, Research and Youth by grant CEEX 5918/2006

References

[1] Peterlongo A, Miotello A and Kelly R 1994 *Phys. Rev. E* **50** 4716
[2] Jordan R and Lunney J G 1998 *Appl. Surf. Sci.* **127–129** 968

Journal of Physics: Conference Series **207** (2010) 012005 doi:10.1088/1742-6596/207/1/012005

[3] Chang J J and Warner B E 1996 *Appl. Phys. Lett.* **69** 473
[4] Amoruso S, Sambri A, Vitiello M and Wang X 2006 *Appl. Surf. Sci.* **252** 4712
[5] Amoruso S,.Sambri A and Wang X 2006 *J. Appl. Phys.* **100** 013302
[6] Hermann J, Boulmer-Leborgne C and Hong D J. 1998 *Appl. Phys.* **83** 691
[7] Amoruso S, Bruzzese R, Velotta R, Spinelli N, Vitiello M and Wang X 2005 *Appl. Surf. Sci.* **248** 45
[8] Amoruso S, Armenante M, Berardi V, Bruzzese R, Pica G and Velotta R 1996 *Appl. Surf. Sci.* **106** 507
[9] Lunney J G, Doggett B and Kaufman Y 2007 *Journal of Physics: Conference Series* **59** 470
[10] Hansen T N, Schou J and Lunney J G 1999 *Appl. Phys. A* (Suppl.) **69** S601-4
[11] Hendron J M, Mahony C M O, Morrow T and Graham W G 1997 *J. Appl. Phys.* **81** 2131
[12] Gurlui S, Sanduloviciu M, 2006 *AIP Conf. Proc.* **812** 279
[13] Gurlui S. Sanduloviciu M, Strat M, Strat G, Mihesan C, Ziskind M and Focsa C 2006 *J. Optoelectron. Adv. Mat.* **8** 148
[14] Swift J D, Shwar M J R 1970 *Electrical probe for plasma diagnostics* (ILLFE-Book)
[15] Gurlui S, Agop M, Nica P, Ziskind M and Focsa C 2008 *Phys. Rev. E* **in press**
[16] Ursu C, Gurlui S, Focsa C, Popa G 2008 *Nucl. Instrum. Meth. B* **in press**
[17] Bulgakova N M, Bulgakov A V and Bobrenok O F 2000 *Phys. Rev. E* **62** 5624

Journal of Physics: Conference Series **207** (2010) 012006 doi:10.1088/1742-6596/207/1/012006

Active species in N_2 and N_2-O_2 afterglows for surface treatments

A Ricard[1], A M Pointu[2], S Villeger[1] and C Canal[3]
[1]LAPLACE , Univ. Paul Sabatier, CNRS, 118 route de Narbonne 31062 Toulouse, France
[2]LPGP, Univ. Paris-Sud, 91405 Orsay, France
[3]Pharmacy and Pharmaceutical Technology Dpt., University of Barcelona, Avda. Joan XXIII s/n 08028 Barcelona, Spain

E-mail: ricard@laplace.univ-tlse.fr

Abstract. Production of active species is studied in N_2 and in N_2-O_2 afterglows of electrical discharges at low and atmospheric gas pressures. They are produced in microwave discharges in a large range of gas pressures from a few Torr to 100 Torr and in corona discharges at atmospheric gas pressure. The active species in N_2 afterglows are the N-atoms which are in the range of a few percents in the afterglows. The effect of O_2 molecules in low percentages in low pressure N_2 microwave plasmas and as impurity in corona N_2 discharges is specially analysed. The interaction of N and O-atoms with surfaces is studied for bacteria decontamination and for transmission of N-atoms though porous membranes. The processes of bacteria decontamination in N_2-O_2 afterglows are described for low pressure microwave and atmospheric pressure corona discharges. Transmission of N-atoms through porous membranes is studied at medium pressure (10–100 Torr) microwave afterglows.

1. Introduction

In the field of surface modification by plasmas, flowing post-discharges of reactive gases have been recently studied to transform sensitive surfaces at gas temperatures as low as 300 K avoiding the interaction of charged particles (electrons and ions) on the treated surfaces.

In such a way, remote plasma reactors have been implemented for microelectronic applications to obtain oxidized and nitrided silicon thin films [1], for polymer activation to increase surface adhesion [2], for plasma nitriding to enhance hardness and resistance to corrosion of metal surfaces [3], for metal surface cleaning [4] and more recently, for bacterial sterilization [5,6]. In these previous works, flowing plasmas were produced at low gas pressures (1–10 Torr), for an optimum production of active species and to obtain a large diffusion of the species into the post-discharge chamber.

There is also a tendency to work with plasmas at higher pressures, up to the atmospheric gas pressure to avoid any pumping system.

Direct current (DC), radiofrequency (RF) [7] and microwave [8] glow discharges have been used for low pressure plasmas. Dielectric barrier discharge (DBD) [9] and corona [10] are suitable for atmospheric discharges.

In the present paper, production of N and O-atoms in N_2-O_2 flowing post-discharge reactors is presented from low (a few Torr) to high pressures and up to the atmospheric gas pressure.

Journal of Physics: Conference Series **207** (2010) 012006 doi:10.1088/1742-6596/207/1/012006

2. Experimental setup

2.1. The microwave plasma afterglow

The experimental setup used for surface treatments (here bacteria decontamination [6]) in afterglow conditions is reproduced in figure 1. In the first part of the reactor, a discharge is produced in a quartz tube of 5 mm internal diameter by a 2.45 GHz microwave surfatron cavity at powers up to 200 Watt. This small discharge tube is connected to a tube of 18 mm i.d. and up to 30 cm length. The reactor consists of a treatment chamber in pyrex of 15 cm diameter and 20 cm height. The gas flow can vary from 0.1 to 3 slm. The gas pressure in the reactor can be adjusted between 1 and 50 Torr by means of a throttle valve located above the primary pump. Bacterial samples are placed in the center of the post-discharge chamber on a steel holder which can be heated up to 120 °C.

The absolute N and O-atom densities in N_2-O_2 gas mixtures were obtained by the NO titration as detailed in [11].

The optical emission was collected from the afterglow by means of an optical fiber, connected to a Jobin-Yvon 270M spectrometer with a CCD detector.

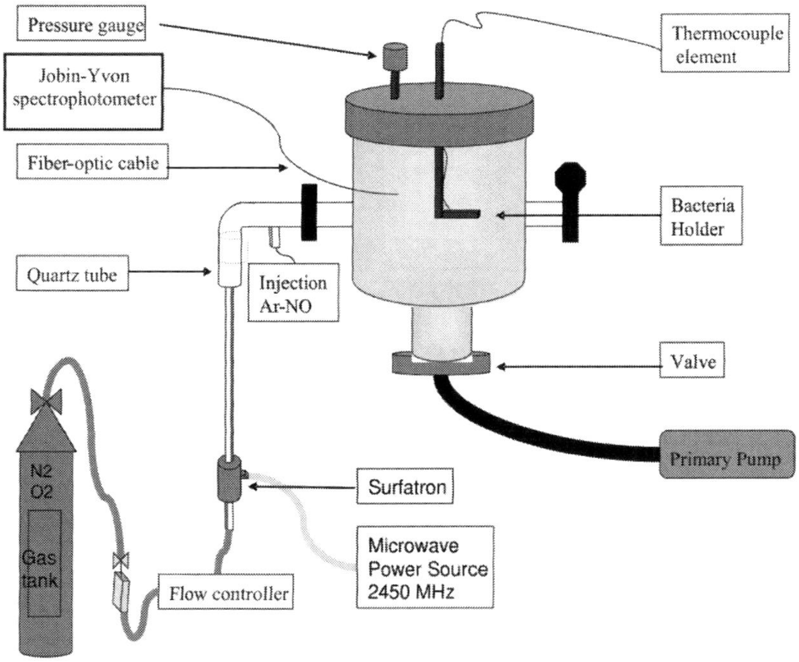

Figure 1. Post-discharge reactor used for bacteria inactivation (LAPLACE – Toulouse)

2.2. The corona discharge reactor

An atmospheric pressure nitrogen afterglow was used as described in [10]. It flows at about 40 slm from successive fast pulsed corona discharges produced between needles with a voltage of some 10 kV, a repetition rate of some 10 kHz and a mean power near 15 W. When contained in a manuril® tube (6 mm inner diameter), it produces external fluorescence over more than 10 meters. Such light emissions have been attributed to transported nitrogen atoms.

A high impedance regime of each corona discharge before it is blown up by the gas flow is associated to an energetic electron distribution function able to give a high dissociation efficiency to the discharge. Moreover, the gas temperature is around 300 K.

Thus, in views of the pursued applications, we shall study a cylindrical post discharge, both by measuring the "active" species using spectroscopic means and by testing their biocidal effects.

A typical experimental set up is shown in figure 2.

High purity N_2 gas (99.995 %) is introduced upstream the discharge and, in parallel, a small proportion of O_2 can be added as controlled impurity. Two flow controllers adjust corresponding flows respectively to 40 slm for N_2 and to 0–50 sccm for O_2. After the discharge region, the afterglow is established in a quartz tube of 8 mm internal diameter, ending in the atmospheric air. The emission spectroscopy device (ROPER SCIENTIFIC) is composed of an ACTON spectrometer (0.25 nm resolution with 1200 lines) and a CCD PIXIS-100 detector working from 200 to 1080 nm.

The spectrometer entrance slit is coupled to one end of an optical fiber of 400 μm diameter through a filter used to eliminate higher interference order. The other end of the fiber, covered with a quartz lens, is mobile in parallel with the discharge axis. Spectral acquisition time is around 20 s.

When necessary, biocidal tests are made in the downstream part of tube, inserting a 250 cm^3 plastic box containing the biological species (figure 2).

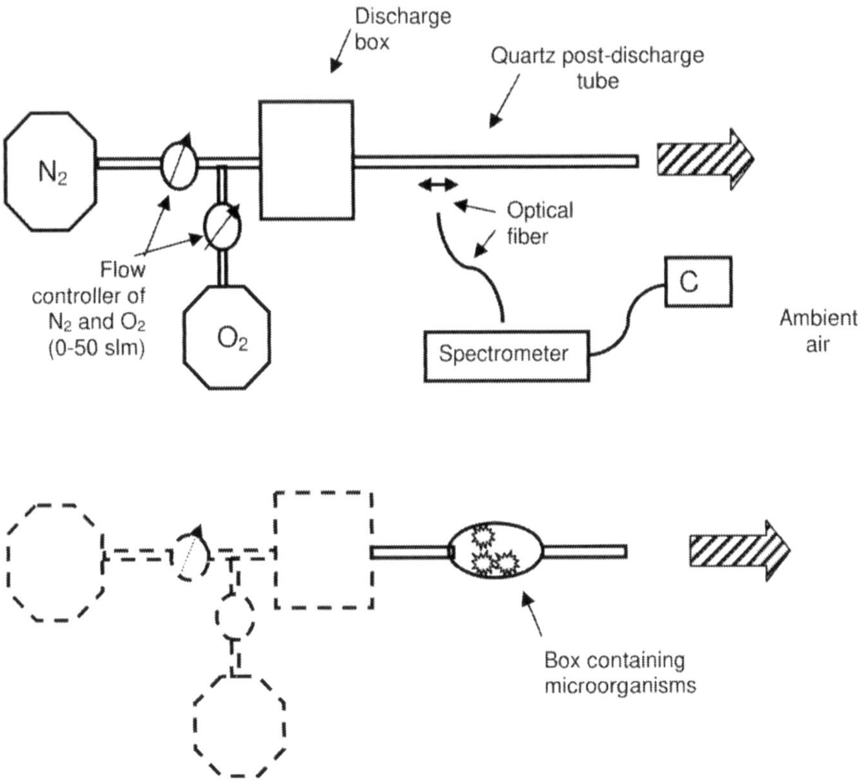

Figure 2. Corona discharge reactor in LPGP-Orsay.

3. Production of excited species in the afterglow

3.1. Afterglows of N_2 and N_2-O_2 flowing microwave discharges

In pure N_2 afterglow, the N-atoms recombine in the gas volume as follows [11,12]:

(a) $$N + N + N_2 \rightarrow N_2\,(B, v'=11) + N_2.$$

Reaction (a) produces the N_2-first positive system (FPS) emission from the N_2 (B, v' = 11) to the N_2 (A, v") states, with emissions of 540.7 (v" = 6), 580.4 (v" = 7) and 625.3 (v" = 8) bands. The 580.4 nm being the most intense of these three bands is considered as the signature of N-atoms in the afterglow.

As the radiative frequency of N_2 (580 nm) is low (5×10^4 s^{-1}), it is demonstrated [11] that the quenching of N_2 (B, 11) by N_2 is the dominant process at pressures above 1 Torr, so that the intensity of N_2 (580 nm) is proportional to the square of N-atom density:

$$I\,(580\text{ nm}) = k\,[\text{N}]^2. \qquad (1)$$

The constant k is calibrated by NO titration as explained in [11].

The homogeneity of the afterglow has been analyzed [13] for several pressures in the reactor of figure 1 by directly connecting the 5 mm quartz tube to the reactor chamber at constant flow rate (1 slm) and microwave power (100 W). It has been found that at a pressure of 4 Torr the afterglow is homogeneous and the same atom density is obtained in several parts of the reactor. At higher pressures, an afterglow jet is observed with an increasingly marked difference in the repartition of N-atoms.

When a few percent of O_2 are introduced into the N_2 microwave discharge there is a production of N and O-atoms by electron collisions on N_2 and O_2 and also UV emission, by the following reaction:

(b) $$N + O + N_2 \rightarrow NO(B) + N_2,$$
$$\text{with } NO(B) \rightarrow NO(X) + h\nu\ (NO_\beta).$$

The emission of NO_β bands are found from 300 to 400 nm.

3.2. Afterglows after N_2 corona discharges with O_2 in impurity

In DBD and corona discharges at atmospheric gas pressure, reaction (a) is always present to produce the specific afterglow characteristic of N-atoms.

With an impurity of oxygen from air or with the addition of a low O_2 quantity, as indicated in figure 2, it is observed the following specific emissions [10]:

- the NO_γ bands produced by excitation transfer from $N_2(A)$ metastable molecules :

(c) $$N_2(A) + NO \rightarrow N_2 + NO(A),$$
$$\text{with } NO(A) \rightarrow NO(X) + h\nu\ (NO_\gamma).$$

- the NO_β produced by reaction (b).

The NO molecules and the O-atoms in reactions (b) and (c) are produced in the plasma by dissociation of O_2 by electron collisions and by reactions of N-atoms with O_2, as follows:

(d) $$N + O_2 \rightarrow NO + O.$$

Reaction (d) is efficient at high discharge temperatures (1000–1500 K) [14].

There is also formation of the $N_2O(^1S)$ excimer by reactions of N_2 with $O(^1S)$ metastable atoms:

(e) $$O(^1S) + N_2 + N_2 \rightarrow N_2O(^1S) + N_2,$$
$$\text{with } N_2O(^1S) \rightarrow N_2O(^1D) + h\nu,\ 557.7\text{ nm}$$ (band emission also present in the aurora).

The $O(^1S)$ metastable atom can be excited in the afterglow by collisions of $N_2(A)$ metastable molecules on O-atoms :

(f) $$N_2(A) + O \rightarrow O(^1S) + N_2.$$

The $N_2O(^1S) \rightarrow N_2O(^1D) + h\nu$, 557.7 nm band is thus a signature of $N_2(A)$ metastable molecules.

In the following sections we describe how the emission spectroscopy in the N_2 (N_2-O_2) afterglows can be related to the mechanisms of bacteria inactivation by N and O(^{1}S) atoms, and by UV emission (section 4), and to the transmission of N-atoms through porous membranes (section 5).

4. Bacteria decontamination

In recent years there has been an increasing interest in plasma technology for bacteria sterilization because of its safe and needs lower temperatures then other treatments [15,16]. Such a plasma process would be useful for complex medical devices that are humidity sensitive and that need to be sterilized at temperatures below 70°C.

4.1. Process in N_2-O_2 and N_2 microwave plasma post-discharges at low gas pressures (1 – 10 Torr)

The inactivation efficiency of N_2-O_2 microwave flowing post discharges on several bacteria was previously studied [5,6,17,18]. The optimization of the post-discharge process was performed by varying different parameters, such as gas mixture, source power, and gas pressure. For a better understanding of the sterilization mechanisms in post-discharge environments the absolute concentrations of N and O atoms were determined by NO titration.

It has been demonstrated that the efficiency of bacteria sterilization was related to the N and O-atom density as determined by emission spectroscopy and NO titration [17,18]. It has been found that the N and O-atom densities increased with the microwave power. This is a result of the electron density increase in the discharge, which directly affects the production of N and O. The O-atom density increased with the percentage of O_2 in N_2 (from 1 to 8%), while the N atom density decreased sharply. Such a decrease comes from reactions (b) and (d).

The sensitivity of bacteria (*E-coli*) to the post-discharge was studied in the reactor reported in figure 1. With a N_2-O_2 plasma the sterilizing agents are the N and O-atoms and the UV emission of NO_β, which is produced by the recombination reaction (b).

E-coli bacteria were diluted into LB broth (Lurian Bertani broth) and deposited inside small sterile aluminium cylinders (diameter 5 mm, height 10 mm). The cylinders were placed on the holder as indicated in figure 1. By pumping down to about 10^{-2} Torr, the liquid part of the LB broth was evaporated. It has been checked that about 10% of bacteria remained alive after this operation. Then the dried bacteria and the broth macromolecules were exposed to the N_2-5%O_2 post discharge at 100 watt with a pressure of 5 Torr, and a flow rate of 1 slm. The N_2-5%O_2 gas mixture was chosen because it represents a good compromise to obtain significant values of both N and O-atom density and NO_β UV intensity.

The sample holder can be heated up to 120°C and the holder temperature can contribute to the reduction of the sterilization time. As shown in figure 3, a reduction of 6 log of *E-coli* is observed after treatment times of 2, 25 and 30 minutes with an external heating of the stainless steel holder at 120, 80 and 60 °C, respectively. Moreover, a bacterial reduction of 12 Log was obtained in 5 minutes at 120 °C. An external heating at 120 °C alone, without plasma, only achieved a reduction of 6 log in 30 minutes (not shown in figure 3, but verified).

Such results indicate the synergy of the metal holder temperature and of the active species for an efficient bacteria inactivation in the post-discharge. It seems that the N and O-atoms interact by an etching process on the bacteria membranes. A small increase of the substrate temperature – up to 120 °C in the present experiment – seems to favor desorption of etched products. The increase of N atoms diffusion inside the bacteria as the temperature increased can also be relevant.

The bacteria inactivation has also been investigated in pure N_2 afterglow in standard conditions of 4 Torr, 1 slm and 100 W. The results are reproduced in figure 4 for *E-coli* bacteria diluted inside two different media: LB broth and sterile water, which were then dried in vacuum.

A complete bacteria inactivation was obtained when bacteria were initially (before pumping) in suspension in water at concentrations lower than 10^8 cfu/cm^3. Sterilization (decrease of 6 log) is achieved after 30 minutes of treatment at ambient temperature.

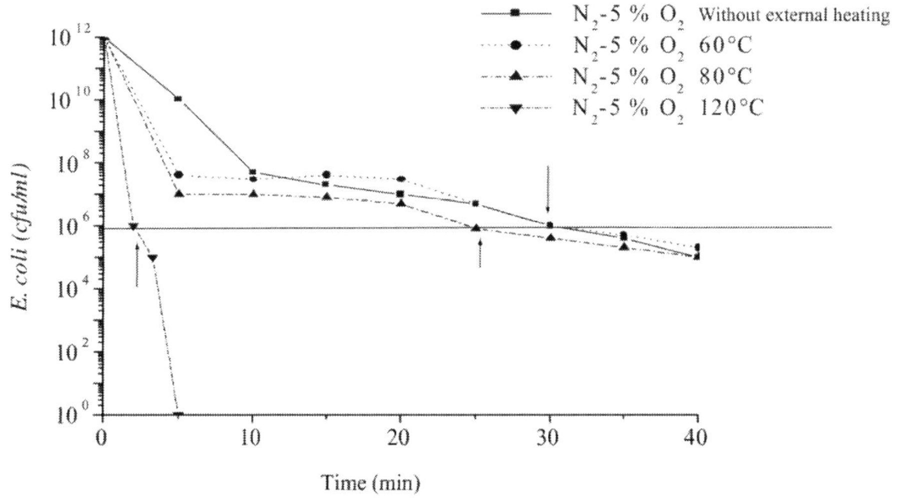

Figure 3. Variations of E-coli survivors versus N_2-5%O_2 afterglow treatment time for several substrate temperatures

Sterilization is more efficient when bacteria are prepared in water, indicating a shadow effect of dried macromolecules from the LB broth.

When bacteria are prepared in LB or in water at high concentrations (10^{10} cfu/cm^3) no sterilization is observed.

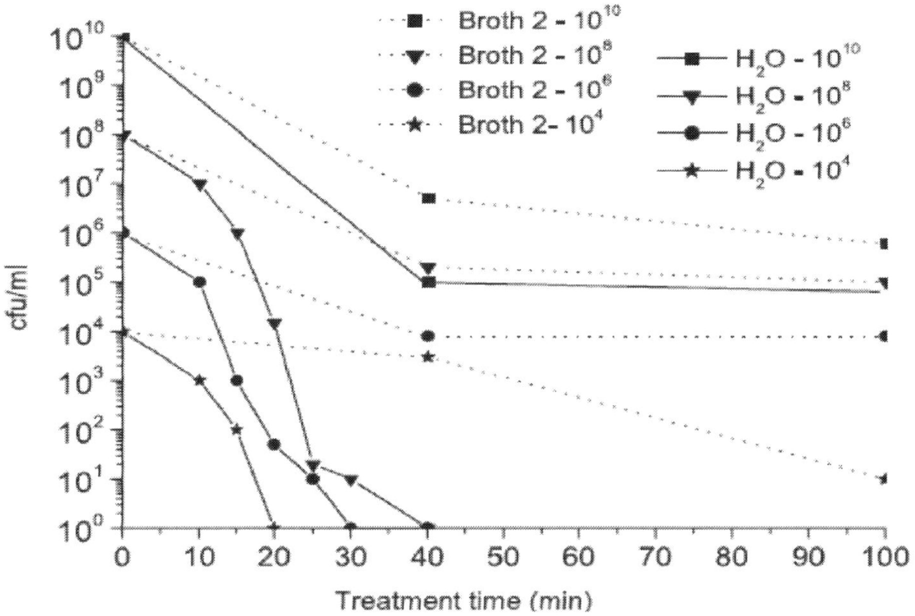

Figure 4. *E-coli* inactivation in N_2 post-discharge versus the treatment time with dilutions of bacteria in Broth 2 (LB) and sterile water [13] 1 slm, 100 W , 4 Torr, 300 K.

By comparing with N_2-5%O_2 post-discharge without external heating (figure 3), the pure N_2 post-discharge is less efficient with the LB broth than the N_2-5%O_2 gas mixture.

However, pure N_2 post-discharges are not oxidizing processes, which is efficient when bacteria are diluted and dried in water media. The etching of bacteria is accelerated by heating the substrate up to 60 °C, which is compatible with a cold sterilization process.

4.2. Process in N_2 corona discharge at atmospheric gas pressure. Effect of O_2 impurities

The N-atom density has been determined by NO titration in [10] as being of 5×10^{14} cm^{-3}, keeping a nearly constant value up to a quantity of 5×10^{-3} O_2 into N_2. By comparing the I_{N2}(580 nm) and I_{N2O*}(557.7 nm) band intensities resulting of reactions (a) and (e, f), it is obtained the order of magnitude of the $O(^1S)$ metastable density with a maximum value of 7×10^{10} cm^{-3} for $(2.5–5)\times10^{-5}$ O_2 into N_2.

Considering the results of spores decontamination (Bacillus Stearotermophilus) as reproduced in table 1, it can be deduced that the best results are obtained with the N_2 – 2.5×10^{-5} O_2 gas mixture when I_{N2O*}(557.7 nm) and thus the $O(^1S)$ density reached a maximum value. The results showed that a small density of $O(^1S)$ was efficient in the sterilization process.

Table 1. Number of Bacillus Stear spores survivors versus exposure times to the corona post-discharge. Effect of addition to N_2 of small O_2 quantities [10].

	[O2]/[N2] ratio		
Exposure time (min)	0	2.5×10^{-5}	2.5×10^{-4}
0	21400	22200	18600
15	1195	130	350
30	595	105	250

5. Transmission of N-atoms through porous membranes

The interaction of N-atoms with nonwoven membranes was investigated on nylon (5 µm) made by co-solvent evaporation from a nylon extruded film and polypropylene (PP) prepared from polypropylene fibers by a melt-blown process, which were exposed as shown in figure 5 to the N_2 flowing (1 slm) microwave (100 Watt) post discharge produced at low gas pressures: from 1 to 30 Torr [19,20,21]. The N atoms transmission through these membranes was measured from the intensity variation of the N_2 (580 nm) afterglow, which results of N atom recombination (reaction (a)), before and after the membranes.

As shown in figure 5, the afterglow emission was collected by a quartz optical fiber which can be fixed in positions 1 or 2 (P_1 and P_2, before or after the membrane) and sent to the entrance slit of the spectrometer as in figure 1.

In order to study the transmission of N-atoms, the nonwoven membranes were set-up across the 18 mm i.d tube, as detailed in figure 5, after the 5 mm i.d discharge tube and before the reactor inlet.

By considering equation (2), the N atom transmission (T_N) through the membranes can be calculated as follows:

$$T_N = \frac{[N]_{P2}}{[N]_{P1}} = \frac{\sqrt{I_{P2}}}{\sqrt{I_{P1}}} \quad (2)$$

After the PP membranes it was observed that the 580 nm intensity was lower than before the membranes and that it kept a nearly constant value up to 30 minutes of post-discharge [21]. This indicates a constant N density after the membranes, which results in a constant T_N value for afterglow times up to 30 minutes.

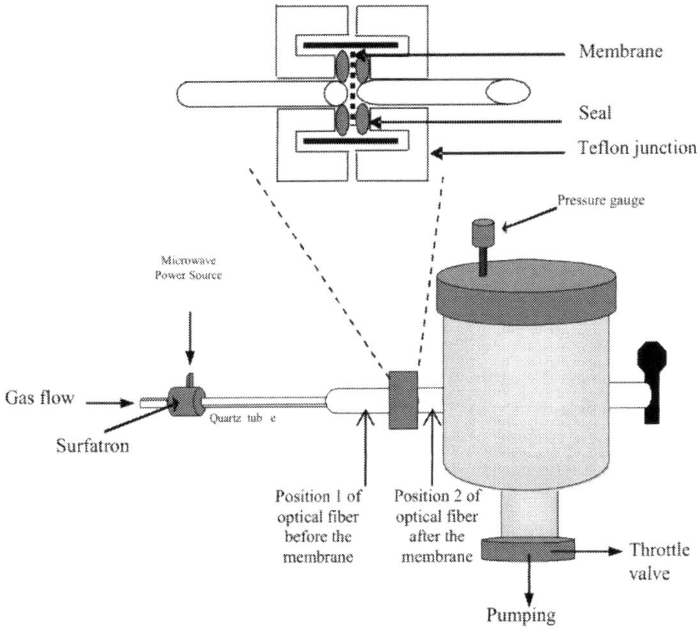

Figure 5. Microwave flowing afterglow reactor. Detail of the membrane insertion in a Teflon junction in the 18 mm id. quartz tube [19].

The mean value of N atom transmission calculated was $T_N = 0.43 \pm 0.03$. This constant N atom transmission through the polypropylene membrane indicates that no significant alterations are produced by the afterglow on the main structure or on the porosity of the membranes, as confirmed by SEM [21].

With the nylon 5 μm membrane, T_N increased from 0.1 to 0.5 after 30 minutes of afterglow time. The increase with treatment time of N atom density after the membrane indicated that the post-discharge treatment produced important alterations on the membrane. In a previous work [19], SEM revealed an increase of the nylon pore size through etching reactions of N atoms with the nylon surface.

Investigations with other membranes also found that T_N increased after 30 min of post-discharge from 0.5 to 1 with PDVF (pores 1 μm) and from 0.33 to 1 with polysulfones (pores 1–3 μm) [20].

Previous results on the treatment of polymer fibers with plasmas pointed out to a preferential etching effect on low crystallinity regions (amorphous regions) [22]. It is known that the drawing of the fibers during the spinning process to produce fibers induces orientation of the polymer chains in the fiber axis resulting in increased crystallinity. According to the fabrication process of the nylon and polypropylene nonwovens, as well as to the SEM micrographs, it can be speculated that the melt-blown spinning process of PP produces a polymer with higher crystallinity than the nylon obtained from a non-oriented polymer film, which is more easily etched by N atoms.

In addition to that, the chemical structures of PP and nylon have to be taken into account in what regards the etching effect. The presence of amide bonds in the nylon molecule provides a point with higher reactivity towards the N atoms than the aliphatic C-C bindings present in the PP molecule, justifying the intense etching effect (and increasing T_N) observed in nylon membranes.

6. Conclusion

Two flowing N_2-O_2 post-discharges have been studied and applied to bacteria sterilization.

The first one is produced by a microwave plasma at reduced gas pressure (1 – 50 Torr) and middle flow rates (0.1 – 3 slm). The second one is a Corona discharge at atmospheric gas pressure and high flow rate (40 slm).

The afterglow has been analysed by emission spectroscopy with detection of N and O atoms whose absolute density has been obtained by NO titration.

The main kinetic reactions in the N_2-O_2 afterglows are reported, showing that the most relevant active species for bacteria sterilization are the N and O-atoms and UV emission at reduced gas pressure with in addition the $O(^1D)$ and $N_2(A)$ metastable atoms and molecules in the Corona post-discharge at atmospheric pressure.

In the microwave post-discharge reactor at reduced gas pressure, it is reported a synergetic effect in the bacteria destruction of gas temperature and density of N and O–atoms with UV emission. Also, it has been found that the macromolecules from the broth, surrounding the bacteria, reduced the bacteria destruction.

In the Corona post-discharge at atmospheric pressure, the $N_2(A)$ and $O(^1D)$ metastables interacted in addition to the N and O atoms in the bacteria (spores) inactivation.

The transmission of N-atoms through membranes, in view to preserve the sterilization of medical instruments, have been measured in the microwave post-discharge reactor. The best results (transmission of 43 %) were obtained with the polypropylene membrane without a marked etching effect after 30 minutes of treatment.

References

[1] Kulich W 1993 *Surf. Coat. Technol.* **59** 193
[2] Normand F, Granier A, Leprince P, Marec J, Shi M K and Clouet F 1995 *Plasma Chem. Plasma Process.* **15** 173
[3] Ricard A, Oseguera-Pena J, Michel H and Gantois M 1990 *IEEE Trans. Plasma Sci.* **18C** 940
[4] Gaillard M, Raynaud P and Ricard A 1999 *Plasmas Polym.* **4** 241
[5] Moreau S, Moisan M, Tabrizian M, Barbeau J, Pelletier J, Ricard A and Yahia L'H 2000 *J. Appl. Phys.* **88** 1166
[6] Villeger S, Cousty S, Ricard A and Sixou M 2003 *J. Phys. D* **36** L60
[7] Chapman B 1980 *Glow Discharge Processes* Wiley
[8] Moisan M and Pelletier J 1992 *Microwave excited plasma* Elsevier
[9] Massines F, Segur P, Gherardi N, Khamphan C and Ricard A 2003 *Surf. Coat. Tech.* **8** 174–5
[10] Pointu A M, Ricard A, Dodet B, Odic E, Larbre J and Ganciu M 2005 *J. Phys. D* **38** 1905
[11] Boudam M K, Saoudi B, Moisan M and Ricard A 2007 *J. Phys. D* **40** 1694
[12] Czerwiec T, Gavillet G, Belmonte T, Michel H and Ricard A 1996 *J. Phys III* (Paris) **6** 1205
[13] Cousty S, Villeger S, Sarrette J P, Ricard A and Sixou M 2006 *Eur. Phys. J. AP* **34** 143
[14] Gordiets B and Ricard A 1993 *Plasma Sources Sci. Tech.* **2** 158
[15] Cariou S, Gauthier M and Glénat M C *Decontamination, bionettoyage, desinfection, sterilisation* Edit. Masson (4th ed. 2000)
[16] Block S S *Desinfection, sterilization and preservation* Edit. LWW (5th ed. 2001)
[17] Villeger S, Ricard A and Sixou M 2004 *Eur. Phys. J. AP* **26** 203–8
[18] Villeger S, Sarrette J P and Ricard A 2005 *Plasma Process. Polym.* **2** 709
[19] Villeger S, Sixou M, Durand J and Ricard A 2006 *J. Phys. D* **39** 3826
[20] Villeger S, Sixou M, Durand J and Ricard A 2007 *Plasma Process. Polym.* **4** 150
[21] Ricard A, Gaboriau F and Canal C 2008 *Optical spectroscopy to control a plasma reactor for surface treatments.* Surf. Coatings Tech.
[22] Canal C 2005 *PhD Thesis - Universitat Politècnica de Catalunya* http://www.tesisenxarxa.net/TDX-0426107-123209/index.html

Journal of Physics: Conference Series **207** (2010) 012007 doi:10.1088/1742-6596/207/1/012007

Study of Dielectric Barrier Discharges in Nitrogen and Afterglows at Atmospheric Pressure

F Clément[1], E Panousis[1], A Ricard[2], E Lecoq[1], J-F Loiseau[1], B Held[1]

[1] Université de Pau – LEGP, IPREM - 64000 Pau, France
[2] Université Paul Sabatier – LAPLACE - 31000 Toulouse, France

Abstract. This article is devoted to the study of Dielectric Barrier Discharges (DBD) and afterglows obtained at atmospheric pressure in nitrogen. Electrical and optical analysis of the DBD reactor are carried out in order to characterize specific activated gaseous species, which can travel with low energy dispersion for long distances in tubes. A hypothesis for the explanation of the presence of long-lived species in such afterglows is finally explained as related to a solitary wave effect.

1. Introduction

Dielectric Barrier Discharges (DBD) are currently used in order to produce non-thermal plasmas at atmospheric pressure. Introduced for the first time by Siemens [1], DBDs consist in depositing on, or between HV and grounded electrodes, one or two dielectrics in order to avoid arc transitions. Depending on the gas and on the HV power supply used, such discharges are usually multi-filamentary but can exhibit diffuse characters [2].

In this work is shown a DBD reactor working in nitrogen and supplied by a pulsed-like HV generator. After a brief description of the set-up, experimental diagnostics allow characterizing the excited gas and a particular attention is given to the afterglows obtained by this way. At last a particular explanation of the formation of luminous afterglows in tubes at atmospheric pressure is given supposing a solitary wave effect.

2. Experimental set-up

Figure 1 shows a schematic representation of the experimental set-up. It can be seen that it is possible to connect an adaptor at the exit of the DBD reactor in order to channel the gas and obtain afterglows at atmospheric pressure.

The DBD reactor has been conceived in a coaxial configuration with a gap of 1mm and a discharge area of 1cm in length. The inner electrode is supplied by a pulsed-like HV generator and the outer is grounded. An alumina coating has been applied on the HV electrode in order to cover it by a dielectric with constant permittivity ($\varepsilon = 9$). Rounded profiles have been machined on the grounded electrode in order to avoid point/edge effects (local enhancement of the Electric Field) between the extremities of the inner electrode where no dielectric coating is applied and the grounded electrode as it is shown in figure 1 (see the sketch of the DBD reactor).

The HV generator is a bipolar pulsed-like supply which delivers high voltage pulses with residual oscillations as it can be seen on figure 2a. Nevertheless, this HV power supply allows adjusting numerous parameters as:

- the frequency in the [8–180 kHz] range,

- the HV amplitude in the [0–8 kV] range,
- the rise times in the [0.2–0.6 kV/ns] range,
- the pulse duration in the [1–5 μs] range.

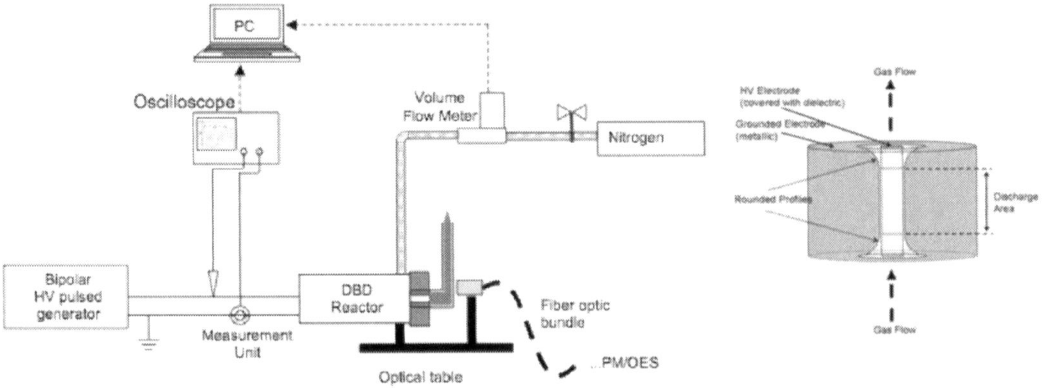

Figure 1. Experimental set-up and sketch of the DBD reactor.

The applied voltage $v_m(t)$ is measured by a 1/1000 Tektronik P6015A voltage probe while the total current $i_m(t)$ flowing in the circuit is measured by a rogowski-type current probe (Tektronix CT-2). The electrical signals are visualized by means of a Tektronix 3054B oscilloscope (500 MHz, 5 Gs/s). By using an electrical equivalent of the system (figure 2b) it can be deduced from the measurements of the delivered voltage $v_m(t)$ and of the total current $i_m(t)$ the gap voltage $v_g(t)$, the discharge current $i_z(t)$, and the plasma power $p(t)$.

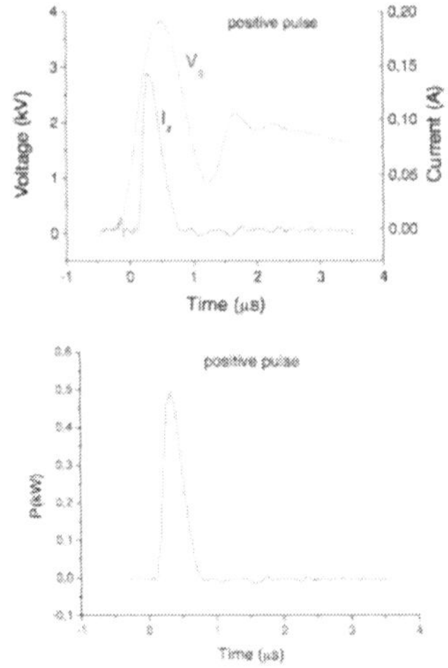

Figure 2a. Electrical voltage, current, and power waveforms for the positive pulse of the bipolar HV pulsed-like generator.

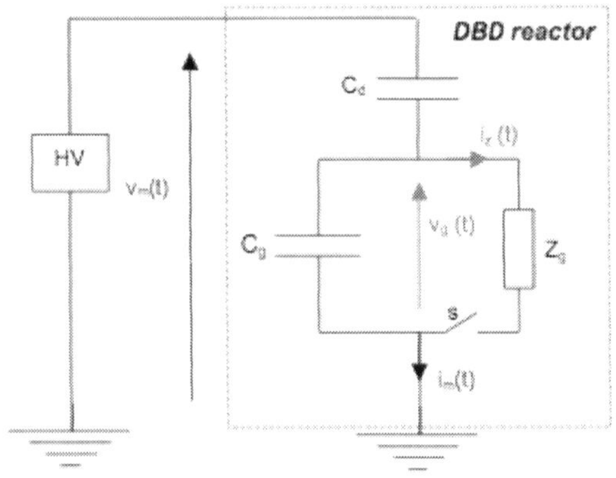

Figure 2b. Electrical equivalent of the DBD reactor (C_d: dielectric capacity, C_g: gas capacity, Z_g: dissipative component in the gas).

This way, the mean plasma power can be deduced during one period of the electrical signals and related to the discharge surface: the mean plasma power density can thus be adjusted in the [1–50 W/cm^2] range.

The gas used is nitrogen and is delivered by a cryogenic vessel (purity 99.995 %) in the axial direction of the electrodes. Due to the configuration of the experimental device, the gas is evacuated in the working room so that the gas is composed of oxygen impurities. The gas flow is controlled by a gas flowmeter at the inlet of the reactor and can be varied from 1 to 100 slm.

Optical analysis of the discharge and of the afterglow have been conducted with a CCD equipped JY-1000M spectrometer with a 1200 gr/mm grating blazed at 500 nm. In each case, a fiber optic is used and the light is focalized by means of an optical lens.

3. Optical Emission Spectroscopy

The analysis of the inside of the discharge by emission spectroscopy indicates mainly the presence of the Second Positive System (SPS) of the molecular nitrogen, as shown in figure 3.

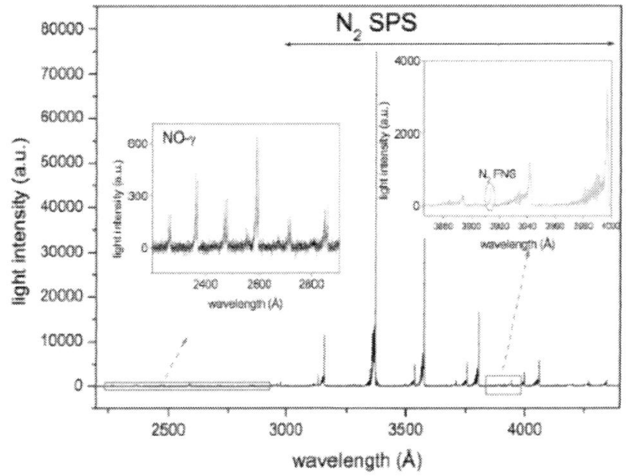

Figure 3. OES diagnostic inside the nitrogen DBD.

The N_2 SPS emissions are due to the following reaction:

$$N_2(C) \rightarrow N_2(B) + h\nu \qquad (1)$$

and its production is characteristic of electronic collisions in the gap with molecular nitrogen. This known result [3–5] is confirmed with the following figure 4 where the correlation can be established between the main emission of the SPS system at 337.1 nm and the mean plasma power. The higher the applied voltage the highest the mean plasma power and, consequently, the highest the electronic collisions in the gas which finally produce higher concentrations of the molecular state $N_2(C)$.

Moreover on figure 3 it can be seen the presence of the First Negative System of the molecular nitrogen which is characteristic of ionic charges in the gap [3–5] and confirm the interpretation of the N_2 SPS.

Finally in the UV part of the spectrum, figure 3 indicates the presence of NOγ emissions which may be due to the oxygen impurities naturally contained in our atmospheric DBD reactor but which are characteristic of the presence of the metastable excited stated $N_2(A)$ of molecular nitrogen by the following mechanisms [3–5] :

$$N_2(A) + NO \rightarrow NO(A) + N_2 \qquad (2)$$

$$NO(A) \rightarrow NO(X) + h\nu \,. \qquad (3)$$

The addition of a metallic adaptor at the exit of the DBD reactor allows to connect a quartz glass tube of several mm in diameter and to channel the excited gas (figure 1). It is remarkable that by this way it is possible to obtain luminous afterglows of several meters at atmospheric pressure and in numerous materials like polymers, glass, quartz and metallic too. Photo 1 exhibits a nitrogen afterglow obtained in a silicone supple tube of 1 meter in length.

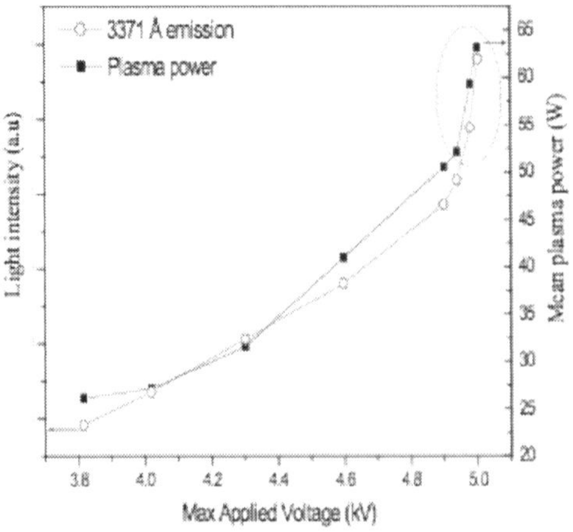

Figure 4. Correlation between the mean emission of the N_2 SPS at 337 nm and the mean plasma power.

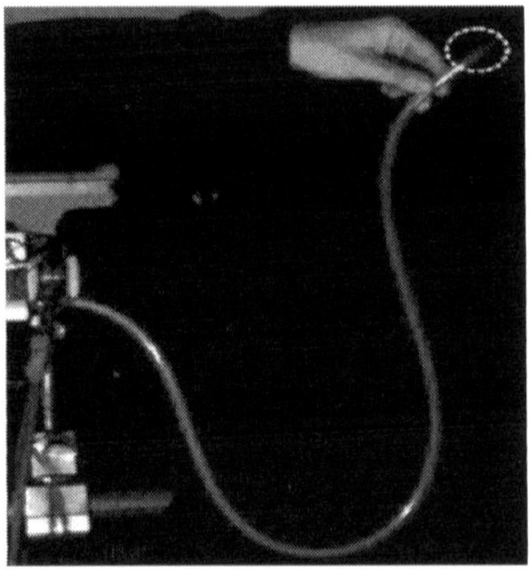

Photo 1. Nitrogen afterglow obtained at atmospheric pressure.

It can be noticed that this afterglow is produced at low temperature and consequently can be applied for useful applications of biological interest like decontamination and sterilization in order to obtain killing effect on parasites, bacteria, fungi, etc [6–10].

The diagnostic of the afterglow by optical emission spectroscopy allows identifying the species contained in such channel tubes. Figure 5 is a spectrum obtained in a quartz tube of 10 mm in diameter and 50cm in length. It is shown like in figure 3 the N_2 SPS which is characteristic of the reaction (1). Nevertheless in afterglows the gas is in remote conditions far from the discharge so that it can be considered that no charges are still present in order to produce another plasma. The mechanism supposed to obtain the N_2 SPS is thus the pooling of the metastable $N_2(A)$ as described in reaction (4) [11]:

$$N_2(A) + N_2(A) \rightarrow N_2(C) + N_2. \quad (4)$$

This is confirmed by the fact that the N_2 First Negative System is now not observed in such conditions.

In the UV part of the spectrum appears again the characteristic emission bands of the NO-γ system but also of the NO-β system which is an indicator of the presence of nitrogen and oxygen atoms:

$$N + O + N_2 \rightarrow NO(B) + N_2 \quad (5)$$

$$NO(B) \rightarrow NO(X) + h\nu \quad (6)$$

Specific characteristic bands denoting the presence of impurities like NH and CN components can also be detected. The formation of such molecules can be explained by the presence of water and by carbon contamination in the set-up from the DBD reactor to the quartz tube.

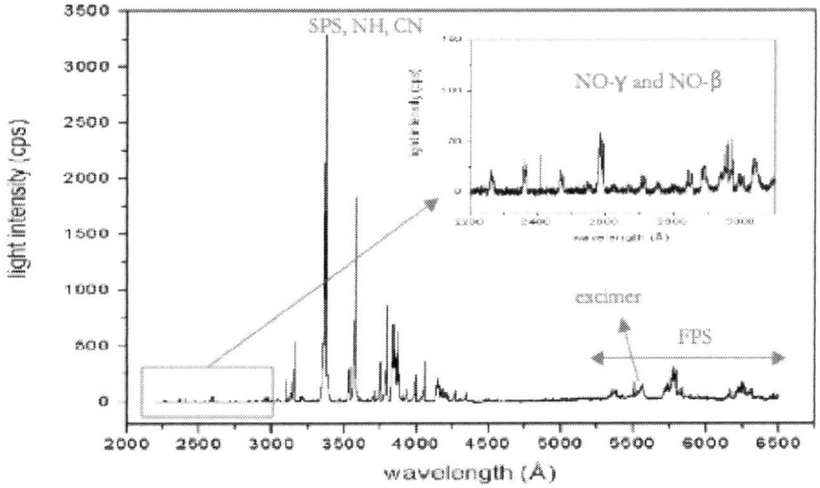

Figure 5. OES diagnostic of the nitrogen afterglow.

But in comparison with figure 3, new bands appear which have been identified to correspond to the emissions of the First Positive System of molecular nitrogen [3–5]:

$$N + N + N_2 \rightarrow N_2(B,11) + N_2 \quad (7)$$

$$N_2(B,11) \rightarrow N_2(A,7) + h\nu \,. \quad (8)$$

FPS emissions allow determining the nitrogen atom concentration in the afterglow as it has previously been achieved by the NO-titration method [12–13]. In our conditions the atomic nitrogen concentration has been estimated to be comprised between 4×10^{14} and $10^{15}cm^{-3}$ and it can be deduced the atomic oxygen concentration which is comprised here between 10^{10} to $10^{11}cm^{-3}$. In comparison with the work of Pointu et al. [14] using another atmospheric pressure reactor, it seems we obtain similar concentrations. Finally figure 5 indicates the presence of the aurora green band characteristic of the excimer $N_2O(^1S)$ as described in reactions (9) and (10):

$$O(^1S) + 2N_2 \rightarrow N_2O(^1S) + N_2 \quad (9)$$

$$N_2O(^1S) \rightarrow N_2 + O(^1D) + h\nu \,. \quad (10)$$

This excimer band progressively appears in the quartz tube denoting a transient phenomenon before the establishment of a steady state regime. This phenomenon has been previously described [15] and seems to be due to a progressive treatment of the tube walls allowing the channelling of the afterglow with low energy dispersion and explaining why it is possible to obtain so long afterglows at atmospheric pressure.

By using an analogy with hydrodynamics solitary laws obtained for the description of the blood flow in vessels of human bodies or for the description of the 'Mascaret', it is possible to describe the afterglows by a solitary wave effect. ('Mascaret' is a natural phenomenon observed in a few rivers and corresponds in the formation of a spectacular wave with constant amplitude moving against the river's flow). In order to analyse the phenomenon, it has been observed that the transient regime of afterglow formation depends of the nature of the tube, of its dimensions, and of the surface state of the internal walls. X-photoelectron spectroscopy (XPS) on the internal walls of the quartz tube have shown an increase of the O/C ratio by a factor ~4. Consequently it may be assumed that the transient phase progressively treats the tube walls in order to form a new gaseous channel called 'soliton tube', as shown in figure 6, in which the excited species may travel with very low energy dispersion. By this way surface recombination of the excited state are probably minimized and only the volume recombination mechanisms could participate to the extinction of the luminous afterglow. Nevertheless afterglows exhibit luminous activity after several meters in the tube. Thus in our conditions, volume recombination is not so consequent. The explanation suggested is that the gaseous channel formed in the tube allows conserves an important quantity of metastables like $N_2(A)$ which glide without being destroyed.

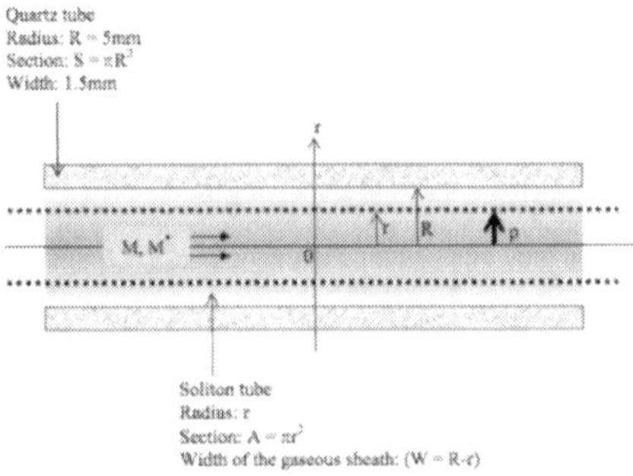

Figure 6. Formation of the "soliton tube" (M, M*: gaseous species, excited or not, p: gas pressure).

By this way it is possible to have an analytical solitary description of the excited gas flow in the tube, as indicated in equation (11):

$$\frac{\partial A_1}{\partial \eta} + \frac{3}{2} A_1 \frac{\partial A_1}{\partial \chi} - \frac{1}{2} \frac{\partial^3 A_1}{\partial \chi^3} = 0 \cdot \qquad (11)$$

In this Korteweg-de-Vries type equation, dimensionless parameters correspond to: A_1 the section of the soliton tube, η to a temporal dimension and χ to a spatial dimension. The low non linear term of this equation compensates the low energy dispersion and consequently may permit to form such systems like afterglows at atmospheric pressure.

9. Conclusion

In this work DBD and afterglows obtained at atmospheric pressure in nitrogen have been experimentally studied. Specific nitrogen and oxygen excited species have been identified and quantified and the obtained results are in agreement with similar studies [14]. These analysis have been completed with a soliton model in order to describe the afterglows behaviours with low energy dispersion and conservation of the long-lived excited species like the metastable state of nitrogen $N_2(A)$.

References

[1] Kogelschatz U et al 1999 *Pure Appl. Chem.* **71** (10) 1819–28
[2] Kogelschatz U 2002 *IEEE Trans. Plasma Sci.* **30** (4) 1400–08
[3] Herzberg G 1989 *Molecular Spectra and Molecular Structure: Spectra of Diatomic Molecules* vol 1 (USA: Krieger Publishing)
[4] Hartmann G 1977 *PhD Thesis, Paris-Sud University – France*
[5] Ricard A 1996 *Plasmas Réactifs* SFV France
[6] Fridman G et al 2007 *Plasma Process. Polym.* **4** 370–5
[7] Moisan M et al 2001 *Int. J. Pharm.* **226** 1–21
[8] Laroussi M 2002 *IEEE Trans. Plasma Sci.* **30** (4) 1409–15
[9] Kieft I E et al 2006 *IEEE Trans. Plasma Sci.* **34** (4) 1331–6
[10] Moreau S et al 2000 *J. Appl. Phys.* **88** (2) 1166–74
[11] Piper L G 1988 *J. Chem. Phys.* **88** 231–9
[12] Ricard A et al 2008 *Eur. Phys. J-Appl. Phys.* **42** (1) 63–6
[13] Boudam M K et al 2007 *J. Phys. D: Appl. Phys.* **40** (6) 1694–711
[14] Pointu A M et al 2005 *J. Phys. D: Appl. Phys.* **38** 1905–9
[15] Clément F et al 2008 *J. Phys. D: Appl. Phys.* **41** 085206 11pp

Parametric study of afterglows issued from Dielectric Barrier Discharges in nitrogen and air and applied on *Aureobasidium* fungi

E Lecoq[1], C Leclaire[2], F Clément[1], G Orial[2], E Panousis[1] and A Ricard[3]
[1]LEGP - IPREM - Université de Pau, 64000 Pau, France
[2]LRMH - 29 rue de Paris, 77420 Champs-sur-Marne, France
[3]LAPLACE - Université Paul Sabatier, 31000 Toulouse, France

Abstract. This article is devoted to the study of afterglows issued from two Dielectric Barrier Discharge reactors driven by two distinct High Voltage generators. The obtained afterglows are applied on *Aureobasidium pullulans* fungi in view of a curative treatment. In a first part, the electrical characteristics of the discharges obtained with both systems are studied and in a second part are presented the results of the fungi treatment efficiency.

1. Introduction

Last years, biological decontamination by plasma discharges or by plasma afterglows has shown very promising results [1–3]. In most of the cases, treatments are applied on bacteria such as *Escherichia coli* or *Bacillus subtilis*, but any study about the inhibition of fungi via plasma treatments was found. Thus, our study consists in applying afterglows issued from Dielectric Barrier Discharges (DBDs) on the *Aureobasidium pullulans* fungi, which is one of the so-called "blue-stain fungi", responsible for the bluish discoloration of wood. These discolorations do not alter wood mechanical properties but can cause heavy economical losses in sawmills or also to deteriorate some historic monument materials and in that way to destroy parts of our cultural heritage.

The main objective is, for a first step, to set up the experimental device and the characterisation means and to find the best parameters for the wanted curative treatments. Consequently, the efficiency of afterglow decontamination is initially assessed directly on fungal cultures, set on a glass slide, in order to isolate it from the wood substrate and to try to understand the mechanisms involved.

2. Experimental set-up

Figure 1 shows a schematic representation of the electrical set-up. The gases used for the study are mixtures of nitrogen and oxygen, the oxygen percentage being comprised between 0 and 20%. Gas flows are controlled by gas flowmeters at the inlet of the reactor, and the total flow value was kept constant at 80 slm for the whole study. The applied voltage is measured by a 1/1000 Tektronik P6015A voltage probe while the total current flowing in the circuit is measured by a Rogowski-type current probe (Tektronix CT-2). The electrical signals are visualized by means of a Tektronix 3054B oscilloscope (500 MHz, 5 Gs/s). The oscillograms obtained permit the calculation of the average plasma power over a period of the applied HV signal.

In the study, two distinct systems of DBD reactors and HV generators were used and are described in the following parts. In each case, an adaptor is added to the reactor in order to collect the gas exiting from the reactor and to guide it in a quartz tube of 1 cm in diameter. This way, the afterglow is applied on the sample which is placed at 1 cm from the exit of the tube. The sample moves thanks to a custom-made conveyor belt so that the treatment is dynamic and the entire surface can be treated. The duration of the treatment is then defined by the velocity of this conveyor belt.

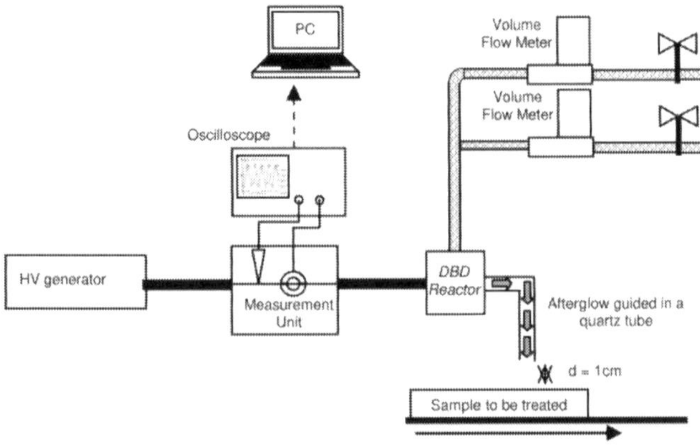

Figure 1. Scheme of the experimental set-up.

2.1 Industrial reactor driven by a quasi-sinusoidal HV generator

The first DBD reactor is an industrial one and is schematically represented on figure 2. It presents a cylindrical coaxial geometry. The HV signal is applied on the inner electrode which is coated with a dielectric, while the outer electrode is grounded. The gas enters through the upper slit, circulates around the inner electrode and exits through the lower one forming a rectangular afterglow which is channelled in the quartz tube. The surface of the discharge is 56.4 cm^2.

This reactor is electrically supplied by a generator which delivers quasi sinusoidal voltage and current waveforms in a frequency value of about 125 kHz in our conditions. An external trigger has been added to this generator in order to chop the signal and to form voltage ON and voltage OFF phases enhancing the variation of the duty cycle. A more detailed description of this generator and of the chopping system can be found in [4].

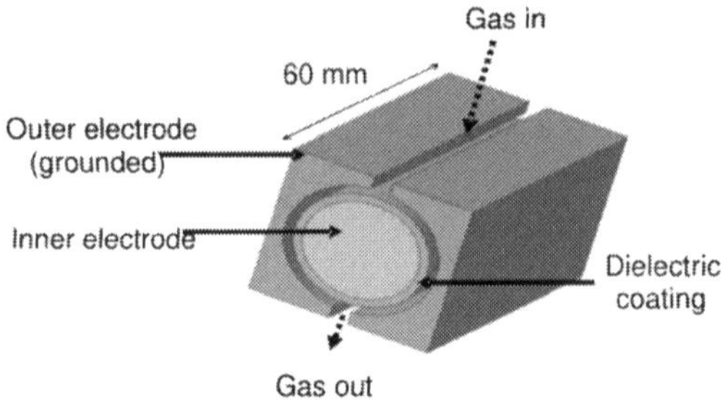

Figure 2. Schematic representation of the industrial reactor used for the first system.

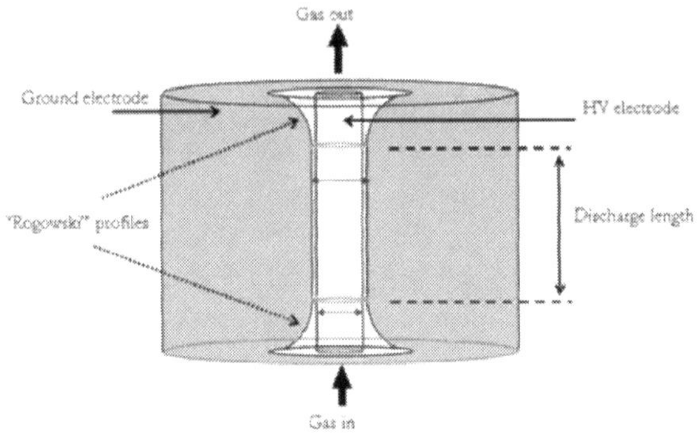

Figure 3. Schematic representation of the custom-made reactor used for the second system.

2.2 Custom-made reactor driven by a pulsed HV generator

For the second system, a custom-made reactor presented on figure 3 was used. As for the industrial one, it is a reactor with a cylindrical coaxial geometry. The coating (alumina) is on the inner electrode, where the high voltage is applied, and the outer electrode is grounded. The gas flow is axial and, again, an adaptor enhances the guidance of the afterglow in the quartz tube. Rogowski profiles have been produced on this reactor in order to avoid parasite discharges. The surface of the discharge obtained with this reactor is equal to 4.15 cm^2.

Journal of Physics: Conference Series **207** (2010) 012008 doi:10.1088/1742-6596/207/1/012008

This custom-made reactor is electrically supplied by a generator which delivers positive and negative impulses. The actual signal presents residual oscillations, but it is nevertheless possible to form ON and OFF phases which allow working in pulsed-type conditions. For a detailed analysis of this pulsed HV generator the reader can refer to [5].

3. Treatment electrical conditions

The treatment electrical conditions were defined for each system in a way to have the highest power density and a temperature of the afterglow inferior to 40° Celsius. Those electrical conditions were kept constant for the whole study and the studied parameters were the percentage of oxygen in the gas mixture and the treatment duration.

With the sinusoidal HV generator, the parameter on which we can play is the total delivered power. It was fixed to 900 Watts. In order to keep the temperature inferior to 40° Celsius, the duty cycle was adjusted to 10%. With the pulsed-type generator, we can vary the frequency and the applied voltage. To be in agreement with our conditions described previously, the frequency value was adjusted to 120 kHz and the maximum applied voltage was around 4 kV.

In order to study the electrical properties of the discharge, the applied voltage and the total circuit current are measured and visualised on an oscilloscope and from Kirchhoff's equations the conduction current (I_z) and the gap voltage (V_g) are calculated and the plasma power can be determined. Figures 4 and 5 represent the typical conduction current, gap voltage, and plasma power waveforms obtained for each system.

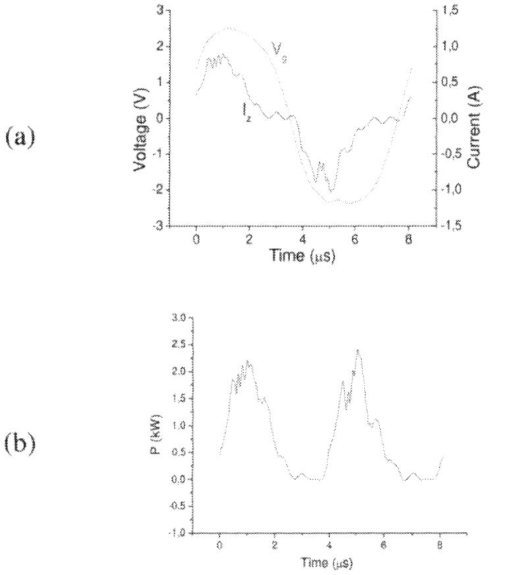

Figure 4. Waveforms of (a) gap voltage and conduction current, (b) plasma power obtained with the first system (on one period of the applied voltage).

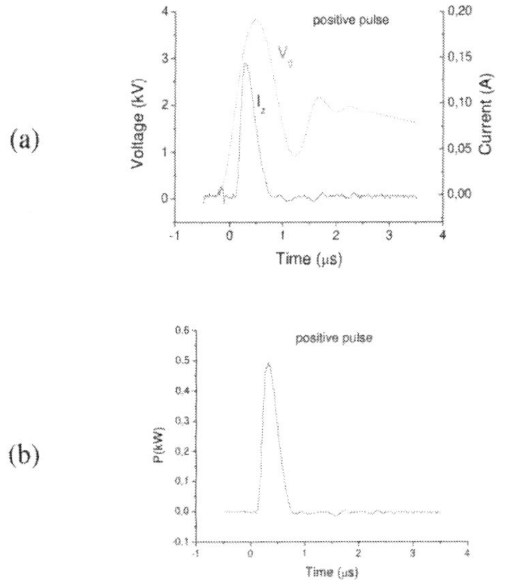

Figure 5. Waveforms of (a) gap voltage and conduction current, (b) plasma power obtained with the second system (for the positive pulse of the applied voltage, the negative one is similar).

Mean plasma powers are, respectively, around 800 and 40 W for the first and the second system but in view of a comparison between the two systems, the surface of the discharge has to be taken into account. That is why mean plasma power densities were estimated following the formula (1):

$$<P_{dens}>=\frac{<P>}{S}, \quad (1)$$

where <P> represents the mean plasma power in watts calculated on one period of the applied voltage and S is the surface of the discharge in cm². Thus, plasma power densities were estimated to be around 15 W/cm^2 for the first system and 10 W/cm^2 for the second one in those electrical conditions.

In view of the application, the parameter which will be representative is the mean energy density injected during the treatment. Energy densities were calculated taking into account the duty cycle r_c, the mean plasma power density $<P_{dens}>$, and the treatment duration t as in formula (2):

$$< E_{dens} >= r_c \times < P_{dens} > \times t \ . \tag{2}$$

For the industrial reactor driven by sinusoidal HV generator, the duty cycle is equal to 10%. Conversely, for the custom-made reactor driven by pulsed HV generator we can consider that the duty cycle is equal to 1 because the mean plasma power density is calculated on one period of the applied voltage, applied 100% of the time. Figure 6 represents the value of mean energy densities injected for both systems as a function of treatment duration.

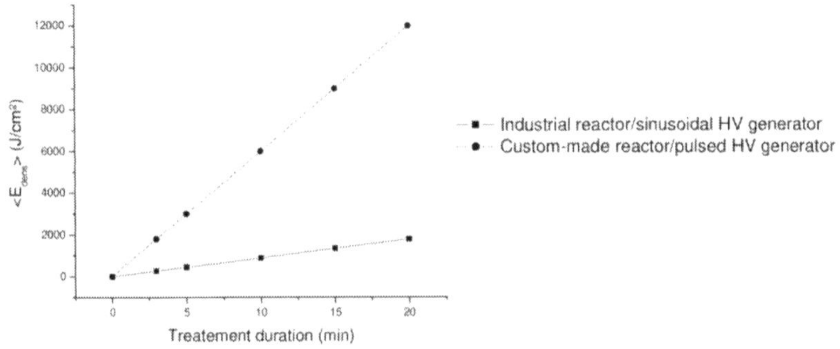

Figure 6. Mean energy densities obtained with each system as a function of the treatment duration.

It can be observed that the mean energy density injected for a given treatment duration is much higher with the custom-made reactor driven by pulsed HV generator than with the industrial reactor driven by sinusoidal HV generator.

4. Ozone measurements

As already mentioned, treatments have been conducted with gas mixtures containing between 0 and 20% of oxygen. As a result, when oxygen is introduced in our system (<1%), the afterglow is not luminescent and optical emission spectroscopy can not be used to study the species present in the afterglows. However, ozone measurements can be conducted, using the absorption properties of this molecule. The objective is here to compare the ozone production as a function of the oxygen percentage for both systems, knowing that ozone can be one (but not the only one) of the chemical species responsible of living cell damaging.

4.1 Experimental set-up

The gas exiting from the quartz tube is brought in a measurement unit (l = 5 cm). A mercury lamp is used and a filter enhances the selection of the 254 nm wavelength, for which the ozone absorption cross-section is the highest. The transmitted intensities are measured at the exit of the measurement unit via a photomultiplier. The ozone number density is calculated using the ozone absorption cross-section at 254 nm and the Beer-Lambert law.

4.2 Results

When the discharge is ignited, we can observe a progressive increase of the ozone density in the measurement cell. After a few minutes, a plateau is reached. Figure 7 represents the ozone densities measured after having reached this steady state as a function of oxygen percentage with both systems.

For the first system, the ozone production first increases with the oxygen proportion, but after 10%, it seems that a plateau is reached. For the second system, the ozone production seems to be constant whatever the oxygen proportion is but the obtained densities are lower compared to the maximum values obtained with the first system.

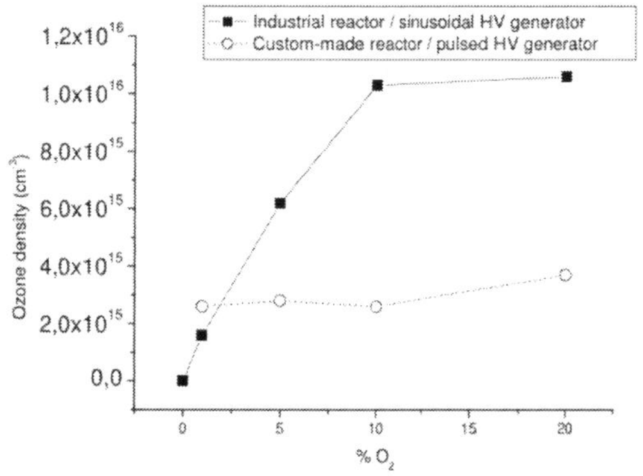

Figure 7. Maximum ozone densities reached as a function of oxygen percentage in the mixture with both systems.

5. Samples preparation

Samples were prepared from ten-day fungi cultures of *Aureobasidium pullulans* which were mixed with 20 ml of sterile water. Then, 100 µl of this liquid culture were spread on microscope slides and let to dry at ambient temperature before being exposed to the afterglow.

6. Characterisation means

After treatments, two characterisation means were used to observe the effect or the afterglow on the fungi: fluorescence microscopy and synthetic culture medium.

6.1 Fluorescence microscopy

The fluorescence microscopy is a technique that can give information about cells viability. Cells are stained with fluorescent dyes: fluorescein diacetate (FDA) and propidium iodide (PI). When observed with a fluorescent microscope, these dyes will, respectively, fluoresce green and red. FDA is used as a viability probe since enzymatic activity is required to activate its fluorescence and because cell-membrane integrity is necessary for its intracellular retention. On the contrary, PI is used for the detection of non-viable cells: it can be taken up only by cells which are dead or have their membrane damaged. Consequently, when cells are observed with fluorescent microscope, viable and non-viable cells, respectively, fluoresce green and red [6]. It is important to remark that this technique can only be used as a qualitative one.

6.2 Synthetic culture medium

The synthetic medium culture is used to observe how fungi can develop themselves and to grow after having been exposed to the afterglow. After treatments, microscope slides are rubbed with a sterile swab which was applied on malt agar plates. After incubation at 24°C during 4 days, colonies are formed and counted in colonies forming unit (CFU). By this way, it is possible to compare the growth of treated fungi with controls.

7. Results

Four runs of experiment were conducted. Table 1 represents the results obtained for the first run. Treatments were conducted with the industrial reactor driven by quasi-sinusoidal HV generator. Treatment duration was constant (15 minutes) and the energy density was consequently equal to 1350 J/cm^2. Gas mixtures containing 0, 5 and 20% of oxygen were tested. Results show that after a treatment conducted with a gas mixture containing 5% of oxygen, few viable cells can be observed with fluorescent microscopy and that any colonies are formed on synthetic culture medium, denoting a good efficiency of the treatment.

Table 1. Results of fluorescence microscopy and synthetic medium culture obtained for the first run of treatments.

Gas mixture N_2/O_2	100/0	95/5	80/20	Control
Fluorescence observations	Presence of viable and non-viable cells	Few viable cells visible	Presence of viable and non-viable cells	Presence of viable and non-viable cells
Synthetic medium cultures	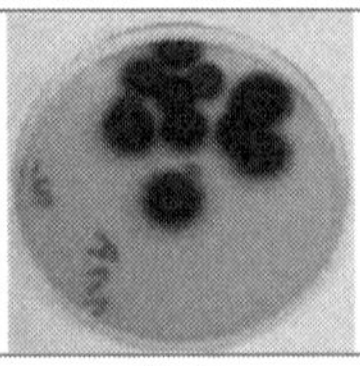	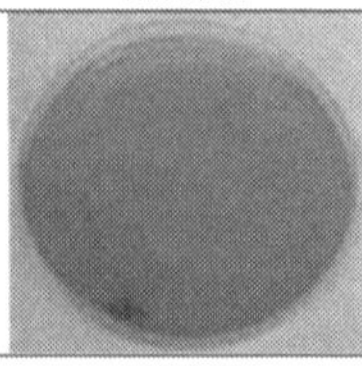	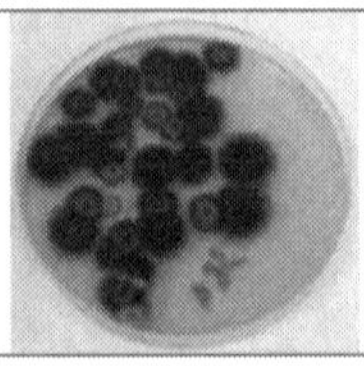	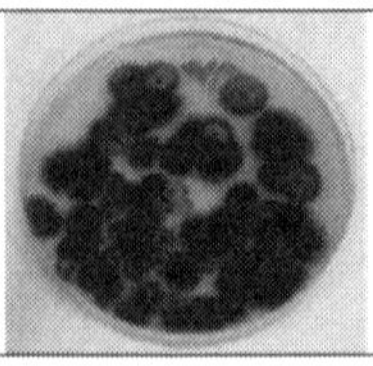
Inhibition rate	92%	100%	75%	–

For the second run, the same system was used. Treatment durations were shorter and equal to 1 and 5 minutes (E_{dens} = 90 and 450 J/cm^2). Five gas mixtures were tested. Figure 8-a represents the CFU number obtained for each gas mixture and treatment duration as compared to the untreated reference. For 1 minute treatments, whatever the oxygen content in the gas mixture, no differences can be observed between the treated samples and the references. However, it seems that after a 5 minute treatment and for gas mixtures containing 1, 5 or 10 % of oxygen, there is a partial inhibition of fungi.

The third run was conducted in the same conditions but for treatment duration and an energy density equal, respectively, to 20 minutes and 1800 J/cm^2. It can be seen on figure 8-b that the treatment conducted with 5% of oxygen seems again to be the most efficient.

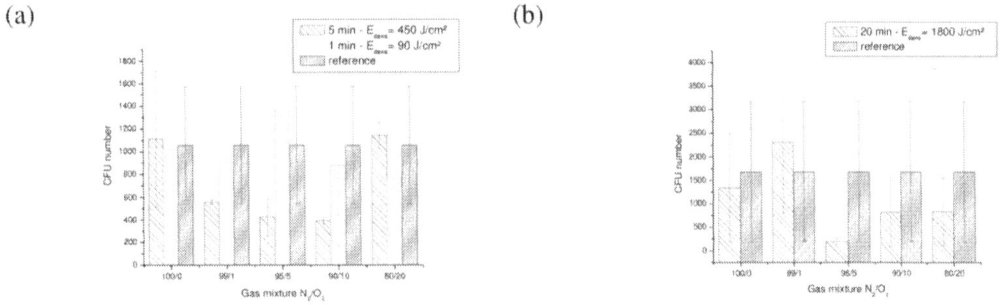

Figure 8. Results of synthetic medium culture obtained for the (a) second and (b) third run.

The fourth run was conducted with the custom-made reactor driven by the pulsed HV generator for the same gas mixtures conditions. Treatment durations were equal to 3 and 20 minutes and, consequently, energy densities were around 1800 and 12000 J/cm^2. The results presented on figure 9 seem to show that 3 minutes of treatment are not sufficient in those conditions to inhibit the fungi growth. Nevertheless, after 20 minutes of treatment, for gas mixtures containing 0 and 5 percent of oxygen a partial inhibition can be observed.

8. Discussion

It can be seen on figures 8 and 9 that the error bars are quite large. In fact, there is a low reproducibility which might be due to a problem encountered during slide preparation. The liquid culture of fungi is not homogeneous because of the presence of spores and hyphae. As a result, the quantity of fungus deposited on the slides is difficult to control. The protocol is now being improved in order to avoid this problem.

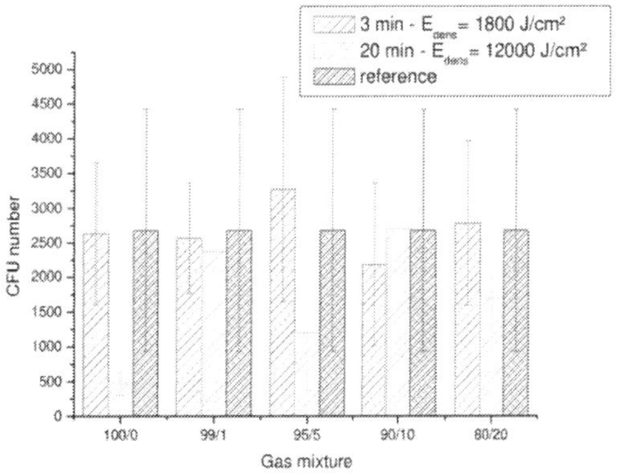

Figure 9. Results of synthetic medium culture obtained for the fourth run.

However, the results show some tendencies. It seems that treatments conducted with a gas mixture containing 5 % of oxygen would be the most efficient and that a minimum of 5 minutes of treatment is required to observe a partial inhibition of fungi.

The electrical study of the two systems showed that the injected energy densities are higher for the second one for an equivalent time of treatment. Nevertheless, for the moment no distinction can be made concerning the efficiency of one system to the other regarding to fungi treatment.

Ozone measurements revealed that ozone production is different for the two systems and also that optimal treatments are not necessary obtained in conditions where ozone densities are the highest, denoting that other species play also a role in fungi treatment.

9. Conclusion and perspectives

To conclude, it can be said that a new curative fungi treatment, conducted in afterglow conditions at atmospheric pressure has been developed. Some of the first results are really encouraging, but the protocol has to be improved in order to draw reliable conclusions.

The influence of other parameters has to be studied and some other characterisation means could be considered in order to go further in the study. For example, scanning electron microscopy could give information about the effect of the afterglow on the fungi at the cell scale. The afterglows have also to be analysed more precisely in order to determine the present species and to try to understand the mechanisms involved during the treatment.

Acknowledgements

This work is financially supported by the French National Research Program ANR-Plasmapal and by a French National Partnership (CIFRE convention: ANRT-Beynel Manustock Society-LEGP).

References

[1] Moreau S, Moisan M, Tabrizian M, Barbeau J, Pelletier J, Ricard A and Yahia L'H 2000 *J.Appl. Phys.* **88** 1166
[2] Moisan M, Barbeau J, Moreau S, Pelletier J, Tabrizian M and Yahia L'H 2001 *Int. J. Pharm.* **226** 1–21
[3] Laroussi M and Leipold F 2004 *Int. J. Mass. Spectrom.* **233** 81–6
[4] Panousis E, Clement F, Loiseau J-F, Spyrou N, Held B, Larrieu J, Lecoq E and Guimon C 2007 *Surf. Coat. Technol.* **201** 7292–302
[5] Panousis E, Clement F, Lecoq E, Loiseau J-F and Held B 2008 Experimental investigation of an atmospheric pressure Dielectric Barrier Discharge (DBD) under bi-polar pulsed high voltage excitation *Proc. GD08 Cardiff*
[6] Jones K H and Senft J A 1985 *J. Histochem. Cytochem.* **33** 77–9

Journal of Physics: Conference Series **207** (2010) 012009 doi:10.1088/1742-6596/207/1/012009

Environmentally Friendly Plasma-based Surface Engineering Technologies

A. Drenik, M. Mozetic, A. Vesel, U. Cvelbar
Jozef Stefan Institute, Jamova 39, 1000 Ljubljana, Slovenia

Aleksander.drenik@guest.arnes.si

Abstract. Weakly ionised reactive plasmas are characterised by a very low degree of ionization, which rarely exceeds the order of 10^{-4}, and by a very high degree of dissociation, which often reaches values above 50%. Thus the most numerous plasma species are free neutral atoms originated from the dissociation of the source gas molecules. Neutral atoms are chemically very reactive species, which makes such plasma suitable for material processing. At the same time the neutral atoms have a low kinetic energy and therefore they cannot penetrate into the bulk material, so their effect is restricted to the topmost atomic layers of the material surface hence weakly ionised, reactive plasmas are suitable for surface engineering. Here we present examples of weakly ionised plasma applications as environmentally friendly alternatives to processes that otherwise utilise aggressive chemicals and produce toxic waste.

1. Introduction

Surface engineering technologies, in which the chemical composition on the surface material is changed while the bulk of the material is left intact, are a growing filed in industry. They can either represent the final stages in an industrial processing or an important step that makes final stage processes possible or more efficient, thus aiding to more environmentally friendly processes. Surface activation, for example, leads to improved adhesive properties of materials, which reduces the amount of paint needed to coat a surface or enable the metallization of composite materials, which, in turn, allows them to be used as substitutes for metals in certain industrial applications, thus eliminating the need for metal.

However, the surface processing itself may not always be environmentally friendly as inducing chemical changes on the surface of a material requires a sufficiently reactive medium. This, as wet chemical treatment, uses acids, lye, solvents or liquid halogenated hydrocarbons, which presents a burden to the environment for several reasons, starting from the consumption of energy and other resources tied to the production and the purification of such chemicals, health hazards during transport, storage and the industrial process itself and to the problem of waste disposal after the process has been completed [1]. In this paper we present the possibility to replace chemical surface engineering technologies by alternatives that do not include the use of said aggressive chemicals.

2. Weakly Ionised Plasma

The plasma used in the processes described in this paper is often referred to as a weakly ionised one. Usually it is created at pressures ranging from 1 to 100 Pa using RF or MW discharges. It belongs to the class of cold plasmas, meaning that the temperature of the electrons is well above the temperatures of other plasma species. Whereas the temperature of electrons is typically between 4 and 5 eV, the ions and the gas remain at room temperature or just slightly above. As the name suggests, the degree of ionisation is relatively low, typically it does not exceed 10^{-4}. However, the degree of dissociation is rather high and often reaches 50 % or even more [2].

As dissociation means breaking up of source gas molecules into neutral atoms, the neutral atoms are thus the most populous plasma species and their characteristics govern the total plasma characteristics. Their main properties are that they are chemically very active while their kinetic energy is small enough so that they cannot penetrate into the bulk of the processed material and thus their effect is restricted to the topmost atomic layers [3]. Therefore, weakly ionised plasmas can be used to induce chemical changes on the surface of a processed material while leaving its bulk properties intact and as such can be a very suitable tool for surface processing.

3. Plasma Based Surface Engineering Technologies

3.1. Surface activation

Surface activation stands for changing the chemical composition of the surface with the intent to modify its adhesion properties. Surface energies of nonpolar plastic materials are raised to levels which ensure a satisfactory adhesion before further processing such as painting, coating, laminating, printing, lacquering, gluing, etc. The increase of surface energies is a combined product of three effects:

- Removal of weak boundary layers.
- Cross-linking of surface molecules.
- Generation of polar groups.

Oxygen atoms remove the surface layers with the lowest molecular weight and oxidise the uppermost atomic layer of the polymer. The oxidation is responsible for the creation of polar groups, which are directly related to the adhesion properties of the surface. Furthermore, oxygen atoms combined with UV radiation help to break up bonds and promote a three-dimensional cross-linking of molecules close to the surface [4–12].

Plasma-based surface activation is already a relatively wide spread application in industry. Examples range from the automobile industry (rear-view mirror housings and bumpers are activated to meet the strict requirements for paint adhesion and polypropylene instrument consoles are activated prior to foaming with polyurethane) to the production of medical equipment (plastic parts of disposable syringes are pre-treated before they are glued to the barrel of the needle and polystyrene containers used in biomedical research are hydrophilised to facilitate wettability by liquids).

3.2. Selective etching

The idea behind selective etching is that different materials have different erosion yields under the same plasma conditions [13]. Therefore, in the case of a two-component material in principle it should be possible to find such conditions for which one of the components is eroded by the plasma atoms while another one is not. Whether a component of the material will be etched depends solely on the chemical properties of the component. In the case of inorganic particles in a polymer matrix, it is possible to achieve a perfect selectivity. Also, since the neutral atoms are propagated through space only by diffusion, such etching is isotropic, meaning that even surfaces with complex shapes will be etched (almost) as well as perfectly flat surfaces.

3.2.1. *Study of pigment dispersion in powder coatings.* Powder coating (PC) is a method of applying coatings to surfaces without using solvents. The coating is applied as a free flowing powder and heated to cause it to flow and adhere to the surface. The powder may be a thermoplastic or a thermoset polymer. It is normally used to create a hard finish on metals. Powder coating is mainly used for coating metals, such as "white goods", aluminium extrusions, and automobiles, although some other materials (like MDF – medium-density fibreboard) also can be coated in this way.

The most common polymers used are polyester, polyester-epoxy (known as hybrid), fusion-bonded epoxy, and acrylics. The majority compound of a powder coating mixture is the binder polymer, which amounts to around 70 mass %. The pigment particles typically represent from 0.1 to 10 mass %. The rest is the filler (usually salts such as $CaCO_3$ or $BaSO_4$), which amounts from 10 to 20 mass % depending on how many different pigments are used.

Powder coating is popular because compared to other coating technologies (procedures) the emission of volatile organic compounds is much smaller, making it an environmentally much friendlier and more acceptable technology.

One of the technological issues of powder coating is the dispersion of the pigment particles. When they are not dispersed well enough – meaning that they form relatively large clusters – the visual properties of the coating are affected (reduced gloss and increased haze values) as well as the viscosity of the final PC system. Needless to say that a good pigment dispersion is desired.

The dispersion of a pigment in powder coatings is usually incomplete. It is influenced by the properties of the pigment itself, by the binder and, to a large extent, by the extrusion process. Several properties of the coating formulation influence the dispersing process, namely the viscosity of the binder, its ability to wet the pigment surface, the attractive forces between the primary particles and the production process, i.e., pre-mixing, extrusion, and grinding. Sufficient pre-mixing of the ingredients is especially important for materials that are used in small quantities. Pigment particles are broken down to primary particles during the extrusion process, which typically lasts for 20 – 30 s. Differences in the extrusion equipment and extrusion shear force also influence the final state of the pigment.

In the past years different approaches for improving the dispersion have been used, which include the use of pigment predispersion in the coating formulation, different additives to improve wetting of the pigment surface, and modifications in equipment design. However, for a truly successful improvement of pigment dispersion, an effective method of analysis of dispersion in needed. Unlike classical coatings where the degree of dispersion can be easily observed in situ, the degree of pigment dispersion in powder coatings can only be evaluated at the end of the extrusion process or at the end of manufacture of the powder coatings [14].

While there are certain indirect methods for evaluating the state of the pigment (using colorimetry, etc.) a direct, quantitative method would be preferred.

One of the few suitable methods for quantitative determination of the state of pigment dispersion in powder coatings is the use of scanning electron microscopy (SEM). However, a scanning electron microscope by itself is not enough. Since the microscope makes only an image of the surface, that would not reveal any appreciable information of the degree of dispersion. As mentioned, the pigment particles amount to only a few mass percent of the powder coating, so there are obviously very few visible at the surface that is mostly represented by the polymer binder. Therefore, before the SEM can be used, the pigment particles must be exposed. That means that the binder must be removed from the surface.

A very efficient way of removing the binder and thus exposing the pigment particles is by etching with weakly ionized oxygen plasma. The oxygen atoms etch away the surface layers of the binder while leaving the pigment particles intact. This happens because the binder is made of polymers that oxygen atoms easily transform to volatile products while pigments are mostly metal oxides and as such do not react with oxygen [15].

3.2.2. *Metallization of a graphite grain – polymer composite.* A less studious and a more applicable use of selective etching can be found in the metallization of composite materials [16]. Currently metal-coated parts made of polymer and graphite grain composite are being used as a replacement for metal commutators in electric motor industry. Such parts are lighter than metal ones; also they are chemically more resistive to certain compounds, which make electric motors built with composite parts more suitable for use in environments where aggressive chemicals are present, e.g. gasoline pumps in automobile motors where the gasoline corrodes the copper that is used for conventional commutators.

The standard metallization procedure consists of etching with ultrasound and NaOH, surface activation with palladium, and chemical deposition of nickel. Only then the galvanic deposition of nickel is possible. While galvanic deposition is a fairly inexpensive and efficient process, the chemical deposition and palladium activation are not. Weakly ionized oxygen plasmas successfully replace the classical procedure. The oxygen atoms not only etch the polymer matrix as seen on figure 1, which replaces the NaOH etching, but also activate the carbon particles, hence eliminating the need for palladium baths. Thus weakly ionized plasma technology offers an environmentally friendlier and a less expensive alternative to the classical procedure.

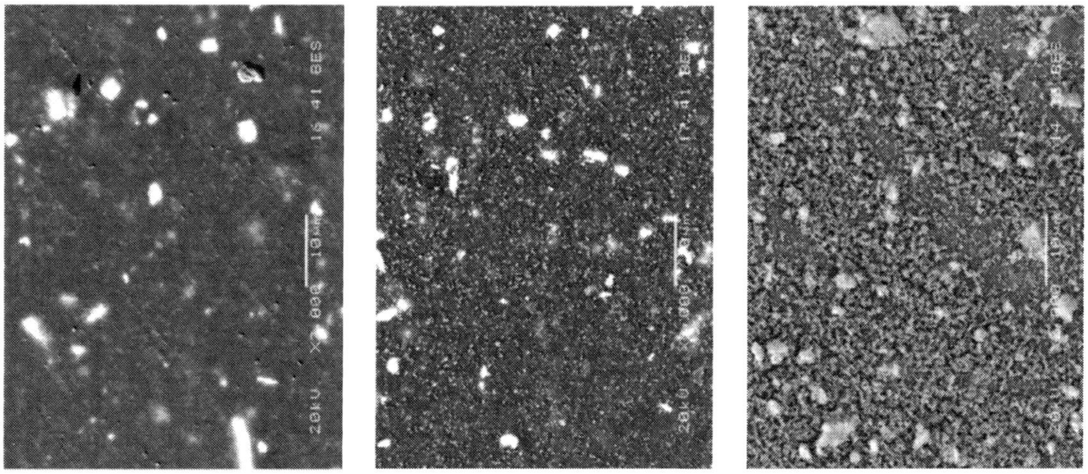

Figure 1. Selective plasma etching of a powder coating (epoxy polymer) sample by weakly ionized, highly dissociated oxygen plasma. The SEM images represent samples (from left to right):) untreated, plasma treated for 60 s, and plasma treated for 400 s.

3.3. Surface cleaning

In terms of plasma processing, cleaning can mean both the removal of organic impurities (greases, oils, waxes, and solvents) by oxygen plasmas or the removal of corrosion products (such as oxide layers) by hydrogen plasmas. The first process is not much different from etching, as described in the previous section. Oxygen atoms transform organic molecules on the surface into volatile compounds such as H_20, CO, etc, which detach from the surface and are pumped away. This procedure is widely used in electronics industry where organic residues on metal surfaces strongly hinder the mechanic and electric properties of joints created on those surfaces.

Hydrogen cleaning is employed in order to remove corrosion products from metal surfaces. H atoms form products such as H_2O, H_2S, HCl, etc, which are desorbed from the surface and are pumped away [17]. This procedure is also widely used in industry [18,19], but also in other applications such as the treatment of archaeological artifacts [20–22]. Figure 2 shows Auger electron spectroscopy depth profiles of a copper surface prior to cleaning, after wet chemical treatment, and after combined oxygen and hydrogen plasma treatment.

3.4. Ashing

Ashing is a procedure used prior to analysis such as atomic adsorption spectroscopy (AAS) which is used to test organic matrices for traces of inorganic compounds. As concentrations of said inorganic compounds can be below the level of detection, the organic matrix has to be removed until so little material remains that the concentration of inorganic compounds is high enough for a reliable detection [23].

Journal of Physics: Conference Series **207** (2010) 012009 doi:10.1088/1742-6596/207/1/012009

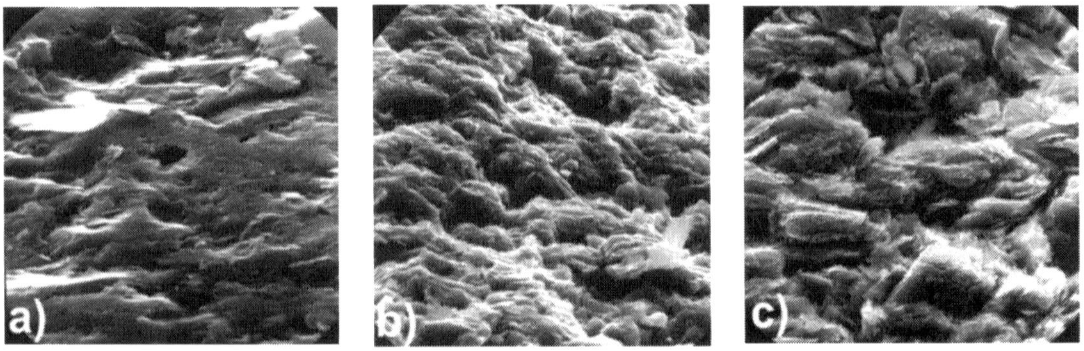

Figure 2. Selective etching of a polymer-graphite composite. SEM images of: a) untreated sample; b) weakly treated sample; c) heavily treated sample. In the first picture it is apparent that the surface consists mostly of the polymer matrix (smooth edges) while on the right hand picture we can see by the increased presence of sharp edges that the carbon particles have been revealed.

Figure 3. Green leaf before (left) and after (right) ashing in oxygen plasma. During the ashing all of the water and the organic matter has been removed yet the structure of the leaf has been very well preserved.

In ashing by oxygen plasma oxygen atoms from the plasma react with the surface material, transforming the organic matrix into volatile molecules (H_2O, CO, etc). Unlike wet ashing, which utilises acids, not only does ashing by plasma cause less damage to the environment but it also eliminates the danger of introduction of other elements into the analysed material during the process of ashing.

4. Conclusion

We have presented weakly ionised plasma as a tool in surface engineering technologies. Due to its high concentration of neutral atoms it is a chemically reactive medium that can be used to induce chemical changes on the surface of the processed material without causing any changes in the bulk of the material. We have presented examples of surface engineering technologies in which plasma processing can offer an alternative to traditional, wet cleaning methods. As plasma treatment eliminates the need of aggressive chemicals, it is an environmentally less damaging procedure.

References

[1] Šorli I, Petasch W, Kegel B, Schmid H and Liebl G 1996 *Plasma Processes Part II: Applications in Electronics MIDEM 26* **2** 113–21
[2] Drenik A, Cvelbar U, Vesel A and Mozetič M 2005 *Inf. Midem 35* 85

[3] Šorli I, Petasch W, Kegel B, Schmid H and Liebl G 2006 *Plasma Processes Part I: Plasma Basics, Plasma Generation MIDEM 26* **1** 35–45
[4] Ostrikov K 2005 *Rev. Mod. Phys.* **77** 489
[5] Dhayal M, Forder D, Parry K L, Short R D and Bradley J W 2003 *Surf. Coat. Technol.* **173–4** 872
[6] *Polymer surface modification: Relevance to adhesion* vol 2 ed Mittal K L 2000 (The Netherlands: VSP)
[7] Esrom H, Seebock R, Charbonnier M and Romand M 2000 *Surf. Coat. Technol.* **125** 19
[8] Belmonte T, Pintassilgo C D, Czerwiec T, Henrion G, Hody V, Thiebaut J M, Louriero J 2005 *Surf. Coat. Technol.* **200** 26
[9] Hegemann D, Brunner H and Oehr C 2003 *Nucl. Instr. and Meth. B* **208** 281
[10] Eon D, Raballand V, Cartry G, Cardinaud C, Vourdas N, Argitis P and Gogolides E 2006 *J. Vac. Sci. Technol. B* **24** 2678
[11] Occhiello E, Morra M, Morini G, Garbassi F and Humphrey P 2003 *J. App. Poly. Sci.* **42** 551
[12] Hody V, Belmonte T, Czerwiec T, Henrion G and Thiebaut J M 2006 *Thin Solid Films* **506–7** 212
[13] Mozetič M 2003 Controlled oxidation of organic compounds in oxygen plasma *Vacuum* **71** 237–40
[14] Kunaver M, Klanjšek-Gunde M, Mozetič M and Hrovat A 2003 The degree of dispersion of pigments in powder coatings *Dyes Pigments* **57** 235–43
[15] Kunaver M, Mozetič M and Klanjšek-Gunde M 2004 Selective plasma etching of powder coatings *Thin Solid Films* **459** 115–7
[16] Cvelbar, Uroš, Mozetič and Miran *Auswählendes Ätzen von Polymermatrix-Verbundstoffen, Munchen: Deutsches Patent-und Markenam*
[17] Mozetič M 2001 *Vacuum* **61** 367–71
[18] DeBoer S J, Dalal V L, Chumanov G and Bartels R 1995 *Appl. Phys. Lett.* **66** 2528
[19] Montgomery J S, Barnak J P, Silvestre C, Hauser J R and Nemanich J R 1995 in *Ultraclean semiconductor processing technology and surface chemical cleaning and pasivation* ed Liehr M, Henys M and Hirose M (Pittsburg: Mater Sci. Soc.) p279
[20] Veprek S, Elmer J T, Eckermann C, Jurčik-Rajman M 1986 *J. Electrochem. Soc.* **31** 29
[21] De Graaf M J, Severens R, Van de Sanden M J F, Schram D C, Meijers H J M and Kars H 1993 *J. Nucl. Mater.* **200** 380
[22] Patscheider J and Veprek S 1986 *Stud. Conserv.* **31** 29
[23] Šorli I, Petasch W, Kegel B, Schmid H and Liebl G 1996 *Plasma Processes, Part III: Surface Activation and Ashing MIDEM 26* **3** 179–86

Diaphragm Discharge in Liquids: Fundamentals and Applications

Frantisek Krcma, Zdenka Stara, Jana Prochazkova

Faculty of Chemistry, Brno University of Technology, Purkynova 118, 612 00 Brno, Czech Republic

E-mail: krcma@fch.vutbr.cz

Abstract. Paper describes results of our research on DC diaphragm discharge creation in water solutions of electrolytes from the viewpoint of its principles, properties and applications. The thermal theory of discharge ignition in bubbles of evaporated solution has been confirmed by both high speed camera and sound records. Static current-voltage characteristics revealed a significant dependence of discharge breakdown on the solution conductivity. Determined breakdown parameters lay in the range of 900–1300 V and 40–100 mA, respectively, for NaCl solution conductivity varying from 150 to 1300 $\mu S \cdot cm^{-1}$. Time resolved electrical characteristics showed a simultaneous appearance of current, voltage and sound oscillations as well as a light emission. Plasma diagnostics by optical emission spectroscopy confirmed formation of reactive species (hydroxyl, hydrogen and oxygen radicals) and excitation of metallic atoms presented in the solution. Moreover, rotational temperature was calculated from the obtained OH spectrum. Experiments focused on the diaphragm discharge applications are outlined in the paper. Results of hydrogen peroxide production, organic dye decomposition and treatment of humic acid solutions are compared from the viewpoint of experimental conditions.

1. Introduction

Electrical discharges generated in liquids are in an intensive research focus for a few last decades. Many scientists deal with various types of discharge configuration and several kinds of applied high voltage. The most frequent electrode geometry used for discharge creation is point-to-plane or coaxial [1–3] together with the application of DC pulsed voltage of about 20 kV. Other experiments in liquids have been carried out with capillary [4] or diaphragm discharge [5]. Discharge generation using AC voltage is also possible [6].

The diaphragm discharge, presented in this paper, is formed in reactor containing two electrodes (cathode and anode) separated by diaphragm in which the pin-hole is made. The DC high voltage is applied on the electrodes. Consequently, the high electric field between two planar electrodes is concentrated just in the pin-hole. If the electric field is sufficiently high, the discharge is formed in the bubbles of evaporated water in and/or near the pin-hole made in the diaphragm. Two kinds of plasma streamers of the opposite polarity are created in two parts of reactor separated each other by the insulating wall containing diaphragm, as it is shown in figure 1. The difference in properties of plasma channels propagating towards the electrodes is not only in their shape, but foremost in the velocity of electrons accelerated by the applied electric field and the energy dissipation in channels [7]. For better understanding of streamer propagation, there is an analogy between the generation of the diaphragm discharge and corona discharge in point-to-plate electrode configuration. The pin-hole behaves as a point electrode of both polarities each one for two separated parts of the reactor. In the part with the plane cathode electrons are accelerated towards the positively charged pin-hole (like to the point anode) and the remaining positive space charge at the end of the streamer is further enhancing the electron velocity. Thus created streamers represent long plasma channels. This form of plasma is called in text as the "positive discharge". Contrary to this, electrons propagating from the negatively charged pin-hole (like from the point cathode) towards the plane anode are dragged by the positive space charge and their velocity decreases. The final shape of negative streamers is represented by shorter bush-like plasma channels having a spherical shape and we called them the "negative discharge" [8], (see figure 1).

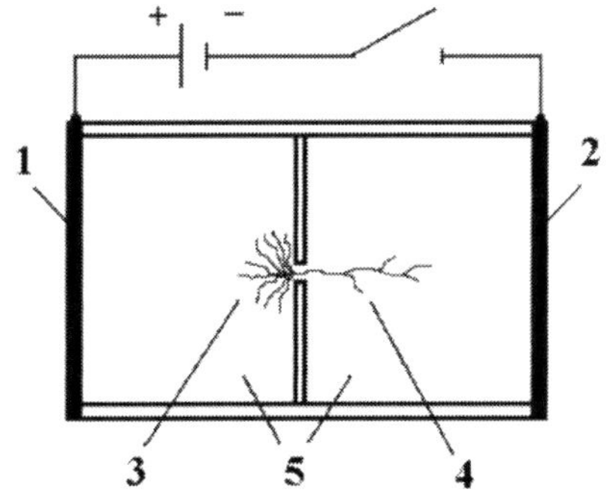

Figure 1. Simplified scheme of diaphragm discharge formation: 1 – anode, 2 – cathode, 3 – negative plasma channels, 4 – positive plasma channels, 5 – electrolyte solution.

Electrical discharges belong to the so-called Advanced Oxidation Processes (AOP´s) that can be utilized for different applications. Their main importance is in the formation of highly reactive species (especially radicals) and production of UV radiation. Subsequently, initiated processes lead to the excitation, ionization and decomposition of particles and compounds contained in treated solution. Thus methods based on electrical discharges are often used in water treatment, surface modification or other applications.

2. Experimental

2.1. Discharge reactor

A specially constructed batch discharge reactor was used in the study of diaphragm discharge formation, properties and applications (figure 2) [9]. The main chamber was divided by an insulating wall into two spaces with one electrode (stainless steel or platinum) in each part. A changeable diaphragm made of inert material (PET, thickness of 0.25 mm) was fixed in the centre of the barrier. Both electrode spaces were connected by a small orifice in the diaphragm (initial diameter of 0.2 mm) where the discharge was ignited. DC high voltage source providing constant power supply up to 300 W was used. Water solutions containing supported electrolyte (NaCl, NaBr, $NaNO_3$, $Na_2HPO_4{\cdot}12H_2O$ or Na_2SO_4; their concentration varied according to the needed conductivity) were split per 1.5 liters into both parts of the reactor. Addition compounds such as organic dyes (textile: Direct Red 79, Direct Blue 106, Direct Yellow 29, food: Acid Blue 74, Acid Red 14; initial concentration of 12–30 $mg{\cdot}l^{-1}$) or humic acids (HUMIN-P 775, initial concentration of 35 $mg{\cdot}l^{-1}$) were added when degradation effects of the discharge were investigated. Both mixing and cooling systems were used to keep experimental conditions homogeneous and constant during the whole process.

Figure 2. Photograph of diaphragm discharge reactor (right) and DC high voltage source (left).

2.2. *Plasma diagnostic and analytic methods*

Breakdown moment of the diaphragm discharge as well as its further behaviour was studied from electrical characteristics (static and time resolved). For that purpose, two and four channel digital oscilloscope Tektronix TDS 1012B and TDS 2024B were used. Signals of discharge voltage (using the HV probe Tektronix P6015A), current, sound and emitted light were recorded. Plasma diagnostics was carried out by optical emission spectroscopy using the spectrophotometer Jobin Yvon Triax 550 in the range of 200–700 nm. Subsequently, rotational temperature was calculated from recorded OH radical spectra.

Detection of hydrogen peroxide generated by the discharge was done by colorimetric method using a specific titanium reagent. In the case of decomposition experiments, concentration of dyes and humic acids was determined by absorption spectroscopy. Moreover, fluorescence spectroscopy was carried out to obtain a humic index of treated humic solutions.

3. Results and discussion

3.1. *Breakdown and formation of diaphragm discharge in electrolyte solution*

According to the thermal theory of discharge initiation in liquid [8], breakdown appears in bubbles of evaporated water solution which is strongly heated by the applied electric field. To confirm this hypothesis in our device, series of photos from high speed camera was correlated with current and sound time resolved records. Obtained pictures showed on the bubble formation [10]. Moreover, determined amplitudes of sound closely corresponded to recorded current peaks (see figure 3). This sound effect could be related to the discharge creation represented by a rapid current increase.

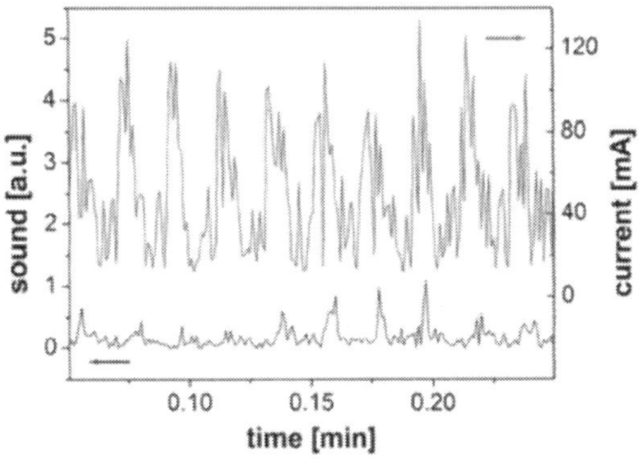

Figure 3. Current and sound records of the diaphragm discharge in 4 mM NaCl solution.

3.2. Electric characteristics

Ignition of DC diaphragm discharge in electrolyte solutions highly depended on the initial conductivity of the liquid phase. Too low conductivity (adjusted by concentration of dissolved electrolyte) meant low amount of charge carriers and subsequently, weak current was insufficient for liquid heating and bubble creation. At the higher conductivities, the system resistance was lower and current was higher. Also the voltage drop in liquid was significantly smaller and thus the discharge breakdown was observed at lower voltage and higher current. Based on these assumptions, breakdown parameters (mean voltage and current) were determined from VA characteristics measured in electrolyte solutions of initial conductivity 50–1400 $\mu S{\cdot}cm^{-1}$. A comparison obtained for two electrolytes (NaCl and KCl) is given in figure 4. It is obvious that in the middle range of tested conductivities (300–1300 $\mu S{\cdot}cm^{-1}$) breakdown voltage was slightly decreasing with the increasing conductivity while breakdown current significantly increased with more or less the same trend in both electrolytes [10]. The extreme values of tested conductivities had a remarkable effect on breakdown appearance foremost due to the consequent discharge instability.

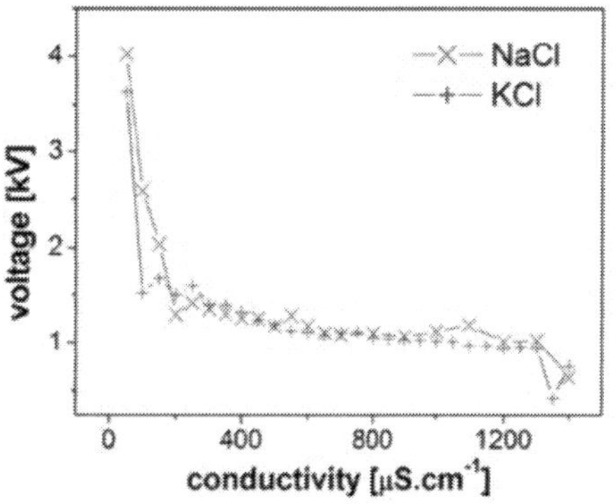

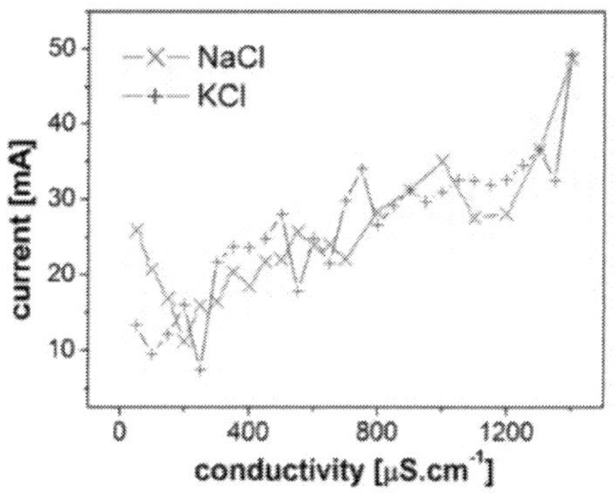

Figure 4. Breakdown values of voltage (left) and current (right) as a function of solution conductivity. Comparison is given for two electrolytes (NaCl and KCl).

An example of characteristics time evolution obtained in NaCl solution during the breakdown moment is demonstrated in figure 5. Increasing the input power into the reactor, voltage and current were slightly increasing and no light emission was recorded until the breakdown moment. Getting over this value, intensive peaks of current appeared together with light flashes while voltage mean magnitude remained more or less constant.

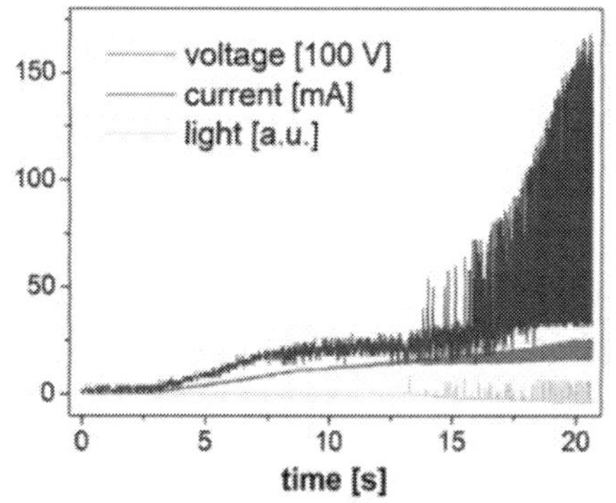

Figure 5. Time resolved characteristics of diaphragm discharge. Given result represents simultaneous records of voltage, current and light signal during the breakdown moment in 4 mM NaCl solution.

3.3. *Optical diagnostics*

Plasma properties were measured by optical emission spectroscopy in the range of 200–1000 nm. Strong emission of OH radicals as well as other reactive species (H and O atomic lines) was determined in the spectrum. Emission of metallic atoms coming from the dissolved electrolyte was observed, too. Figure 6 shows the dependence of sodium atomic line intensity on the concentration of NaCl electrolyte in the solution. The result that the Na intensity was directly proportional to the Na concentration in the solution can be applied as a determination method in the analytical chemistry [11]. Integral emission intensity of OH radicals remained constant over the whole tested concentration range. Additionally, rotational temperature was calculated from OH spectrum and its obtained magnitude was 500–900 K depending on the solution conductivity, applied power and electrolyte kind.

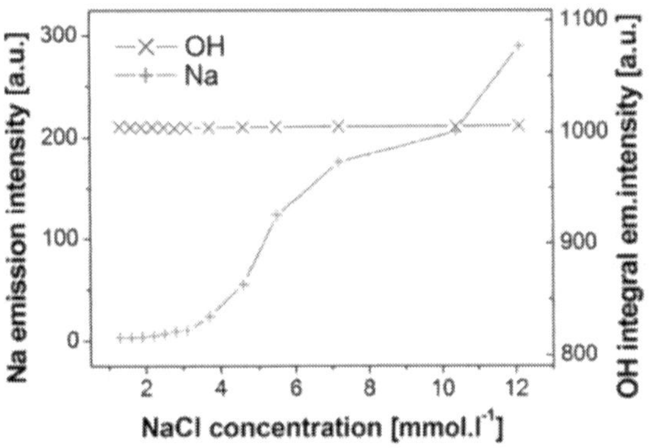

Figure 6. Emission intensity of Na atomic line (589 nm) and integral OH intensity over 305–315 nm as a function of electrolyte concentration (NaCl, input power of 160 W).

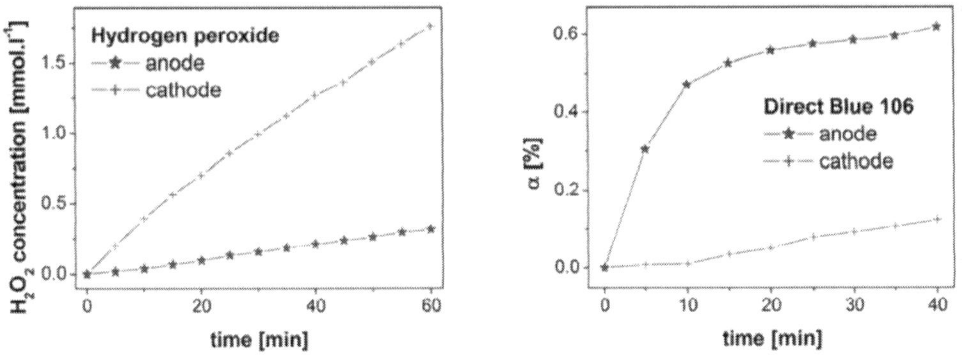

Figure 7. Left: Production of hydrogen peroxide by the diaphragm discharge in both electrode spaces (input power of 200 W, 4 mM NaCl). Right: Decomposition rate of the dye Direct Blue 106 achieved by the diaphragm discharge in both electrode spaces (input power of 170 W, 4 mM NaCl).

3.4. *Example of applications*

Diaphragm discharge generated in water solutions by constant DC voltage can be applied in more fields of utilization. The possibility of an excitation source for determination of metallic atoms in the solution has been already mentioned above [11]. Next useful process studied by our work is production of hydrogen peroxide [12]. Figure 7 (left) demonstrates its formation in both electrode spaces of the reactor. The second picture in figure 7 (right) represents the next discharge application for organic dye decomposition (the used textile dye Direct Blue 106). It is evident that electrode polarity played a significant role in both processes. While H_2O_2 was formed with higher efficiency in the cathode space, decoloration of the dye went faster in the anode space [13]. Plasma treatment of humic solutions led to the enhancement of aliphatic components and decrease of aromatic substances [14,15].

4. Conclusion

Paper describes diaphragm discharge generated using constant DC voltage up to 4 kV in water solutions containing supported electrolytes. Confirmation of thermal theory as well as electric characteristics is presented and an influence of electrolyte (kind and concentration) is discussed. Plasma diagnostics by OES determined rotational temperature and formation of the main radicals (OH, H, O). Application of the discharge in dye and humic acid decomposition, hydrogen peroxide formation and metallic atoms determination is suggested, too.

Acknowledgments

This work was supported by the Czech Science Foundation, projects No. 202/03/H162 and No. 202/07/P371.

References

[1] Sun B, Sato M, Clements J S 2000 Oxidative processes occurring when pulsed high voltage discharges degrade phenol in aqueous solution *Environ. Sci. Technol.* **34** (3) p 509–513
[2] Sugiarto A T, Ito S, Ohshima T, Sato M, Skalny J D 2003 Oxidative decoloration of dyes by pulsed discharge in water *J. Electrostatics* **58** p 135–145
[3] Sunka P, Babicky V, Clupek M, Lukes P, Simek M, Schmidt J, Cernak M 1999 Generation of chemically active species by electrical discharges in water *Plasma Sources Sci. Technol.* **8** (2) p 258–265
[4] De Baerdemaeker F, Simek M, Schmidt J, Leys C 2007 Characteristics of ac capillary discharge produced in electrically conductive water solution *Plasma Sources Sci. Technol.* **16** p 341–354
[5] Monte M, De Baerdemaeker F, Leys C, Maximov A I 2002 Experimental study of a diaphragm discharge in water *Czech. J. Phys.* **52** p D724–D730
[6] Manojlovic D, Ostojic D R, Obradovic B M, Kuraica M M, Krsmanovic V D, Puric J 2007 Removal of phenol and chlorophenols from water by new ozone generator *Desalination* **213** p 116–122
[7] Kuzhekin I P 1995 *Proc. 9th Int. Symp. High Voltage Eng. (Graz)* p 8073–1
[8] Lukes P 2001 Water treatment by pulsed streamer corona discharge *Ph.D. Thesis* Prague
[9] Stara Z 2006 Study of chemical processes in electrical discharges in liquids *Ph.D. Thesis* Brno
[10] Stara Z, Krcma F, Prochazkova J 2008 *Proc. 23rd SPPT (Prague)* p 185
[11] Prochazkova J, Stara Z, Krcma F 2006 Optical emission spectroscopy of diaphragm discharge in water solutions *Czech. J. Phys.* **56** p B1314–B1319
[12] Stara Z, Krcma F 2004 The study of H_2O_2 generation by DC diaphragm discharge in liquids *Czech. J. Phys.* **54** p C1050–C1055
[13] Stara Z, Krcma F, Nejezchleb M, Skalny J D 2008 Influence of solution composition and properties on organic dye removal in DC diaphragm discharge in water *J. Adv. Oxid. Technol.* **11** p 155–162
[14] Stara Z, Krcma F 2006 Treatment of humic acids solutions by DC diaphragm discharge *Czech. J. Phys.* **56** p B1351–B1356
[15] Halamova I, Stara Z, Krcma F 2008 Effects of diaphragm discharge in water solutions containing humic substances *J. Phys., Conf. Series* (Issue of IWSSPP conf. – submitted)

Plasmachemical Removal of Corrosion Layers from Iron in Pulsed RF Discharge

Vera Sazavska, Frantisek Krcma, Tereza Simsova, Nikola Zemanek

Faculty of Chemistry, Brno University of Technology, Purkynova 118, 612 00 Brno, Czech Republic

E-mail: krcma@fch.vutbr.cz

Abstract. Plasmachemical treatment of archaeological artefacts by RF hydrogen low pressure plasma is a relatively new technique developed during the 1980s. The corrosion removal process is very complex and its mechanism is not fully understood up to now and also optimal conditions are not known well. To study the process condition influence, the model corroded samples were prepared on iron with defined roughness using HCl, HNO_3 and H_2SO_4 corrosion agents. The corrosion layers before and after plasma treatment were analyzed by electron microscopy including the study of corrosion layer profile. The experiment was done in Quartz reactor with outer electrodes creating capacitive coupled RF discharge in pure hydrogen at pressure of 170 Pa. The duty cycle was varied from continuous regime up to 1:25. The plasma treatment duration was 40–120 minutes depending on the duty cycle. The results showed that the application of pulsed discharge regime could maintain the corrosion removal process at lower mean power as well and thus the treated samples were affected by lower heating stress. On the other hand, the process was significantly longer.

1. Introduction

Plasmachemical treatment of archaeological artefacts is a relatively new technique developed during the 1980s [1,2]. Although this technique was used for conservation of artefacts which components are mainly metals, the plasmachemical process in itself was used more or less intuitively [3]. The process is based on using low-pressure hydrogen plasma in which the artefacts are treated for several hours, usually in more steps. The problem is how long period of the plasmatic treatment is optimal and how to characterize its efficiency. The ancient artefacts always contain some corrosion layers on their surface. The studies focused on the composition of the layers show that various oxide and chloride complexes, usually containing crystalline water, too, are the dominant compounds of the layer. Bulk corrosion contains especially chlorides and sulphides while surface corrosion is caused by oxygen and water. Each original archaeological object is unique with specific composition of corrosion layers and thus these objects are not ideal for the detail research of the plasma treatment process. The main problem of plasma chemical removal of corrosion layers is the elevation of object temperature. The treatment above 200 °C caused irreversible changes in the metallographic structure of many metals. This could result in serious mistakes when interpreting the original processing of such metal [4]. The sample temperature is proportional to the plasma applied power. So the best choice should be to keep the power as low as possible. On the other hand, too low power is not sufficient for the creation of reactive particles in plasma in needed concentration. The application of pulsed plasma systems could be a possible solution of these antagonistic tasks. This paper gives our first results obtained during the removal of various corrosion structures artificially prepared on iron.

2. Experimental

The flowing RF capacitively coupled discharge in pure hydrogen was used for our study. Its schematic drawn is showed in figure 1 with more details. The pressure was kept at 170 Pa during all the treatment at hydrogen flow rate of 10.2 Sccm. The integral spectrum of OH radical in the range of 305–315 nm was used as the process monitor according to our previous works [5]. The plasma treatment was stopped at the moment when OH integral intensity decreased down to 10% of its maximal value according to [5]. This spectrum was also used for the rotational temperature calculation. The model corroded samples were prepared on iron with defined roughness using HCl vapour (for one week) and HNO_3 and H_2SO_4 acids (by dipping for 5 seconds and one week storage in desiccator). This allowed study of the influence of discharge conditions on the more or less same corroded samples and thus the results were comparable. The corrosion layers were analyzed by electron microscopy including the study of corrosion layer profile.

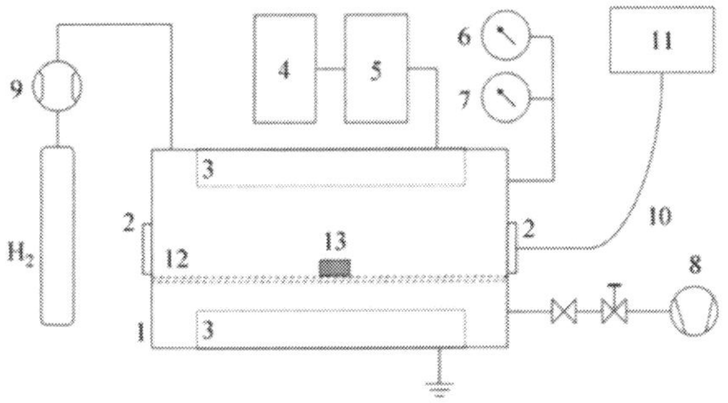

Figure 1. Experimental set-up. 1 – Quartz reactor (i.d. 96 mm, length 900 mm); Quartz windows; 3 – outer copper electrodes; 4 – RF power supply Cesar 136; 5 – automatic matching network; 6 – capacitance gauge; 7 – Pirani gauge; 8 – rotary oil pump; 9 – MFC; 10 – Quartz optical cable; 11 – spectrometer Jobin Yvon TRIAX 550 with CCD; 12 – glass sample holder; 13 – treated sample.

3. Results

The time evolution of OH radical integral intensity during the removal of chlorine based corrosion at different discharge conditions is shown in figure 2. In the case of pulsed discharge at 400 W, a peak at about 13 min after the process beginning is well visible at duty cycle of 1:50 and at 23 min at duty cycle of 1:25. No such peak is visible in the continuous regime. At 500 W this peak has higher intensity and moves to the shorter treatment times (at duty cycle 1:25) and disappears at duty cycle of 1:50. These experimental results show that the corrosion removal is dependent on the total energy supplied into the plasma. The corrosion removal is faster with the energy increase.

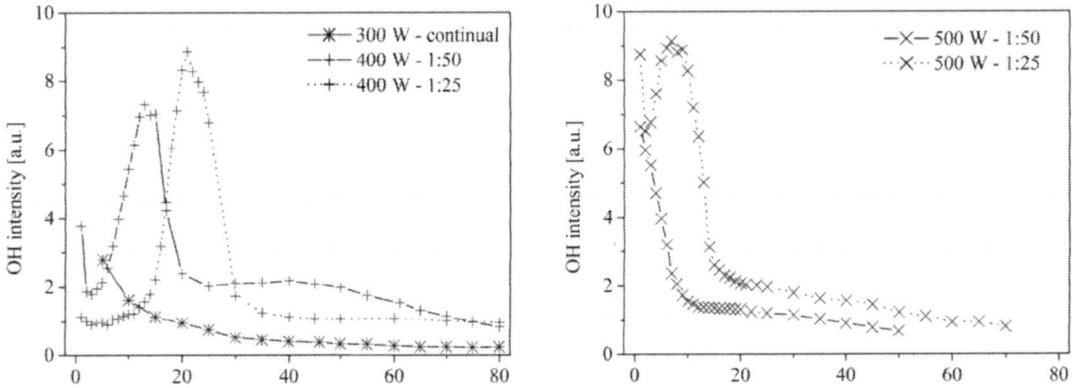

Figure 2. Time evolution of OH integral intensity during the treatment of samples corroded in HCl vapour at different discharge conditions.

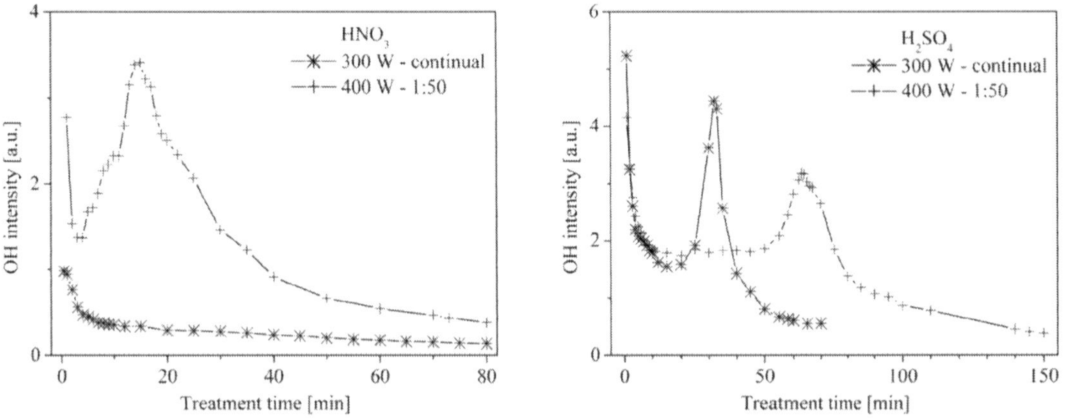

Figure 3. Time evolution of OH integral intensity during the treatment of samples corroded in HNO_3 (left) and H_2SO_4 (right) at different discharge conditions.

Figure 3 shows the similar situation for the other kinds of corrosion. The nitrate corrosion layer removal shows the same behaviour as chlorine ones but the peak of OH intensity is not such sharp and the removal process is slower. Unfortunately, the strong emission of molecular nitrogen second positive system was observed during the treatment in short time interval at the OH radical maximal intensity. The sulphuric corrosion layers are the hardest removable as it is shown in figure 3 – right. Also in this case there is a well visible characteristic peak of OH radical emission but it is observed at significantly later treatment time.

The rotational temperature of plasma was monitored. An example of its time dependence during the treatment is shown in figure 4. Nearly the same results were obtained during treatment of all samples more or less independently on the discharge conditions. It must be noted at this place that this temperature is not the temperature of sample which we have not been able to measure correctly up to now.

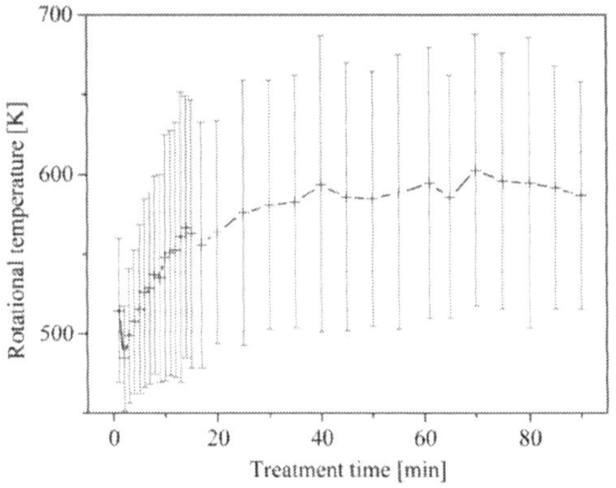

Figure 4. Rotational temperature of hydrogen plasma at 400 W with duty cycle 1:50 during the treatment of sample corroded in HCl vapour.

Finally, figure 5 shows the corrosion layer structure after the plasma treatment and its elementary composition. It is well visible that the chlorine content in the sample surface layers was strongly decreased although it was not removed completely yet.

4. Conclusion

The experimental results show the presence of the characteristic peak in OH radical emission during the plasma treatment of model corrosion layers. Its position during the treatment depends on the corrosion layer kind as well as on the total applied energy into the plasma. Its dependence on the corrosion composition can be dependent on the chemical boundary or on the anion weight. The detail study of this effect is a subject of the contemporary study. The rotational temperature of plasma is slightly increasing during the treatment in the interval 500–600K and it is not dependent on the plasma conditions and corrosion kind. The contents of the corrosion salts in the surface layers of the samples is significantly decreased but not fully removed.

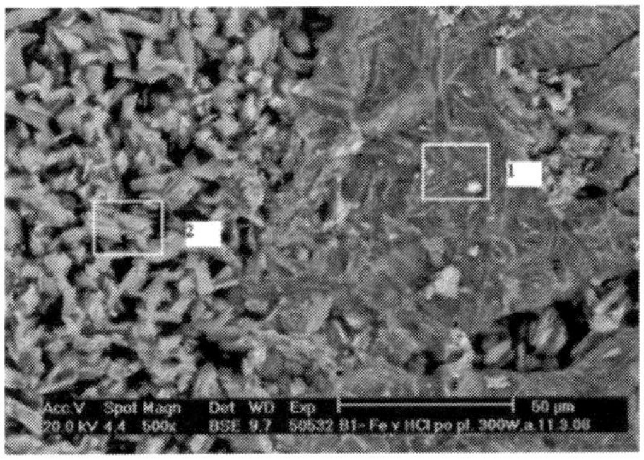

	Non-treated		Treated – position 1		Treated – position 2	
Element	**weight %**	**atomic %**	**weight %**	**atomic %**	**weight %**	**atomic %**
O	11.45	25.70	24.65	52.10	29.78	56.89
Si	0.40	0.51	0.55	0.66	6.71	7.30
Cl	46.39	46.96	5.29	5.05	2.98	2.57
Ca	–	–	0.34	0.28	0.42	0.32
Mn	–	–	1.95	1.20	1.39	0.77
Fe	41.75	26.83	67.22	40.70	50.73	32.15

Figure 5. A result of material analyzes of HCl vapour corroded sample after plasma treatment at 300 W in continuous regime and its comparison with non-treated sample. The sample after treatment was stored for 1 month at ambient laboratory conditions where sample oxidation was not protected.

Acknowledgments
This work was supported by the Czech Ministry of Education, Youths and Sports, Research Plan No. MSM0021630501.

References

[1] Veprek S, Patscheider J, Elmer J 1985 Restoration and conservation of ancient artifacts: A new area of application of plasma chemistry *Plasma Chem. Plasma Proces.* **5** 201–209

[2] Veprek S, Eckmann Ch, Elmer J 1986 Recent progress in the restoration of archeological metallic artifacts by means of low-pressure plasma treatment *Plasma Chem. Plasma Proces.* **8** 445–465

[3] Lieberman M, Lichtenberg A 1994 *Principles of plasma discharges and materials processing* (John Wiley & Sons, New York)

[4] Perlik D 2001 The effect of plasma on metallography and deionization of archaeological iron objects *Proc. Conservator Restorer Symp.* (Ceske Budejovice) 89–95 – in Czech

[5] Raskova Z, Krcma F, Klima M, Kousal J 2002 Diagnostic of plasmachemical treatment of archaeological artefacts *Czech. J. Phys.* **52** D927– D932

Effects of Diaphragm Discharge in Water Solutions Containing Humic Substances

Ivana Halamova, Zdenka Stara, Frantisek Krcma

Faculty of Chemistry, Brno University of Technology, Purkynova 118, 612 00 Brno, Czech Republic

Email: xchalamova@fch.vutbr.cz

Abstract. Preliminary results of research focused on the applications of DC diaphragm discharge in water solutions containing humic substances are presented in this paper. Diaphragm discharge investigated by this work was created in the reactor using constant DC high voltage up to 2 kV that gave the total input power from 100 to 200 W. Presented work investigated decomposition of humic substances by the electric discharge in the dependence of discharge conditions (electrode polarity) as well as solution properties (electrolyte kind, pH). Especially substantial effect of pH on humic acid decomposition has been observed when acidic conditions stimulated the degradation process. Absorption spectroscopy in UV-VIS region together with fluorescence spectroscopy has been used for the detection of changes in humic solutions. Index of humification was calculated from obtained fluorescence spectra and a significant decrease of aromatic components in the humic mixture was determined during the discharge treatment.

1. Introduction

Water pollution is one of the biggest problems of present world. Among these pollutants humic substances are often contained that are otherwise find in soils, ponds and sediments. In the last years, floods are becoming a more frequent phenomenon due to climate changes and tree felling in the surrounding of water flows. Thus, humic acids are more frequent in surface water. Humic acids are not toxic like organic wastes from factories, but they limit water use for purposes in households. In general, humic substances are mixtures of mainly aromatic compounds of high molecular weight (up to 10 000 mass units). Water treatment can be generally realized by three methods: physical, chemical and biological. These methods have disadvantages – insufficient efficiency, production of another ecologically or health risk substances and the resources are too expensive. Therefore interactions of organic compounds with electric discharges are investigated recently. Effects of electric discharges on decomposition of organic dyes [1] or other organic reactants (phenol [2], 4-chlorophenol, etc.) were examined and confirmed. This method is based on the transformation of high electrical energy into an aqueous phase and generation of highly reactive species (hydroxyl radicals, hydrogen peroxide, ozone, etc.) [3]. These species are able to oxidize organic matter and such reactions could lead to destruction of organic pollutants dissolved in water. The main goal is not the full humic acids decomposition but at least the breaking of aromatic ring is necessary for further biological purification.

In this work, degradation of humic substances dissolved in water was investigated by diaphragm discharge, which was generated in a very small orifice in dielectric diaphragm placed between two

Journal of Physics: Conference Series **207** (2010) 012012 doi:10.1088/1742-6596/207/1/012012

planar electrodes. Effects of electrolyte kind and pH value were examined in this work. Aromatic and aliphatic content in humic acid mixture was compared by humification index calculated from fluorescence spectra.

2. Experimental setup

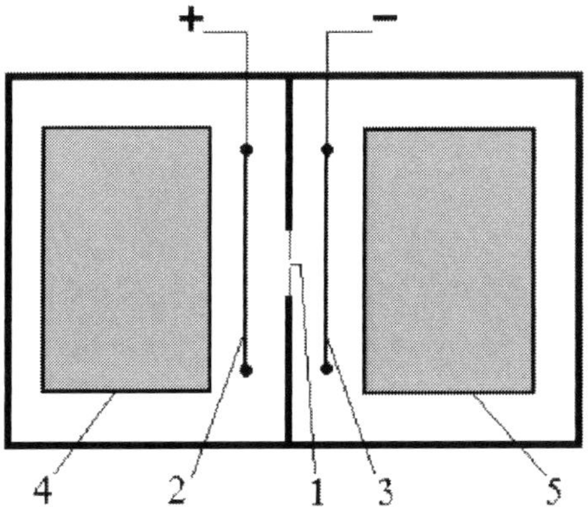

Figure 1. Scheme of the discharge reactor: 1 – diaphragm with pinhole, 2, 3 – electrodes, 4, 5 – cooling boxes.

Diaphragm discharge studied in this work was generated in a batch reactor (figure 1) divided into two parts – cathode and anode space [4]. The DC non-pulsed HV source giving the voltage of about 2 kV was used for the discharge generation. General parameters describing experimental set-up are listed in Table 1. Each part of the reactor contained 2 litres of treated solution containing commercial humic acid HUMIN-P 775 dissolved in demineralised water and electrolyte (NaCl, Na_2SO_4). Solution pH was primary adjusted by the electrolyte kind. As pH significantly changed during all experiments [5], its value was either modified by acid/base adding or it was let without any correction. Analyses of quality changes of treated solutions were carried out by spectroscopic methods (UV-VIS and fluorescence spectroscopy).

Table 1. General parameters of experiments

Discharge power	200 W
Electrodes	planar stainless steel
Dielectric diaphragm	PET, thickness 0.25 mm
Pinhole	central, diameter 0.25 mm
initial solution conductivity	500 $\mu S\ cm^{-1}$

3. Results and discussion

First attempts with humic acid decomposition by DC diaphragm discharge have already revealed a remarkably higher HA destruction in the anode space of the reactor [4]. Moreover, a significant decrease of pH value in the anode space and an increase of pH value in the cathode space were detected during the discharge treatment. Therefore we came up to attempts focused on pH value maintenance at a neutral value by the method of dropping the definite amount of surfactant (acid or base) to the both parts of the discharge reactor during the reaction.

3.1. Change of pH value during the treatment

At first, it was detected how pH value turned during the decomposition in the dependence on the used electrolyte kind (figure 2). Changes of pH value for sodium sulfate and sodium chloride were almost coincident. It decreased from the value of approximately 6.7 to 3.6 in the anode space and increased to

Journal of Physics: Conference Series **207** (2010) 012012 doi:10.1088/1742-6596/207/1/012012

9.8 in the cathode space after 20 minutes of the discharge treatment. 1M H_2SO_4, HCl and NaOH solutions were used as surfactants because of their ion identity with used electrolytes. Injection step of surfactant was determined on every 2 minute.

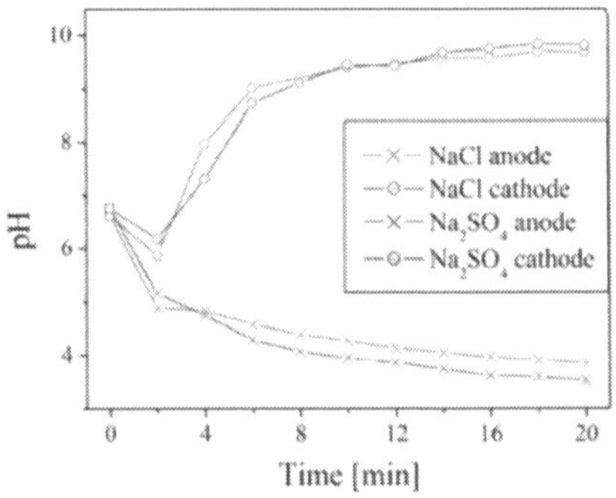

Figure 2. Time evolution of pH value for NaCl and Na_2SO_4 electrolytes by the diaphragm discharge (input power 200 W, initial conductivity 500 $\mu S{\cdot}cm^{-1}$).

3.2. *Correction of pH value*

Experiments with humic matters were carried out by using two different electrolytes – NaCl and Na_2SO_4 (figure 3). In both cases low degradation efficiency of humic acid in the anode space was observed, while in the cathode space a slight degradation improvement was determined in comparison with experiments without pH value correction. It could be assumed that the maintenance of neutral pH value during discharge treatment was inconvenient because it generally lead to lower degradation. On the other hand, it is possible to assume that the low pH value stimulated the degradation process.

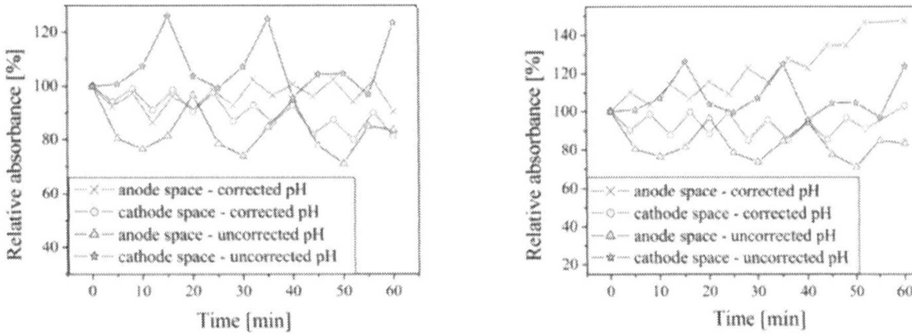

Figure 3. Effect of pH value correction on humic acid decomposition by the diaphragm discharge (initial HA concentration: 35 mg l^{-1}, electrolyte: 4 mM NaCl (left), 2 mM Na_2SO_4 (right), corrected pH 6.7, initial conductivity: 500 $\mu S\ cm^{-1}$).

3.3. *Analysis – humification index (HIX)*

Initial and final samples were analysed by fluorescence emission spectrophotometry, whereas exciting wavelength was adjusted on 280 nm (figure 4). From the ratio of the emission intensity values at 470 and 400 nm, humification index (HIX) was calculated, which indicated the mutual content of aromatic and aliphatic components in the mixture (Table 2). HIXs pointed to decrease of aromatic part at final samples from anode space. Molecules of humic acid were decomposed however, created up to now non-identified by-products distorted the absorption spectra of humic acid.

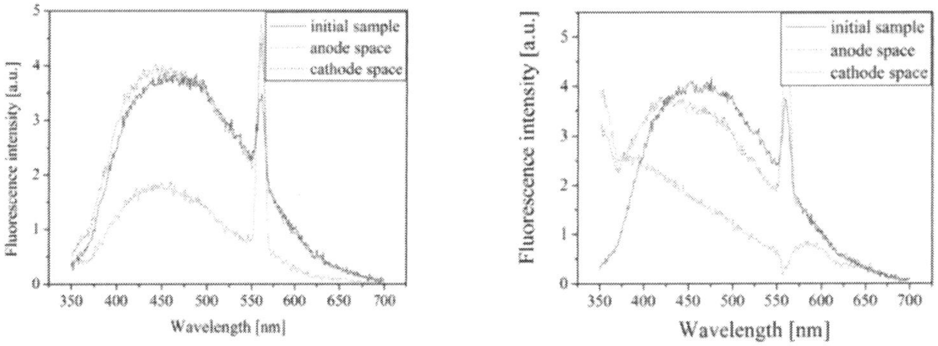

Figure 4. Fluorescence emission spectra of samples before and after treatment with correction pH (initial HA concentration: 35 mg l^{-1}, electrolyte: 4 mM NaCl (left), 2 mM Na_2SO_4 (right), corrected pH 6.7, initial conductivity: 500 μS cm^{-1}, input power: 200 W).

Table 2: Humification index (HIX).

Samples – corrected pH		HIX	Samples – uncorrected pH		HIX
NaCl	initial	1.47	NaCl	initial	1.47
	final – anode	1.28		final – anode	0.55
	final – cathode	1.31		final – cathode	1.27
Na_2SO_4	initial	1.44	Na_2SO_4	initial	1.44
	final – anode	0.64		final – anode	0.37
	final – cathode	1.06		final – cathode	1.40

4. Conclusion

Commercial humic acids dissolved in water solution were treated by diaphragm discharge. The changes of quality were observed by the absorption spectroscopy and fluorescence spectroscopy. The presented work gave the results about the effect of pH value on the decomposition of humic acid mixture. Processes of degradation became more efficient at lower pH values (acidic medium). Fluorescence analysis provided not so misrepresenting results than absorption analysis (UV-VIS). This discrepancy and finding of ideal analytical method will be an object of our further study as well as a detail analysis of final degradation products.

Acknowledgements

This work has been supported by the Czech Science Foundation, project No. 202/07/P371 and by the Czech Ministry of Education, Youths and Sports, Research Plan No. MSM0021630501.

References

[1] Sugiarto A T, Ito S, Ohshima T, Sato M, Skalný J D 2003 Oxidative decoloration of dyes by pulsed discharge in water *J. Electrostatics* **58** 135–45
[2] Lukes P, Clupek M, Sunka P 2005 Degradation of phenol by underwater pulsed corona discharge in combination with TiO_2 photocatalysis *Res. Chem. Intermed.* **31** 285–294
[3] Sun B, Sato M, Clements J S 1997 Optical study of active species produced by a pulsed streamer corona discharge in water *J. Electrostatics* **39** (3) 189–202
[4] Stara Z, Krcma F 2006 Treatment of Humic Acids Solutions by DC Diaphragm Discharge *Czech. J. Phys.* **56** B1351–B1356
[5] Krcma F, Stara Z, Slavicek P 2006 *Proc. ESCAMPIG XVIII (Lecce)* 467

Experimental characterization of the H_β-line profiles in microwave-produced plasmas at atmospheric pressure

J M Palomares[1,*], J Torres[1], M A Gigosos[2], J J A M van der Mullen[3], A Gamero[1] and A Sola[1]

[1]Departamento de Física, Universidad de Córdoba. Campus Universitario de Rabanales C2. E-14071 Córdoba, Spain.

[2]Departamento de Óptica, Universidad de Valladolid. Campus Universitario de la Merced. E-47072 Valladolid, Spain.

[3]Department of Applied Physics, Eindhoven University of Technology. P.O. Box 513. 5600 MD Eindhoven, The Netherlands.

*E-mail: *f02palij@gmail.com*

Abstract. An experimental study on the asymmetry of the Balmer H_β profile in plasmas produced by microwaves at atmospheric pressure is presented. The study is based on the definition of several functions and parameters that quantify the asymmetry and different shape aspects of the profile. The study shows the experimental dependence of these characteristics on the electron density and control parameters such as the gas flow and the hydrogen admixture ratio. The possible use of these newly introduced profile characteristics to plasma diagnosis is discussed.

1. Introduction

The broadening, shifts and asymmetries of spectral lines emitted by plasmas have been widely studied in the past [1,2]. The spectral line studied most extensively in the past is the H_β line of the hydrogen Balmer series especially in relation with its function in plasma diagnosis. Most of the experimental and theoretical studies on the asymmetry of the hydrogen spectral lines has been made up for plasmas with electron densities equal or higher than 10^{23} m^{-3} [3,4,5] for which one expects this asymmetry to be appreciable. The objective of this work is the study and characterization of the Stark asymmetry and dip formation of the line H_β at lower electron densities, *i.e.* in the order of 10^{21} m^{-3}, and to investigate whether the behaviour of the asymmetry and dip-formation follows a tendency that is consistent with the plasma characteristics. This could point towards new possibilities for plasma diagnosis and might give guidelines for the future theory formation.

2. Characterizing the asymmetry of H_β profiles

In Figure 1 the most relevant features of the H_β line is shown. The asymmetry of the H_β line has been described in several studies (a good and early example is that of Wiese [2]) in a qualitative way the *blue peak is higher than the red one*, while *the red wing is higher than the blue one*. Moreover, it is generally accepted that the asymmetry of the profile becomes higher when the electron density is increased. To make the asymmetry in the profile quantitative, relevant definitions of line aspects have to be introduced. The zero position of the intensity, *i.e.* the background level has been determined by adjusting the profile wings to an expression of the type $I \propto (\lambda-\lambda_0)^{-5/2}$ [1]. The height of the line is given by I_{max}, the average maximum of the blue and red peak intensities. The width of the line is described by the full-width at half-maximum (*FWHM*) where the role of the maximum is played by I_{max}. The central position λ_c of the H_β line is now defined as the centre of this *FWHM* (cf. Figure 1). We can now introduce functions that quantify the line asymmetry for the line profile for which we have used the following expressions (cf. [3]):

$$A_1(\Delta\lambda) = \frac{I_r(\Delta\lambda) - I_b(\Delta\lambda)}{I_r(\Delta\lambda) + I_b(\Delta\lambda)} \tag{1}$$

$$A_2(I) = \frac{\Delta\lambda_r(I) - \Delta\lambda_b(I)}{\Delta\lambda_r(I) + \Delta\lambda_b(I)} \tag{2}$$

$$A_3(I) = \frac{\Delta\lambda_r(I) - \Delta\lambda_b(I)}{2} \tag{3}$$

where $I_r(\Delta\lambda)$ and $I_b(\Delta\lambda)$ are respectively the intensities of the red and blue wings of the profile at a certain displacement $\Delta\lambda$ with respect to the central wavelength, while $\Delta\lambda_r(I)$ and $\Delta\lambda_b(I)$ represent the spectral position of these red and blue sides at a given intensity I. From the definitions it follows that the functions $A_1(\Delta\lambda)$ and $A_2(I)$ are dimensionless, while the function $A_3(I)$ has the unit of wavelength; moreover the following equalities are fulfilled: $A_1(\Delta\lambda_{1/2}/2) = A_2(I_{max}/2) = A_3(I_{max}/2) = 0$, $\Delta\lambda_{1/2}$ being the full-width at half-maximum (*FWHM*).

This study reports on the measurements of H_β lines emitted by the *TIA* plasma (the *Torche à Injection Axiale* is a device in use for the production of atmospheric plasmas [6]). With this axial injection torch microwave energy with a power of 1000 W at 2.45 GHz has been coupled into an Ar atmospheric flow in which different amounts of hydrogen gas are introduced. The measurement where taken at a fixed axial position of 1 mm above the nozzle's tip. A detailed description of this setup can be found in [7,8]. In the range of electron densities existing in our plasma (10^{20} m^{-3} – 10^{21} m^{-3}), the asymmetry of the H_β profiles due to the Stark effect is weak. Therefore it is necessary to keep other possible asymmetry sources in mind. The instrumental asymmetry introduced by the optical setup has been calibrated and eliminated of the experimental profiles. The thermal Doppler and van der Waals broadenings are symmetrical and do not introduce asymmetry. The self-absorption is worthless since the H_β line is optically thin in our plasma. The internal structure (inhomogeneities) in the observed plasma zone could be of certain importance, but at present these preliminary results show the presence of Stark asymmetry in plasma with electron density of the order of 10^{21} m^{-3}, even disregarding this structural detail.

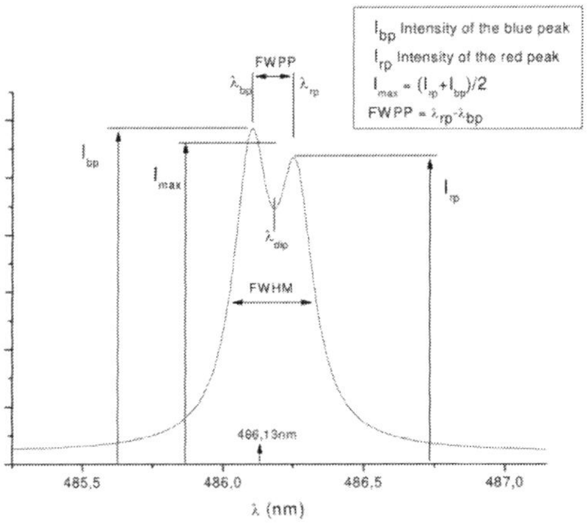

Figure 1. Main characteristic features of the H_β line profile: full width at half-maximum (*FWHM*), spectral separation between dip peaks or full width peak-to-peak (*FWPP*), intensity difference between peaks, dip depth, dip position.

3. Results and discussion

3.1. Influence of the electron density on the asymmetry functions

In figures 2, 3 and 4 the three asymmetry functions corresponding to four H_β profiles at different electron densities are depicted for the Ar *TIA* plasma. The experimental settings are a power of 1000 W, an Ar flow of 5 L/min and a concentration of H_2 in the range of 2 % – 17 %. The general tendency is that increasing the hydrogen admixture ratio reduces the electron density. The values of electron density are calculated using the *FWHM*-Stark broadening of the H_β lines and the computational results provided by Gigosos *et al* [9]. The relative error for the electron density calculated in this work remains less than 10 %. These figures show the asymmetry-functions give a clear response to the variation of n_e and that they follow a coherent behavior (the asymmetry increases with the electron density) in the wings of the profiles ($I < 0.5I_{max}$ and $\Delta\lambda < 0.5\Delta\lambda_{1/2}$). It can be seen that $A_1(\Delta\lambda)$ is much more irregular that the other two. Also, that the wing-side of the $A_2(I)$ is less sensitive to n_e-variations than the function $A_3(I)$.

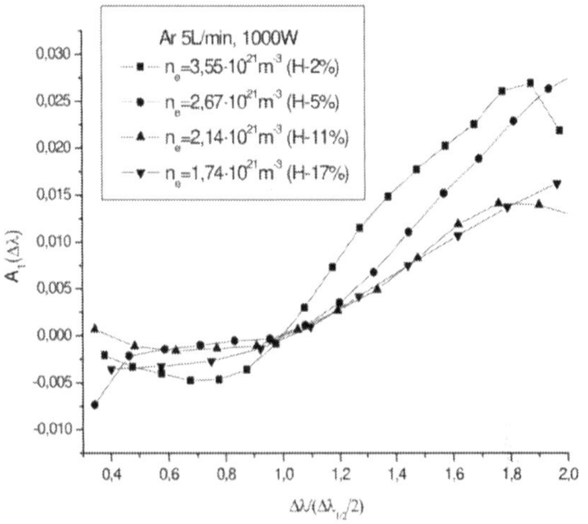

Figure 2. Function of asymmetry $A_1(\Delta\lambda)$ of the H_β profile at 5 L/min of Ar flow, 1000 W of *HF* power and different electron density values.

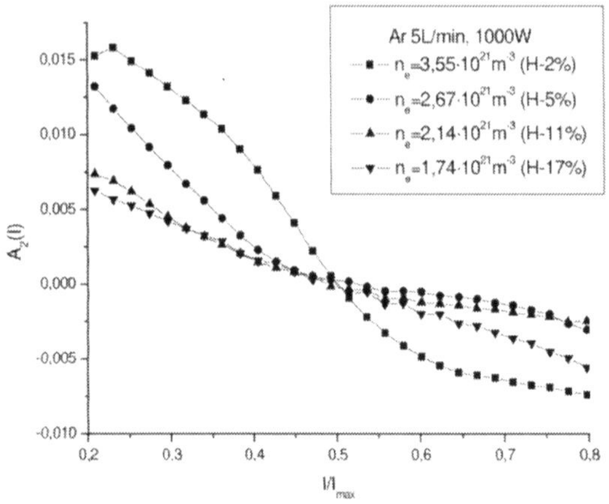

Figure 3. Function of asymmetry $A_2(I)$ of the H_β profile at 5 L/min of Ar flow, 1000 W of *HF* power and different electron density values.

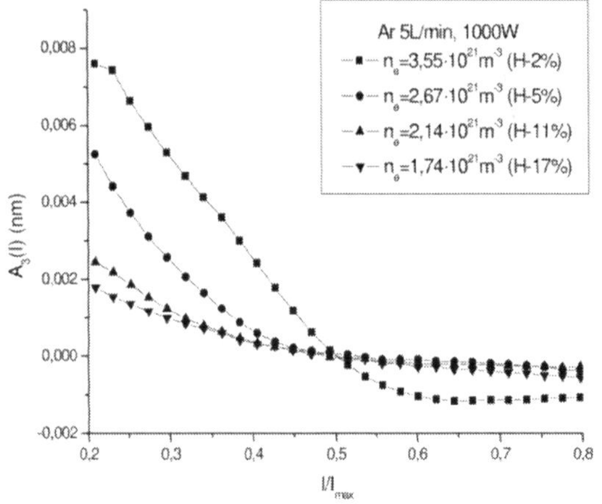

Figure 4. Function of asymmetry $A_3(I)$ of the H_β profile at 5 L/min of Ar flow, 1000 W of *HF* power and different electron density values.

3.2. Influence of the gas flow in the asymmetry functions

The internal structure of the plasma flame provided by the *TIA* strongly depends on the gas flow and it's well known fact that the line shape is affected by the presence of gradients in the discharge and therefore the gas flow could affect the asymmetry and shape of H_β. This has been studied further using the next set of figures that were obtained for 5 L/min and 1 L/min of Ar flow. In figures 5 and 6 we show the central valley and the corresponding $A_3(I)$ function with an H_2 concentration chosen to kept electron density constant. In figures 7 and 8 the concentration of H_2 has been kept constant. Figures 5 and 7 show how the dip changes when the flow is decreased from a normal asymmetry to an inverse or anomalous asymmetry in which the red peak is higher than the blue one. In figures 6 and 8 the corresponding $A_3(I)$ functions show how the anomaly in the asymmetry for the case of lower flow also alters the spectral form of the whole profile from 60 % of I_{max} up. A possible explanation is that the reduction in the gas flow affects the internal structure of the plasma introducing higher in-homogeneities.

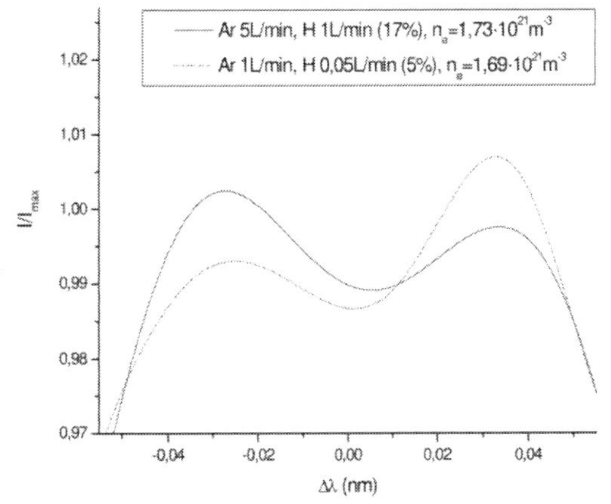

Figure 5. Detail of the central valley at two different gas flux, 5 L/min and 1 L/min, maintaining the electron density = 1.7×10^{21} m^{-3}.

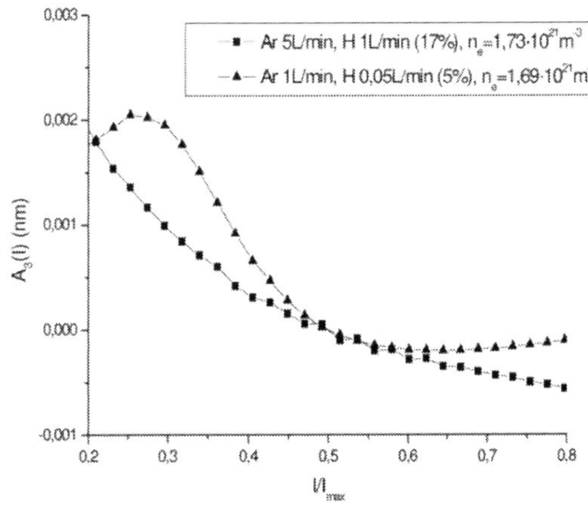

Figure 6. Asymmetry function $A_3(I)$ at two different gas flux, 5 L/min and 1L/min, maintaining the electron density = 1.7×10^{21} m^{-3}.

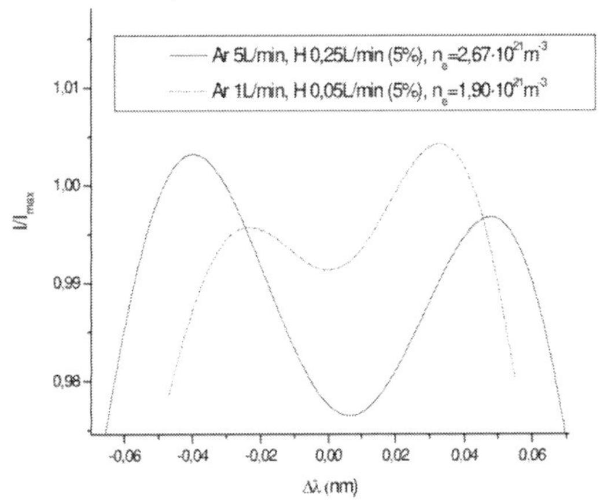

Figure 7. Detail of the central valley at two different gas flux, 5 L/min and 1 L/min, maintaining the H_2 concentration = 5 %.

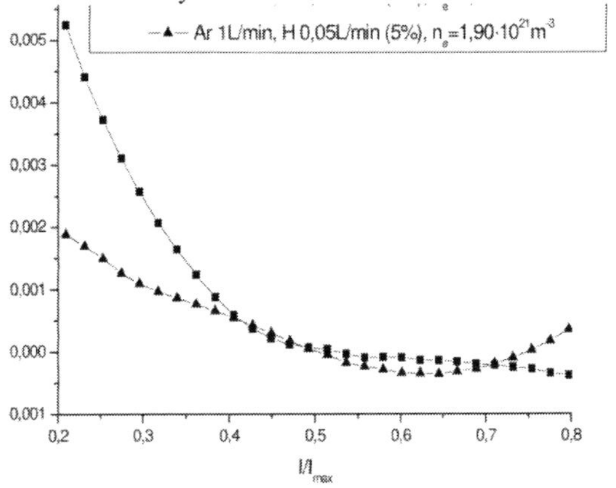

Figure 8. Asymmetry function $A_3(I)$ at two different gas flux, 5 L/min and 1 L/min, maintaining the H_2 concentration = 5 %.

3.3. Specific characterization of the H_β dip shape

In this section other specific parameters have to be constructed in order to study the central valley of H_β. The measurements were done at 1000 W of *HF* power, changing the concentration of hydrogen from 17 % (lower electron density) to 2 % (higher electron density) in plasmas generated with Ar flows of 5 L/min and 2.5 L/min.

3.3.1. Spectral separation between peaks (full-width peak-to-peak, FWPP) and dip depth (DD)

The values of $FWPP = \lambda_{rp} - \lambda_{bp}$ and DD (normalized to the maximum of intensity: $(I_{max} - I_{dip})\times100/I_{max}$) by themselves do not provide any direct information about the asymmetry, however they provide information of the shape and the H_β line structure as the spectral distance between peaks and the dip depth are directly related to the internal distribution or separation of the line-components. In figures 9 and 10 the values obtained for these two parameters as a function of the electron density are depicted. It is shown that these parameters follow an increasing dependence on the electron density which is less pronounced when the gas flux is higher.

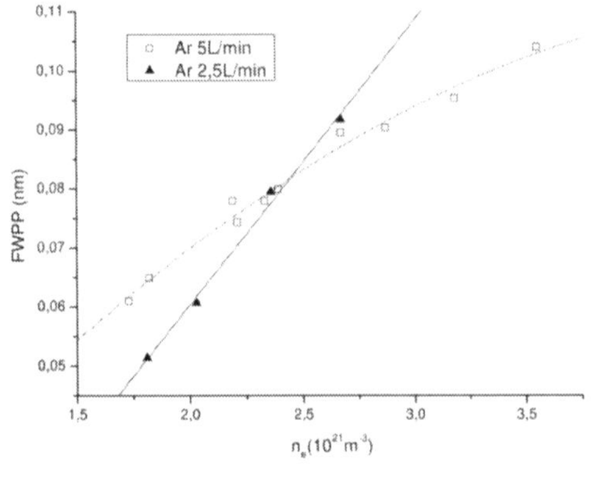

Figure 9. Experimental values of the *FWPP* as a function of the electron density for 5 L/min and 2.5 L/min of Ar flow and 1000 W of *HF* power.

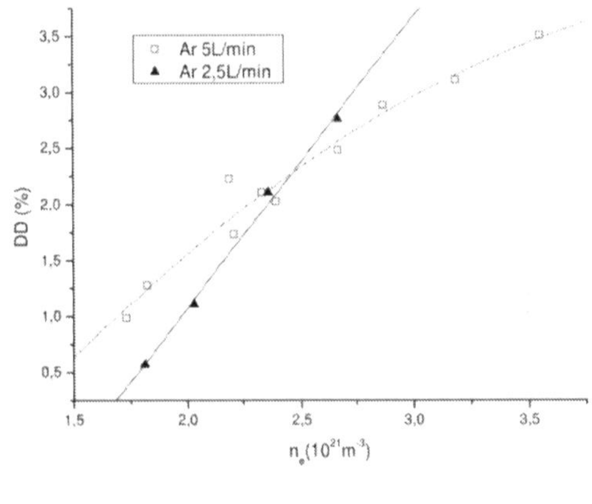

Figure 10. Experimental values of the *DD* as a function of the electron density for 5 L/min and 2.5 L/min of Ar flow and 1000 W of *HF* power.

3.3.2. Intensity difference between peaks (PID) and spectral dip position (DP)

Next, we introduce the values of *PID* (normalized to the maximum intensity: $(I_{bp} - I_{rp})\times100/I_{max}$) and *DP* (normalized to the *FWPP*: $(\lambda_{dip} - \lambda_{bp})\times100/(\lambda_{rp} - \lambda_{bp})$). These parameters give us direct information on the asymmetry in the central valley since they compare characteristic of the blue and

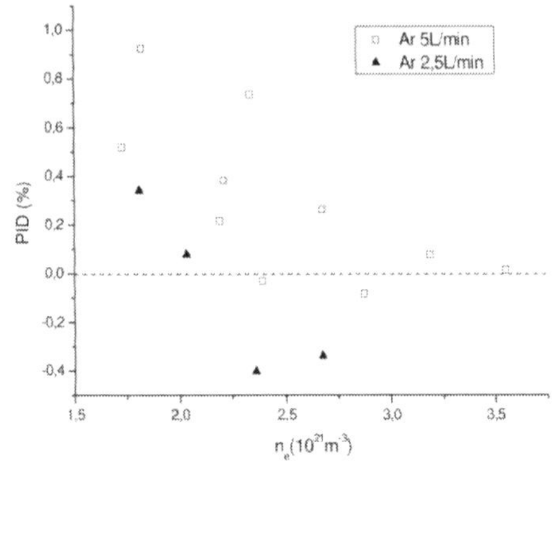

Figure 11. Experimental values of the *PID* as a function of the electron density for 5 L/min and 2.5 L/min of Ar flow and 1000 W of *HF* power.

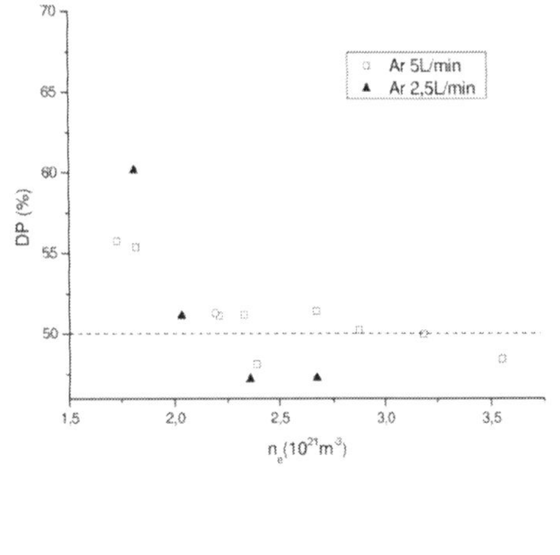

Figure 12. Experimental values of the *PID* as a function of the electron density for 5 L/min and 2.5 L/min of Ar flow and 1000 W of *HF* power.

red sides of the dip. The normal asymmetry of the H_β line is that blue peak is more intense than the red one (*PID* > 0%) and that the dip is closer to the red peak (*DP* > 50 %). Figures 11 and 12 show the experimental values of *PID* and *DP*. These last two parameters show a much larger scatter; thus they do not follow a clear tendency as a function of the electron density. The influence of the internal structure fluctuation is then appreciable in this representation, and could explain the great dispersion of the points, particularly the appearance of some negative values concerning the anomalous or inverse asymmetry (under the dashed line).

4. Conclusions

In this experimental study, the asymmetry concerning the spectral profiles of the Balmer H_β line has been experimentally characterized in high-pressure, microwave plasmas produced by the *TIA* device in which the electron density reaches a value of 10^{21} m^{-3}, which implies a considerably lower value than in previous characterizations (10^{23} m^{-3} or higher). The asymmetry has been studied with the help of some asymmetry functions for the flanks of the profile, including the wings. It has been shown how the asymmetry of the profiles increases with the electron density in the discharge, which is coherent with the known behavior for the studied line. With respect to the gas flux influence, is detected that at low enough gas flux strange or anomalous values of the asymmetry have been measured for which the red side of the profile is more intense than the blue one.

For the central valley or dip, specific functions charactering its shape have been used. The spectral distance between peaks, *FWPP*, and the dip depth, *DD*, exhibit a clear and bare increasing (quasi-linear) dependence on the electron density, less pronounced at a higher gas flux. The other two functions giving direct information about the central valley asymmetry *i.e.* the difference of intensity between peaks, *PID*, and the spectral position of the dip, *DP*, are significantly affected by the fluctuations in the plasma. In some cases, when the gas flux drops enough, it can end up inverting the normal tendency of the profile asymmetry, giving rise to the so-called anomalous asymmetry in which the blue peak of the H_β is less intense than the red one. From a practical point of view, the *FWPP* and *DD* seem to be good candidates for the determination of the electron density since their n_e-dependence shows almost no scatter.

References

[1] Griem H R 1974 *Spectral line broadening by plasmas* Academic Press Inc. (New York, US)
[2] Wiese W L, Kelleher D E and Paquette D R 1972 Detailed study of Stark Broadening of Balmer Lines in a High-Density Plasma *Phys. Rev. A* **6** 1132–1153
[3] Djurovic S, Nikolic D, Savic I, Sörge S, Demura A V 2005 Asymmetry of H-beta Stark profiles in T-tube hydrogen plasma *Phys. Rev. E* **71** 036407-1–036407-15
[4] Sörge S and Günter S 2000 Simulation of shifted and asymmetric hydrogen line profiles *Eur. Phys. J. D* **12** 369–375
[5] Halenka J, Vujicic B and Djurovic_ S 1989 Shift of the peaks of the H-beta spectral-line *J. Quant. Spectros. Rad. Transfer* **42** 571–573
[6] Moisan M, Sauvé G, Zarkrzewski Z and Hubert J 1994 An atmospheric pressure waveguide-fed microwave plasma torch: the *TIA* design *Plasma Sources Sci. Technol.* **3** 584–592
[7] Torres J, Palomares J M, Sola A, van der Mullen JJAM and Gamero A 2007 A Stark broadening method to determine simultaneously the electron temperature and density in high-pressure microwave plasmas *J. Phys. D: Appl. Phys.* **40** 5929–5936
[8] Torres J, Palomares J M, Gigosos M A, Gamero A, Sola A and van der Mullen JJAM 2008 An experimental study on the asymmetry and the dip form of the H_β line profiles in microwave produced plasmas at atmospheric pressure *Spectrochim. Acta B* **63** (in press) (doi:10.1016/j.sab.2008.05.007)
[9] Gigosos M A, González M A and Cardeñoso V 2003 Computer simulated Balmer –alpha, –beta and –gamma Stark line profiles for non-equilibrium plasma diagnostics *Spectrochim. Acta B* **58** 1489–1504

Journal of Physics: Conference Series **207** (2010) 012014 doi:10.1088/1742-6596/207/1/012014

Study of spatial distributions of F, H and CF_2 radicals in DF CCP discharge in Ar/CF_4 and Ar/CHF_3 mixtures

O. V. Braginsky, A. S. Kovalev, D. V. Lopaev, T. V. Rakhimova, A. T. Rakhimov, A. N. Vasilieva, S. M. Zyryanov

Lomonosov Moscow State University Skobeltsyn Institute of Nuclear Physics (MSU SINP), 1(2), Leninskie gory, GSP-1, Moscow 119991, Russian Federation

E-mail: lopaev@dnph.phys.msu.su

Abstract. The comparative research of production and loss of atoms (F, H) and radicals (CF_2), important for plasma processing, is carried out in DF CCP plasma in Ar/CF_4 and Ar/CHF_3 mixtures modeling two etching plasmas (fluorine-rich) and (fluorine-poor) respectively. Electron density, electron temperature and ion energy spectra are measured for single and dual frequency discharge modes. Spatial profiles of radical and atom densities are measured in both mixtures by using spatially resolved actinometry and UV absorption techniques. The opportunity for plasma-surface interaction control using spatial radical and atom density distributions is shown. The influence of high and low frequency powers on balance between surface and volume processes is discussed.

1. Introduction

Fluorocarbon plasma is of great importance for silicon microtechnology [1]. It is widely used for SiO_2 etching. At the present time this process is mostly carried out in dual frequency capacitive discharge (DF CCP) [2, 3]. DF CCP discharge is well known to have possibility of separate control over ion energy and flux by manipulating the rf powers at different rf frequencies [4–7]. This so called decoupling allows precise controlling of the surface processes like etching, because most of them are ion-stimulated. As it's known, highly selective anisotropic etching goes through a thin nm-size fluoro-hydro-carbon film produced on the surface of SiO_2 in the discharge [8, 9]. The complex plasma chemistry leads to the tight interconnection between volume and surface processes [10–13]. Thereby one can obtain information about the surface processes following the spatial distributions of the active particles (such as radicals and atoms) densities and then use it for a feedback to change discharge parameters to sustain the optimal conditions for SiO_2 etching. In this work these spatial distributions were measured using actinometry and UV absorption techniques. Electron temperature, electron density and ion energy spectra were also measured to specify the peculiarities of DF CCP discharge in fluorocarbon plasma.

2. Experiment

Experiments were carried out in quartz chamber of 115 mm diameter at pressure of 45 mTorr in symmetrical electrodes geometry (Al electrodes of 100 mm diameter with the interelectrode

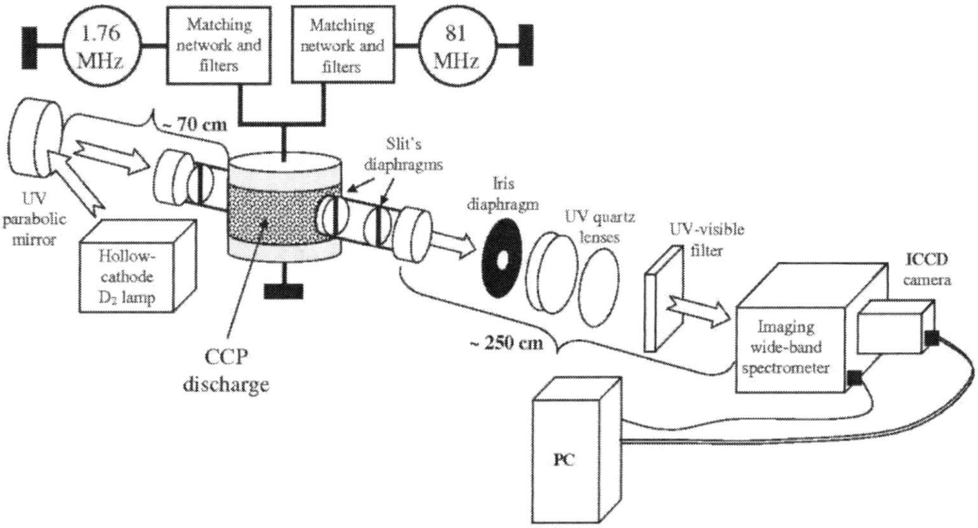

Figure 1. Experimental setup for optical measurements.

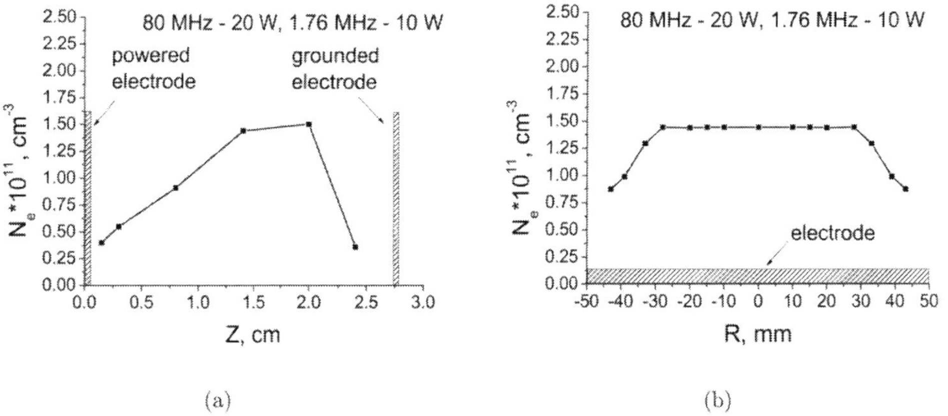

Figure 2. Electron density distributions (a) - axial, (b) - radial.

gap of 28 mm). Both rf frequencies (81 MHz and 1.76 MHz) were applied to the same electrode. Two gas mixtures were used: CF_4/Ar (30%/70%) and CHF_3/Ar (30%/70%). These mixtures model two real situations corresponding to so-called flourine rich and flourine-poor conditions respectively of poly-Si etching and SiO_2 etching through prodicing the fluorocarbon film. The experimental setup for optical measurements is shown in Fig. 1. Besides, to measure plasma parameters, movable Langmuir probe and ion energy analyser at the grounded electrode were also used. Ar^+ ion bombardment allowed carrying out the probe and energy analyser measurements for rather long time before surface contamination by fluoro-carbons.

The axial and radial distributions of the electron density measured by the movable Langmuir probe in CHF_3/Ar (30%/70%) at 45 mTorr are shown in Fig. 2. It can be seen that the discharge was shifted a little toward the grounded electrode, that shows a slight asymmetry of the discharge. Nevertheless along the electrodes the discharge was quite symmetric and uniform.

The electron temperature dependence on high frequency power in single frequency (SF 81 MHz) and on low frequency power in dual frequency (81 MHz + 1.76 MHz) discharge is shown in Fig. 3. Compared to SF in DF CCP discharge the measured electron temperature decreases with the increase of the low frequency power, that is a result of the dramatic changes in the discharge structure. With the addition of the low frequency the number of electrons in the EEDF tail grows [14]. Due to the EEDF nonlocality the rf power is mostly absorbed by high energy electrons, reaching the electrodes sheaths, while in the probe measurements the temperature of the low energy electrons in the discharge center is obtained.

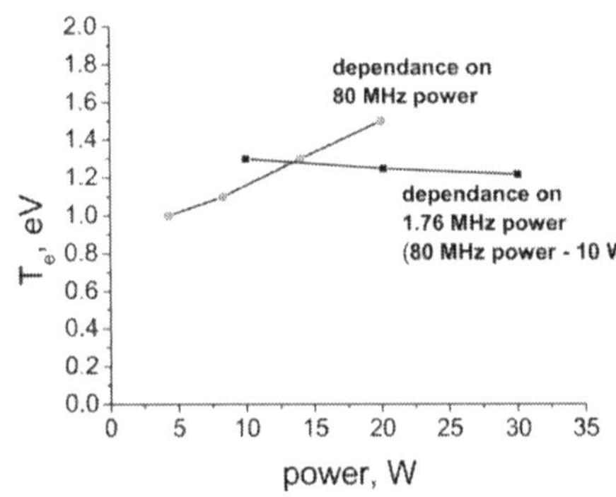

Figure 3. Electron temperature dependency on low and high frequency power in the discharge center (CHF_3/Ar (30%/70%), p = 45 mTorr).

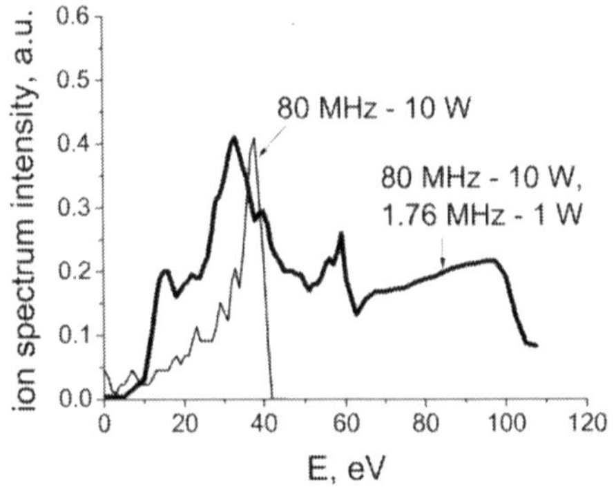

Figure 4. Ion energy spectra at the grounded electrode in single and dual frequency discharges (CHF_3/Ar (30%/70%), p = 45 mTorr).

The ion energy spectra in single and dual frequency discharges in CHF_3/Ar (30%/70%) at 45 mTorr are shown in Fig. 4. The application of high frequency voltage to the electrodes produces only a peaked distribution of ion energy because of thin collisionless electrode's sheath. The distribution maximum is slightly shifted with increasing the high frequency power. With the addition of low frequency the electrode's sheath becomes wider and ion collisions lead to a wide ion energy distribution with the high-energy part of the ion energy spectrum (responsible for ion-stimulated surface processes) being fully controlled by the low frequency power [15, 16].

Two spectroscopy diagnostics were used to obtain the spatial distributions of H, F and CF_2 densities in the discharge. For atoms and CF_2 radicals actinometry technique on Ar atoms was used, while for CF_2 UV absorption spectroscopy was also applied. As it is known actinometry technique is based on that fact, that for certain optical transitions in radicals, the ratio of the radical emission intensity to the actinometer (usually rare gas) intensity is proportional to the radical mole fraction. Whereas the actinometry technique is quite applicable for measuring the absolute values of H and F mole fractions due to possibility of the correct calculations of the atom excitation rate coefficients by electrons, its applicability to CF_2 isn't so obvious. But simultaneous using actinometry and absorption spectroscopy for measuring CF_2 spatial profiles allowed us to obtain reliably both density and true mole fraction of CF_2 radicals. The emission and absorption spectra for two gas mixtures at the pressure of 45 mTorr is shown in Fig. 5 and 6 respectively. It can be seen, that CF_2 emission and absorption signals are much bigger in the CHF_3/Ar mixture. It indicates the higher radical density in CHF_3, that leads to faster polymerization processes and as a consequence to fluoro-carbon film deposition on the SiO_2

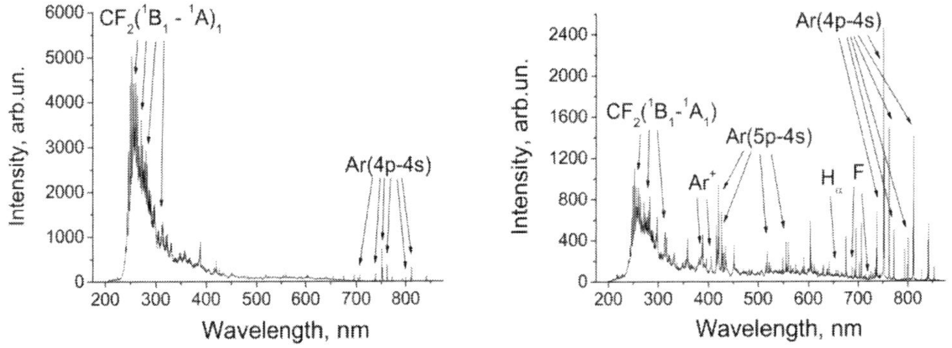

(a) (b)

Figure 5. Emission spectra of the discharges in different mixtures, p = 45 mTorr. (a) - CHF_3/Ar (30%/70%), (b) - CF_4/Ar (30%/70%).

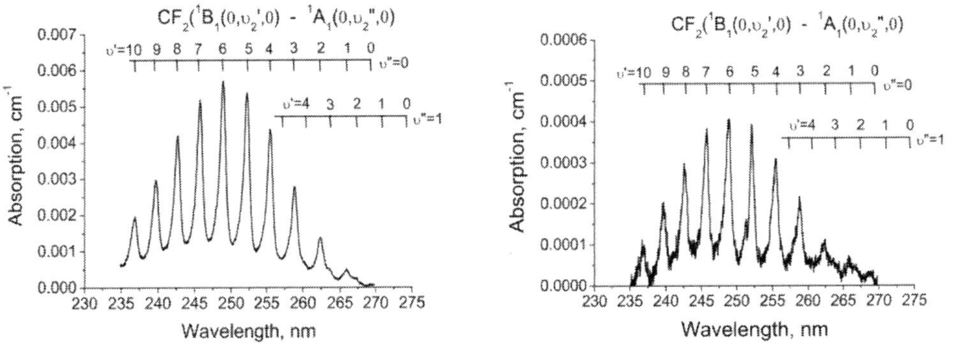

(a) (b)

Figure 6. UV absorption spectra of the discharges in different mixtures, p = 45 mTorr. (a) - CHF_3/Ar (30%/70%), (b) - CF_4/Ar (30%/70%).

Table 1. Sticking coefficients.

	CHF_3/Ar (30%/70%)	CF_4/Ar (30%/70%)
CF_2	$7{\cdot}10^{-3}$	$7{\cdot}10^{-3}$
H	$4{\cdot}10^{-3}$	$7{\cdot}10^{-3}$

surface. So the influence of the surface processes on the volume radicals densities is expected to be stronger in this case.

The spatial distributions of CF_2 radicals and H atoms mole fraction in CHF_3/Ar (30%/70%) and CF_4/Ar (30%/70%) at 45 mTorr are shown in Fig. 7 and 8. Beacuse of the noisy

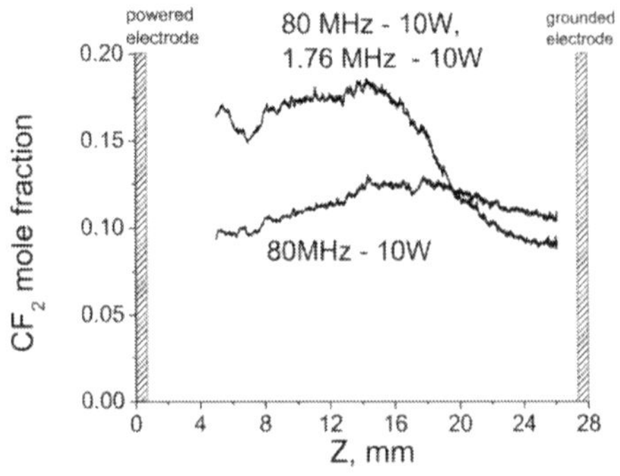

(a)

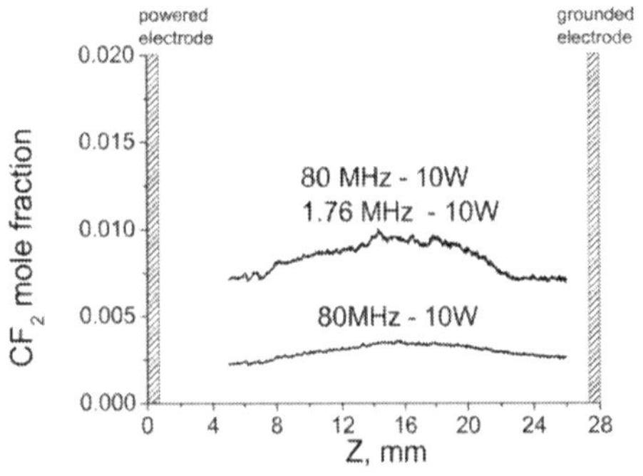

(b)

Figure 7. CF_2 mole fraction axial distributions in different mixtures, p = 45 mTorr. (a) - CHF_3/Ar (30%/70%), (b) - CF_4/Ar (30%/70%).

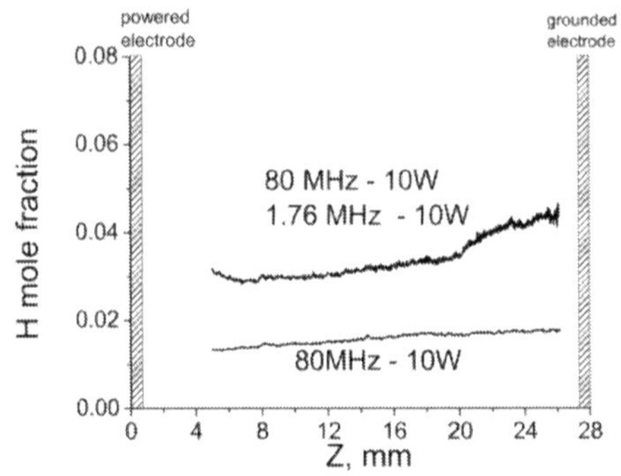

(a)

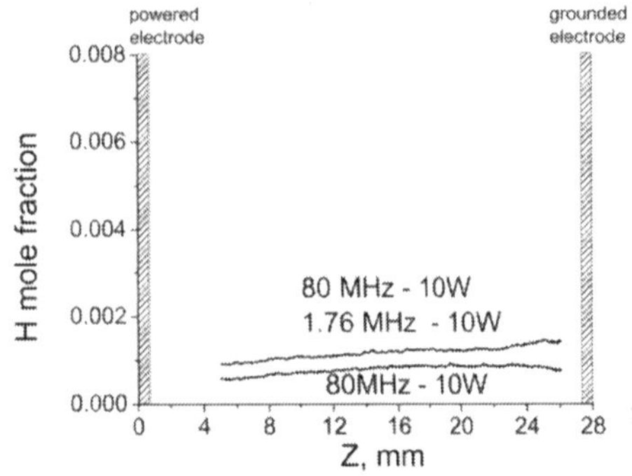

(b)

Figure 8. H mole fraction axial distributions in different mixtures, p = 45 mTorr. (a) - CHF_3/Ar (30%/70%), (b) - CF_4/Ar (30%/70%).

actinometric signals from F atoms, the F spatial density distributions were hardly observed, so they are not used in folowing analysys.

Using the gradients of H and CF_2 density distributions near the electrodes, sticking coefficients (loss probabilities) for H and CF_2 were estimated (Table 1). Due to the thicker fluoro-carbon-film on the electrodes in CHF_3/Ar mixture, the rate of ion stimulated surface processes increases noticeably with increasing the low frequency power, that changes the spatial CF_2 and H density distributions. On the opposite side in CF_4/Ar mixture there seems to be no fluoro-carbon film on the SiO_2 surface, so the increase of the power at the low frequency changes only the radical and atom mole fraction because of the lack of decoupling between two frequencies in this discharge, but the form of the spatial distributions remains the same.

3. Conclusion

It was shown that spatial distributions of atom and radical densities in fluorocarbon DF CCP can give detailed information on the surface processes. Obtaining this information can provide a reliable feedback for plasma processing control. Spatially resolved spectroscopy is a powerful, sensitive, non-intrusive tool for spatial density distribution measurements for many particles simultaneously. In comparison to the existent plasma processing control systems, the additional information on the spatial density distributions of the active particles allows to control not only such global processes as the etching end point, but also the surface processes themselves during etching.

Acknowledgments

This research was supported by the Russian Foundation of Basic Research (08-02-00465-a) and Key Science School (SS-133.2008.2).

References

[1] Lieberman M A and Lichtenberg A J 2005 *Principles of Plasma Discharges and Materials Processing, Second Edition* (New York: John Wiley and Sons)
[2] Tachia S 2003 *J. Vac. Sci. Technol.* A **21** S131
[3] Tsai W, Mueller G, Lindquist R and Frazier B 1996 *J. Vac. Sci. Technol.* B **14** 3276
[4] Kim H C, Lee J K and Shon J W 2003 *Phys. Plasmas* **10** 4545
[5] Boyle P C, Ellingboe A R and Turner M M 2004 *J. Phys.* D **37** 697
[6] Boyle P C, Ellingboe A R and Turner M M 2004 *Plasma Sources Sci. Technol.* **13** 493
[7] Gans T, Schulze J, O'Connell D, Czarnetzki U, Faulkner R, Ellingboe A R and Turner M M 2006 *Appl. Phys. Lett.* **89** 261502
[8] Tatsumi T 2007 *Appl. Surf. Sci.* **253** 6716
[9] Sankaran A and Kushner M J 2004 *J. Vac. Sci. Technol.* A **22** 1242
[10] Booth J P 1999 *Plasma Sources Sci. Technol.* **8** 249
[11] Takada N, Iida T, Shibagaki K and Sasaki K 2004 *J. Vac. Sci. Technol.* A **22** 413
[12] Fendel P, Francis A and Czarnetzki U 2005 *Plasma Sources Sci. Technol.* **14** 1
[13] Hori M and Goto T 2006 *Plasma Sources Sci. Technol.* **15** S74
[14] Schulze J, Gans T, OConnell D, Czarnetzki U, Ellingboe A R and Turner M M 2007 *J. Phys.* D **40** 7008
[15] Lee J K, Manuilenko O V, Babaeva N Y, Kim H C and Shon J W 2005 *Plasma Sources Sci. Technol.* **14** 89
[16] Shannon S, Hoffman D, Yang J G, Paterson A and Holland J 2005 *J. Appl. Phys.* **97** 103304

Hairpin-resonator probe design and measurement considerations

Plamen Dankov[1], Peter Stefanov[2], Valentin Gueorguiev[3] and Tzvetan Ivanov[3]
[1] Faculty of Physics, Sofia University, James Bourchier blvd., Sofia 1164, Bulgaria
[2] ISMA Ltd., Samokovsko shausee 2, Sofia, Bulgaria
[3] Institute of Solid State Physics, BAS, Sofia 1784, Bulgaria

Corresponding Author: Valentin Gueorguiev, e-mail: val@issp.bas.bg,

Abstract. A hairpin-resonator probe is designed to measure the spatial distribution of the electron density in RF plasmas. An effective simulation procedure with the commercial software simulator Ansoft® High Frequency Structure Simulator 8 in eigen-mode option was applied to evaluate the influence of the resonator structures on the resonance frequency $(f_v)_{00p}$ and unloaded Q-factor $(Q_v)_{00p}$ of the resonance modes. Many geometrical constructions were examined and finally, an optimal modification is proposed.

1. Introduction

The hairpin probe is a simple microwave resonator successfully used for electron density measurements in low-pressure gas discharges. This type of plasma probe was introduced by Stenzel in 1976 [1]. Two modifications of the structure in "transmission" or "reflection" regimes have been discussed by Piejak et al. [2]. The transmission-type probe consists of a hairpin resonator, an input loop, and a pick-up loop, while the reflection-type probe includes only a coaxial cable with a magnetic loop at the end that feeds the hairpin structure. We use in our investigation a reflection-type hairpin resonator.

The considered hairpin probe is a quarter-wavelength two-wire resonator with one short-circuited end and one open end and resembles a "hairpin antenna". The structure can be simply excited by a coaxial feed loop with a well-defined proximity to the short-circuited end. The resonant frequency of the hairpin antenna is related to the relative permittivity of the surrounding medium (vacuum or plasma). Therefore, the electron density can be accurately determined from the change between the resonance frequencies without (in vacuum) and with a low-pressure gas discharge. A theoretical analysis of the influence of the plasma sheath around the hairpin antenna has been performed in references [2, 3]. A detailed model of resonator plasma probe was reported in reference [4].

Recently, a number of papers have been published, which are dedicated to the design of different hairpin probes and give results from measurements of plasma parameters in low-pressure inductively-excited industrial plasmas with a variety of gas mixtures used for micro machining of micro devices for modern electronics and Micro-Electro-Mechanics (MEM) applications [5–8].

In this paper we describe experimental results about the hairpin design and its influence on the resonant frequency and return losses.

2. Design and measurement of hairpin probe in air

2.1. Hairpin resonator design and simulations

Outline of the used in the recent investigation hairpin antenna and its coupling with the GHz transmission/receiver coaxial cable is shown in figure 1. The spring section of the antenna wire allows an easy replacement of the hairpin antenna when it is necessary.

Figure 1. Outline of the mechanically stabilized hairpin with a wire spring section in the plasma probe.

The schematic view of the hairpin resonator and the 3D model for the resonator used in the commercial software simulator Ansoft® High Frequency Structure Simulator (HFSS) 8 in eigen-mode option is shown in figure 2. The wire thickness is t, while the distance between the "hairpin" wires is W.

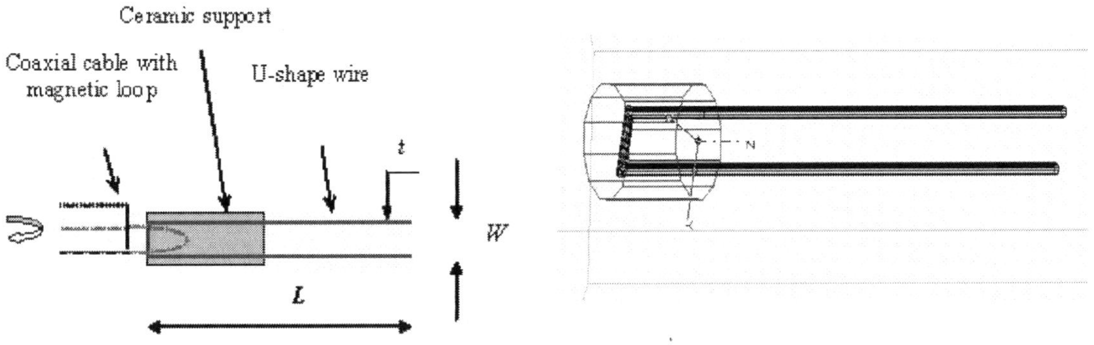

Figure 2. Schematic view of the hairpin resonator wire with length L and cylindrical ceramic support and the 3D model of the resonator used in the Ansoft® HFSS simulator.

The simplest design of the hairpin resonator should be performed on the basis of the assumption for pure TEM-mode excitation in this structure. Therefore, the full hairpin length L should be equal to an integer number of quarter wavelengths ($p\lambda_0/4$; λ_0 – wavelength) of the propagation mode in vacuum. In this case the simple expression for the resonator length L

$$L,\text{ mm} = p\frac{300}{4f_v,\text{ GHz}}, \quad p = 1, 2, 3... \tag{1}$$

is valid, where f_v is the work frequency of the hairpin probe in vacuum, and p is the mode number (usually $p = 1$). Unfortunately, this simple analysis does not take into account the other dimensions of the resonator (e.g. W and t) and the influence of the ceramic support. Because of this, we additionally simulate this open-type resonance structure by the commercial software simulator Ansoft® HFSS-8 in eigen-mode option – see figure 2, in order to evaluate the influence of the other parameters on the resonance characteristics – the resonance frequency, $(f_v)_{00p}$, and the unloaded Q-factor, $(Q_v)_{00p}$, of the resonance modes in the hairpin resonator. Figure 3 presents the electric and magnetic field distribution of the first resonance TEM_{001} in the hairpin resonator with tungsten wires and alumina ceramic support. Some results from the simulations are compared with the experimental data further on (see figure 7).

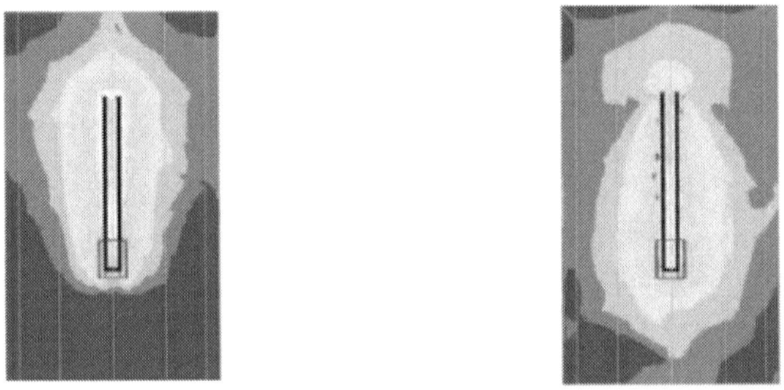

Figure 3. Electric-field and magnetic-field distribution of the first-order resonance obtained by HFSS-8 in eigen-mode option.

First of all, the software simulations show that a practical resonance structure, like this in figure 2, has lower resonance frequencies than those obtained by the simple expression (1). This effect is caused by two main reasons: the influence of the alumina ceramic support (with dielectric constant ~9.8) and the effective "prolonging" of the resonator length due to the so-called "fringing-field" effect in the E-field distribution near to the open resonator edge – see picture in figure 3. It is easy to show that the resonance frequency can be shifted mainly by the variation of the hairpin length L – the resonance frequency decreases with the increase of L (−78 MHz/mm of the resonator length), but the Q factor also decreases for longer resonators. The increase of the distance W between the hairpin wires leads to a small decrease of the resonance frequency (−2.2 % for doubling of W). The thickness of the hairpin wires does not significantly affect the resonance frequency, but the Q factor strongly increases (for example +41 % for doubling of t). A similar strong effect over the unloaded Q-factor increase is observed by increasing the wire conductivity (for example, with +36 % from tungsten *W* to silver Ag), but the choice of the hairpin wire material depends on other reasons (for example, chemical resistance to the plasma gas mixtures).

2.2. *Experimental hairpin probe and measurement results*

The reflection-type hairpin probe is feed by a 50-Ohm coaxial cable with a coupling loop at one end a standard female SMA connector at the other end. The performances of the cable and the connector as well as the transition "connector-cable" between them are very important for the successful operation of the designed plasma probe. We have investigated many combinations between different commercial connectors and cables – soft, semi-flexible, semi-rigid and rigid. These structures are preliminary tested in Time-Domain Reflectometry (TDR) option (250 ns gate length) with a Vector Network Analyzer (VNA) Agilent 88720E. Then the results are converted into the Frequency-Domain (FD) option in a wide frequency range of 0–20 GHz. We have accepted the combinations for which the own return losses *(RL)* are smaller than −25 dB. This requirement will ensure a reproducible response from the hairpin probe at resonance frequencies in the range of 2.5 to 6 GHz.

Next step in the practical realization of the probe is the joining of the experimental hairpin resonator. We investigated several structures with different metal wires (tungsten, soft silver, silver-plated cooper, stainless steel, etc.) with practically equal wire diameter. The other differences in the experimental probes were the distance W between the wires and finally the length L of the each resonator simply controlled by cutting. Both structures, the coaxial feed line and the resonator, are designed to be independent components and we combined several structures to obtain the best

conditions. In this paper, we present a portion of these measurements only: for well-matched semi-flexible cable with female SMA connector and a hairpin resonator with cylindrical alumina ceramic support (diameter ~4 mm and length ~10 mm) with an overlapping between the coaxial input loop and the probe U-sector ~5 mm and with the following other parameters:
Structure S1: tungsten-wire probe with fixed length L_1 = 25.5 mm and fixed width W_1 = 2 mm;
Structure S2: tungsten-wire probe with different length L_2 ~ 25.5–35 mm and fixed width W_2 = 4 mm.

The length L of the structure S2 is changed by cutting, keeping the distance W between the wires.

2.3. Results and discussion

The variation of the resonance frequencies in air due to the change of the geometrical length of the hairpin wires is demonstrated in figure 4. It is easy to be shown that the resonance frequencies satisfy the simple equation for the fundamental frequency given in [9]. To the values of L shown in figure 4 must be added about 5 mm for the length of the wires inside the ceramic body.

The data for the high-order modes show that the influence has more complicated character – there are also parasitic effects caused by a non-effective coupling at a given resonator length (a worsening of the resonance curves and even a vanishing of the resonance peak or the appearance of a lot of nearby parasitic resonance). Thus, an important question appears: which resonance of the hairpin resonator could be most suitable for gaseous plasma diagnostics? We recommend the frequency range below 5–6 GHz. In this range a simple and not very expensive measuring set-up could be realized.

When we measured the resonance structures in the reflection-signal option with a vector or scalar network analyzer, we observed a number of resonance curves in the RL plot. Some of them are connected with the self resonances of the hairpin resonator. Unfortunately, we also have seen a lot of parasitic resonances, especially at higher frequencies, caused by the finite length of the feeding cable, coupling loop, etc. The parasitic resonances do not appear below 3 GHz. In contrast, above 6 GHz, a lot of parasitic resonances are observed and a reliable identification of the self (useful) resonances of the hairpin resonator could not be obtained. A metal cylindrical screen over the outer surface of the ceramic body of the structure could not considerably influence the self resonances, if its end is far enough from the coupling region (>10 mm). The coupling degree between the end of the transmission/receiver cable and the U shaped part of the antenna influences the resonance frequencies stronger (figure 5) compared with the case of other conventional semi-wavelength resonators.

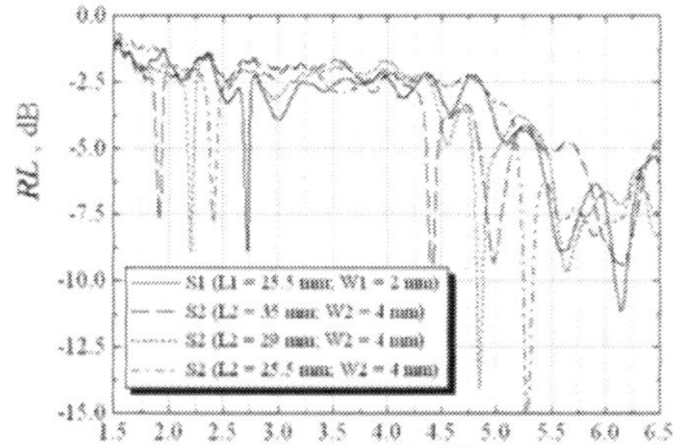

Figure 4. Measured resonance curves of the return losses RL in the structures S1 and S2.

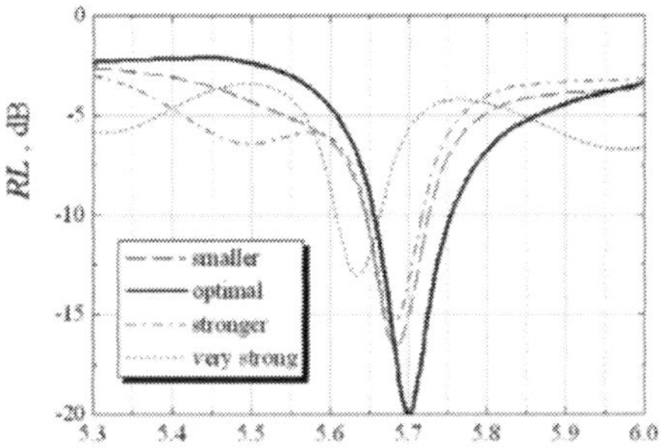

Figure 5. Resonance-curve shift due to an accidental change of the coupling degree.

This is due to the fact that the change of the coupling loop distance to the short-ended part of the hairpin resonator modifies the effective resonator length, and therefore it changes the resonance frequency. The frequency shift is only ~60–70 MHz at 5.7 GHz, but this shift may affect the

Journal of Physics: Conference Series **207** (2010) 012015 doi:10.1088/1742-6596/207/1/012015

preliminary calibration of the resonance structure. From now on, we focused our considerations over the resonances in the considered structures S1 below 6 GHz.

At the curve minimum, where the return losses are L_r (dB), we determine the resonance frequency f_v in vacuum. The corresponding loaded Q factor Q_{Lv} is calculated by the expression $Q_{Lv} = f_v/\Delta f$, where

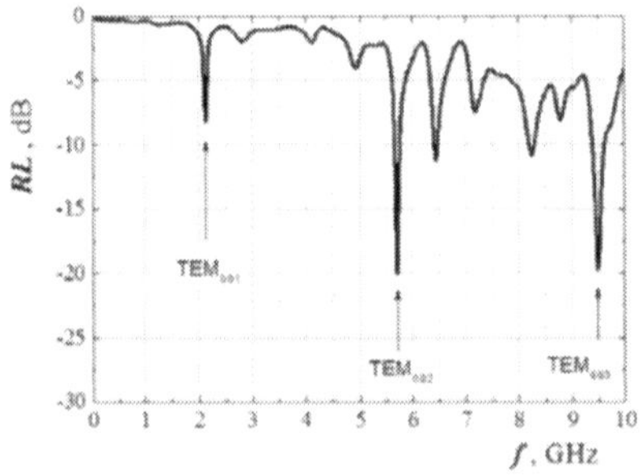

Figure 6a. Wide-band reflection response of the hairpin structure with three excited resonances.

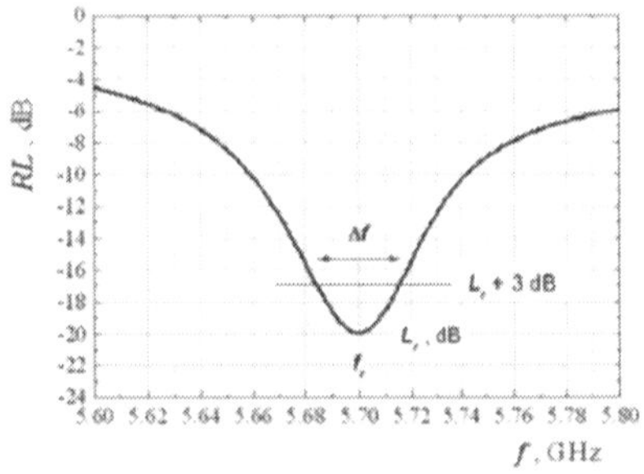

Figure 6b. Narrow-band response of TEM_{002} mode: $f_v \sim 5700$ MHz, $\Delta f \sim 30$ MHz, $Q_{Lv} \sim 190$, $Q_{0v} \sim 211$.

the resonance line-width Δf is determined at level $L_r + 3$ dB of the return losses. Finally, the unloaded Q factor Q_{0v} can determine by the expression

$$Q_{0v} = \frac{Q_{Lv}}{1 - 10^{-|L_r|/20}} \tag{2}$$

At one most effective coupling we have measured three resonance frequencies for structure S1. The corresponding Q-factors were as follows: 1) TEM_{001} mode: $f_v = 2127.5$ MHz with loaded Q-factor Q_{Lv} ~61 (unloaded Q_{0v} ~79); 2) TEM_{002} mode: $f_v = 5700$ MHz, Q_{Lv} ~190 (Q_{0v} ~211); 3) TEM_{003} mode: f_v = 9490, Q_{Lv} ~158 (Q_{0v} ~176). The first two resonances are well excited and without nearby parasitic resonances. Their Q-factors are high enough for similar open-type resonators as the hairpin. The resonances curve of the third mode is partially masked by a nearby parasitic resonance and the Q factor is not high enough. We conclude that the control of the plasma parameters should be performed on both low-order resonances (TEM_{001} and TEM_{002} modes).

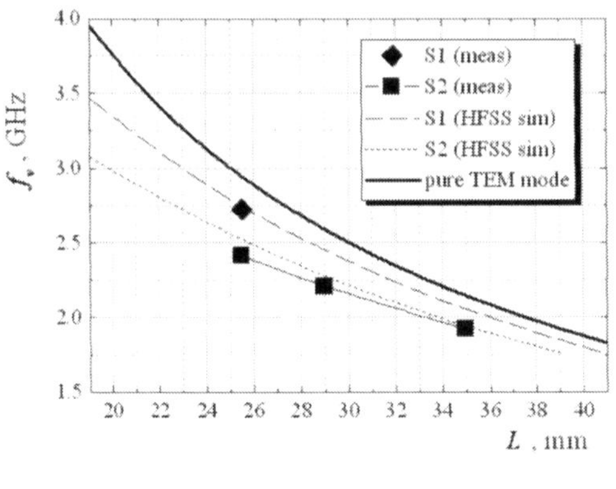

Figure 7a. Dependencies of the resonance frequencies of structures S1 and S2 versus the wire length *L*.

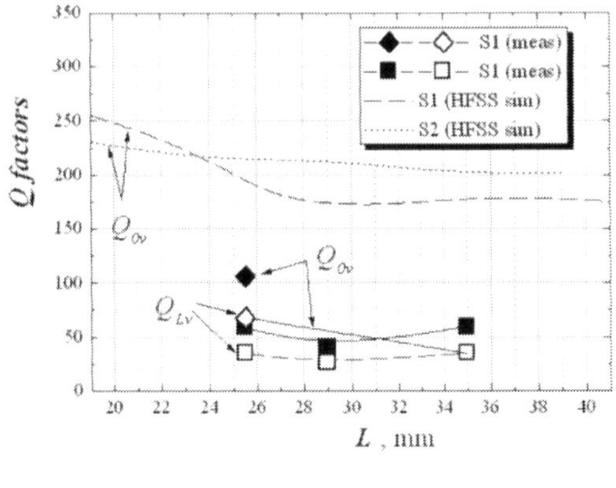

Figure 7b. Dependencies of the unloaded (Q_0) and loaded (Q_L) Q-factors versus the wire length *L*.

Figure 7 (a-b) illustrates the measured and the HFSS-simulated influence of the length of the hairpin wires over the resonance frequency and the unloaded Q-factor of the TEM_{001} mode for structure S1 and S2.). The decrease of the length of the structure shifts linearly the resonance frequency of the TEM_{001} mode to higher values (~ 52 MHz/1 mm). The loaded (measured) Q-factor also changes with the wire length, but changes are relatively small. The comparison between the measured and the simulated resonance frequencies is very good. Unfortunately, the simulated unloaded Q-factor is 2 to 4 times greater due to many reasons. The most important reason is that, we cannot determine well the actual value of the surface resistance of the metal wires due to the poor surface cleanness, oxidation, roughness, temperature change, etc. Thus, the HFSS simulations for the Q factor are only informative for the probe performance.

In the final case (figure 7b), we compare the parameters of the both structures S1 and S2 with fixed length 25.5 mm and different width between the wire ~2 and 4 mm. The narrow hairpin resonator has better performance: resonance frequency ~2.72 GHz and unloaded Q_0 factor 106, while the wide resonator shows worse parameters: resonance frequency ~2.41 GHz and unloaded Q_0 factor 60.

3. Conclusions

An effective simulation procedure with the commercial software simulator Ansoft® High Frequency Structure Simulator 8 in eigen-mode option can be applied to evaluate the influence of the resonator structures on the resonance frequency $(f_v)_{00p}$ and unloaded Q-factor $(Q_v)_{00p}$ of the resonance modes. The coupling between the U shaped hairpin antenna and the transmitter/receiver end of the coaxial cable is essential for reliable and repeatable measurements of the electrons density in RF plasma chambers.

4. Acknowledgments

This work is supported by EU grant NanoPlasma NMP-2004-IST-NMP-3. P. Dankov thanks to Scientific Research Found of University of Sofia "St. Kliment Ohridski" for the partial support concerning hairpin-resonator simulations and measurements in air.

References

[1] Stenzel R L 1976 *Rev. Sci. Instrum* **47** 603
[2] Piejak R B, Godyak V A, Garner R and Alexandrovich B M 2004 *J. Appl. Physics* **95** 3785
[3] Haas F A, Al-Kuzee J and Braithwaite N St J 2005 *Appl. Phys. Letters* **87** 111
[4] Warne L K, Johnson W A, Coats R S, Jorgenson R E and Hebner G A Printed April 2007 *SANDIA REPORT SAN2007-2513 United Release*
[5] Karkari S K and Ellingboe A R 2006 *Appl.Phys. Letters* **88** 101501
[6] Karkary S K, Law V J and Ellingboe A R 2005 *XXVIIth ICPIG Eindhoven the Nederlands* **18–22 July** Topic number 8
[7] Piejak R B, Al-Kuzee J and Braithwaite N St J 2005 *Plasma Sources Sci. Technol.* **14** 734
[8] Dine S, Booth J P, Curley G A, Corr C S, Jolly J and Guillon J 2005 *Plasma Sources Sci. Technol.* **14** 777
[9] Sands B L, Siefert N S and Ganguly B N 2007 *Plasma Sources Sci. Technol.* **16** 716

Ion-bombardment modification of the surface of mirrors fabricated of ZrTiCuNiBe amorphous alloys

V S Voitsenya*, A F Bardamid[1], A I Belyaeva[2], V G Konovalov, K V Kovtun, D I Naidenkova, V I Ryzhkov, A F Shtan', K A Slatin[2], S I Solodovchenko, O V Trembach and A A Vasil'ev

National Science Center "KIPT", 61108 Kharkov, Ukraine
[1]Taras Shevchenko National University, 01033 Kyiv, Ukraine
[2]National Technical University "KPI", 61002 Kharkov, Ukraine

*E-mail: voitseny@ipp.kharkov.ua

Abstract. When preparing mirror samples of amorphous metal alloys some inhomogeneities of the structure became to be seen on the surface. These inhomogeneities were modified during bombardment with ions of deuterium and argon plasma. Besides, a new blister-like type of inhomogeneities was found on the mirror surface in one experiment. In the paper a short description of obtained results are presented.

1. Introduction

Not long time ago the Bulk Amorphous Alloys (BAA) appeared [1,2] which have some properties different from properties of polycrystalline materials that are used at present time in technology and industry. Because of rather short time past since the BAA appeared, not many their characteristics were studied in detail, what limit their use in everyday life. Among characteristics that are not well known yet are those ones important for interaction of BAA with ions, like sputtering rate, gas absorptivity, modification of the near-surface layer, etc.

The looking for the answers these questions is the main objects of this paper. We present here some results of investigations directed to obtaining the data on peculiarities of impact of ions (deuterium and argon) with different energy on properties of mirror samples fabricated of BAA, namely, surface modification under long term ion bombardment, on deuterium uptake, reflectance, and sputtering yield.

2. Samples

The experiments were provided with mirror specimens fabricated of moldings of amorphous materials prepared by:

- Liquidmetal Technology Corporation (**LTC**) in the USA,
- Institute of Solid State Physics, Materials Science and Technologies of National Science Center "Kharkov Institute of Physics and Technology" (**ISSPMT**), and
- Hahn-Meitner-Institute (**HMI**) in Berlin, Germany.

The composition **LTC** and **ISSPMT** samples was – Zr(41.2)Ti(13.8)Cu(12.5)Ni(10)Be(22.5) and the composition **of HMI** samples was a little different – Zr(46.2)Ti(8.2)Cu(7.5)Ni(10)Be(27.5).

The round shape LTC and ISSMT samples were 22 mm in diameter, and the diameter of HMI samples was 5 mm; the thickness of all samples ~2.5 mm. After polishing to the mirror quality with gradual decrease of the abrasive particles, the samples were washed in acetone and saved in air. Before providing any experiment with exposing samples to ions of deuterium or argon plasma, the surface of every sample was cleaned with low energy (50–60 eV) ions of deuterium plasma (typical fluence 5×10^{19} ion/cm^2) to clean the surface from the organic film appeared due to washing in acetone.

The cleaning procedure and experiments with ion impact on mirror samples were provided in the DSM-2 installation described elsewhere [3], with parameters of ECR discharge plasma: $n_e \sim 10^{10}$ cm^{-3} and $T_e \leq 5$eV.

3. Experimental results

Several results on behavior of LTC mirror samples subjected to long-term bombardment with deuterium plasma ions have been already published in [4,5]. Here we repeat some of them for comparison with data obtained in experiments with samples of two other producers.

3.1. Surface of LTC mirror samples

3.1.1. After polishing and cleaning

After polishing and cleaning on surfaces of LTC samples some inhomogeneities became clearly seen [4]. In figure 1a the SEM photo obtained with scattered electrons of similar inhomogeneities is shown. They are a little above the matrix and occupy small part of the whole sample surface (1–2 %), therefore do not affect noticeably the reflectance. The fact they do protrude over the matrix indicates on higher local hardness in comparison to the hardness of the matrix. To clear up the reason of appearance of these inhomogeneities, the SEM photos were made with elastically backscattered electrons and the micro-probe analysis of the composition was provided (by measuring the characteristic X-ray radiation in X-ray spectrometer system attached to TEM). The photo of the same part of the surface shown in figure 1a is reproduced in figure 1b by registration of elastically reflected electrons. The darker color of the inhomogeneities is a qualitative indication on abundance of elements with lower atomic number. This qualitative result was proved by quantitative data obtained with application of microanalysis. It should be noted that, as we know, such inhomogeneities have not been described in literature previously.

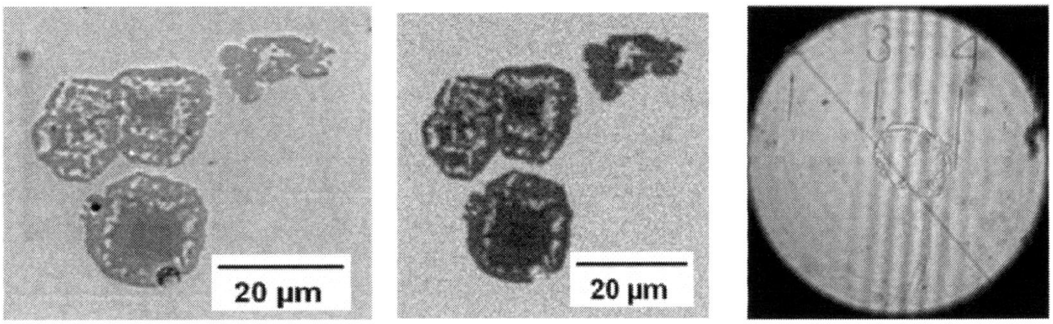

a) b) c)

Figure 1. SEM photos with secondary (a) and backscattered electrons (b) of inhomogeneities on the LTC sample surface; c) photo of interferometer microscope after the layer of ~2 μm was sputtered with deuterium plasma ions.

3.1.2. After long-term ion bombardment.

After long-term ion bombardment the inhomogeneities shown in figure 1 turned into shallow ‘pits’, what indicates that their sputtering rate is higher than the sputtering rate of the matrix, in agreement with lower portion of zirconium measured with micro-probe method. The photo of interferometer microscope in figure 1c was made when the ion sputtered layer thickness (bombardment with deuterium plasma ions) was estimated as ~2 μm. It can be seen that the depth of the 'pit' is ~200 nm. The matrix maintained the initial smoothness of the surface.

When the effect of ion energy (deuterium plasma) was investigated, the new inhomogeneities of the surface microrelief were found out [5]. They were observed only on a rather small area of the part of mirror sample bombarded with 300 eV ions. The photos in figure 2a show the view of these 'blister-like' artifacts in different scales, and figure 2b,c – the reconstruction made from AFM data.

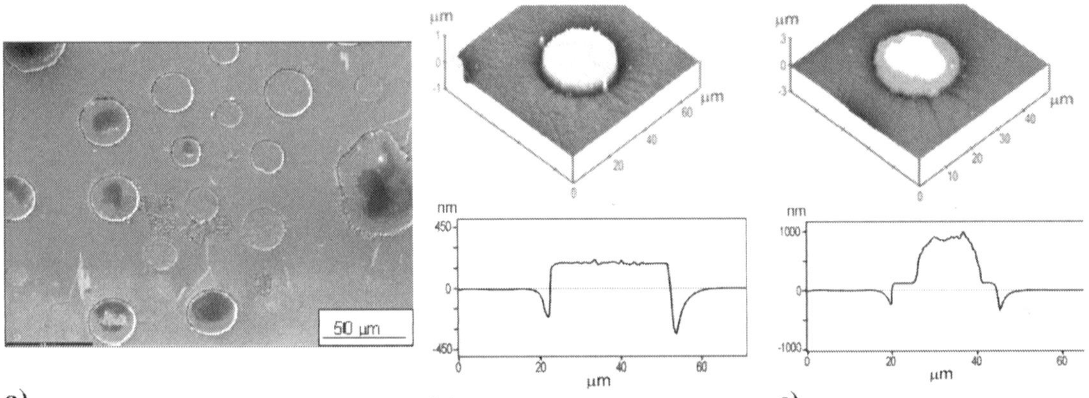

a) **b)** **c)**

Figure 2. SEM photo of blister-like features (a), and the AFM-images of different form 'blisters' and their profiles (b, c).

3.2. Surface of ISSMT mirror samples

3.2.1. After polishing and cleaning

After polishing, washing and cleaning in deuterium plasma, the ISSMT samples viewed in SEM very peculiarly as figure 3a shows. The shallow pits were almost uniformly distributed along the surface. Their depth, according to data of interferometer and atomic-force microscopes, was 60–70 nm. At the same time, the SEM photos made in backscattered electrons, figure 3b, demonstrate a very homogeneous picture without any compositional inhomogeneities, what was supported also by results of microanalysis. So, at present we do not have a non-controversial explanation of appearance of such 'pits'. However, it should be mentioned that because they are not deep and have a quite plane bottom, there is no measurable effect on reflectance at normal incidence in the range where optical measurements were done (250–650 nm).

3.2.2. After long-term ion bombardment

The long-term sputtering of ISSMT amorphous mirror sample with argon ions resulted in decreasing the contrast for observation of the pits that were clearly seen before sputtering (compare figure 3a and figure 3c). This is probably connected with gradual decrease of the depth of pits relatively the matrix level. Important that the surface continues to be smooth and reflectance does not change even after sputtering the layer of 13.4 μm in thickness, what is the experimental confirmation of the hypothesis stated in [6].

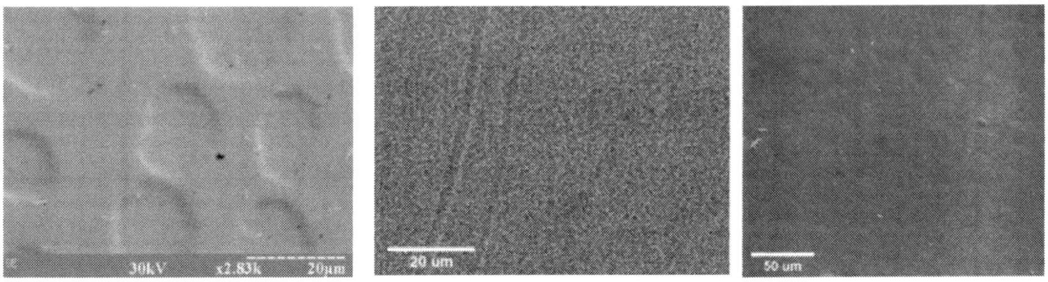

a b c

Figure 3. SEM photos of the polished mirror surface of a ISSPMT sample obtained with scattered electrons (a) and with backscattered electrons (b); SEM photo after layer of ~7 μm was sputter eroded with Ar ions of 1000 eV (c).

3.3. Surface of HMI samples

On the HMI polished samples small inhomogeneities were observed on the surface after polishing and they continued to be seen after ion bombardment. Majority of them are probably similar to that shown in figure 4, but were seen only partly. The shape of inhomogeneity of figure 4 is very much similar to the shape of the crystal, the photo of which was published in paper [7]. Therefore we can suppose that these inhomogeneities are crystals in the matrix and their composition is different in comparison to the matrix of the alloy, as was found in [7].

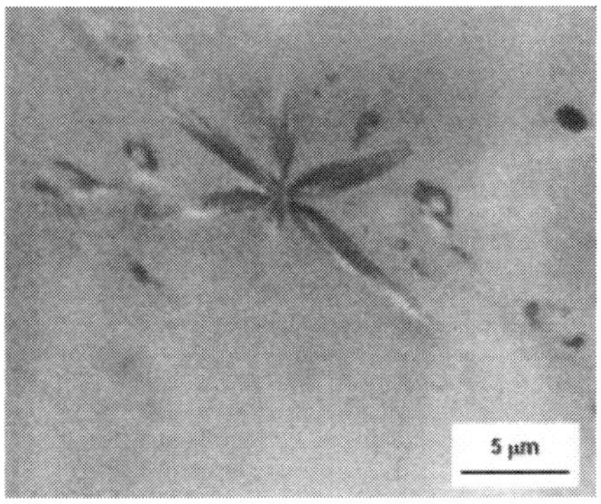

Figure 4. SEM photo of the polished mirror surface of a HMI sample after sputtering with argon ions the layer ~5 μm in thickness.

4. Conclusion

Some inhomogeneities were found out on the surfaces of mirror samples prepared of materials fabricated by 3 producers. The inhomogeneities differed from the main matrix by composition and structure (LMC and HMI samples) and are of a not known nature yet on the surface of ISSMT samples.

In the case of LMC material the inhomogeneities are abundant with elements of lower Z and their sputtering yield is a little higher than the sputtering yield of a matrix. As a result, the shallow pits (~0.2 μm) become to be visible after a layer of ~2 μm in thickness was sputtered by ions of deuterium plasma.

After sputtering by Ar^{+} ions of the layer ~13.4 μm the surface of amorphous ISSMT mirror sample continued to be smooth.

Acknowledgments

The authors acknowledge the financial support of STCU Grant #3668.

References

[1] Johnson W L March 2002 *JOM Journal of the Minerals, Metals and Materials Society* 40
[2] Inoue A and Takeuchi A 2004 *Material Science and Engineering* **A 375-377** 16
[3] Bardamid A F, Gritsyna V T, Konovalov V G, Orlinski D V, Shapoval A N, Shtan' A F, Solodovchenko S I, Voitsenya V S and Yakimov K I 1998 *Surf. Coat. Technol.* **103–104** 365
[4] Bardamid A F, Belyaeva A I, Bondarenko V N, Galuza A A, Kolesnyk O G, Konovalov V G, Naidenkova D I, Ryzhkov I V, Shapoval A N, Skinner C H, Shtan' A F, Solodovchenko SI, Voitsenya V S and Yakimov K I 2006 *Physica Scripta* **T123** 89
[5] Bardamid A F, Voitsenya V S, Lytvyn O S, Lytvyn P M, Konovalov V G, Shapoval A N, Solodovchenko SI and Yakimov K I 2008 *J. Nucl. Mater.* **376** 125
[6] Voitsenya V S, Bardamid A F, Becker M F, Gritsyna V T, Konovalov V G, Shtan' A F, Solodovchenko SI, Orlinskij D V and Yakimov K I 1998 *Rev. Sci. Instr.* **70** #1 Pt. II 790
[7] Wanderka N, Wie Q, Sieber I, Czubayko U and Macht M-P 1999 *Mater. Sci. Forum* **312-314** 369

Application of a multi-pass absorption cell for measurement of sputtered tungsten atoms' density

A Blagoev, A Pashov, A Stanev, I Bozhinova and I Rusinov[1]

Sofia University, Faculty of Physics, 5 James Bourchier blvd., 1164 Sofia, Bulgaria

[1]E-mail: irusinov@phys.uni-sofia.bg

Abstract. A dedicated gas discharge tube equipped with a hollow cathode made of tungsten is setup for sputtering of W atoms. A multi-pass White type absorption cell is designed to increase the sensitivity of the absorption method. The optical system consists of three externally mounted aluminium-coated spherical mirrors with 70 cm radius of curvature. Measurements of the cell relative transmittance in the UV and visible spectral regions are carried out. Time-resolved measurements of sputtered ground-state tungsten atoms' density in 2-pass configuration , as well as assessment of the diffusion coefficient of these atoms in argon at 350 K temperature are made.

1. Introduction

Tungsten has been widely applied in various plasma devices, given its high thermal resistance, excellent mechanical properties and low rate of sputtering due to bombardment with ions and neutral atoms with high energy. These are various gas discharge lighting sources, plasma reactors for layer deposition, as well as fusion devices of type tokamak, stellarator, etc. Under certain conditions, however, tungsten components are still eroded and tungsten atoms enter the plasma. Typical example is the electrode destruction in gas discharge lamps at start-up [1]. This effect has an impact on the lifetime of the lighting sources, on their stability and parameter reproduction. A key problem, concerning controlled fusion research is the interaction of plasma with the reactor walls. The requirements for the materials that come into contact with the burning plasma are complex. In the scope of the international fusion reactor project ITER, the plasma–facing materials to be used are beryllium for the main chamber wall, and carbon and tungsten in the divertor. A lot of tests are made with tungsten on existing experimental devices (ASDEX-Upgrade, Garching, Germany, JET, Culham, UK) [2,3]. A program for gradual covering of the device's walls with tungsten is carried out, and the obtained results are quite optimistic. A potential problem that has to be addressed is plasma pollution with high-Z impurity ($Z = 74$), which leads to radiation cooling. The maximum accepted tungsten concentrations in the tokamak core plasma, beyond which the normal operation of the device is restricted by radiation cooling, are very low (of the order of 10^{-3} % of plasma density). Whether relevant to low-temperature or high-temperature laboratory plasma, tungsten erosion makes it necessary to control the influx of sputtered atoms and their concentration in the near-wall plasma with high enough sensitivity. Time resolved measurements help clarifying the processes in the plasma device and are needed for optimizing its operation regimes.

This work is a part of research, aimed at application of optical methods (absorption and laser induced fluorescence) for sensitive detection of low densities of sputtered tungsten atoms with time resolution. A dedicated experiment, based on multi-pass absorption measurement of sputtered ground- and excited-state tungsten atoms is described.

2. Experimental

The experimental setup (figure 1) is similar to one previously used [4], with the exception of the White type multi-pass cell [5]. It consists of a light source, absorption tube, optical system, monochromator, photomultiplier working in gated photon-counting mode, computerized experiment controller and a 2-channel high voltage pulsed source (last two not shown in the figure). The absorption tube, used for sputtering of the metal by pulsed gas discharge, is made of pyrex glass. It contains a 20 cm in length and 30 mm in inner diameter cylindrical hollow cathode, the inner surface of which is covered by a tungsten sheet. The anode is a small aluminium cylinder placed at one side of the tube, close to the edge of the cathode. The glass tube is sealed at both ends with quartz optical windows. The White optical system is mounted externally and consists of 3 concave spherical mirrors, which have 70 cm radius of curvature. The main mirror has 2 holes allowing the light from the external source to enter the system and to leave it after a number of passes, produced by successive reflections from the mirrors. An additional spherical mirror is used as a condenser, introducing into the cell a converging beam from the external light source. The latter is a commercial tungsten hollow cathode lamp (Hamamatsu, filling gas neon). Both the absorption tube and the lamp are operated in synchronized pulse-periodic mode. The typical periods of pulse repetition are chosen in the range 25–30 ms. The absorption tube pulses have widths in the range 2–3 ms with 300–400 mA average current. The lamp discharge current is 40–50 mA and its duration – 500–600 μs. Figure 2 shows typical emission spectra of both tubes.

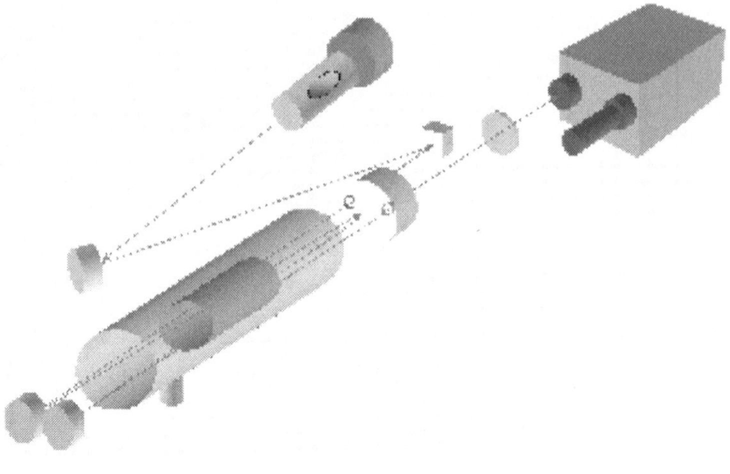

Figure 1. Experimental setup.

The experimental arrangement provides time-resolved measurements of the relative absorption of a given spectral line in the absorption tube volume with extended effective absorption path. The number of passes through the volume can be varied by proper alignment of the cell mirrors.

Several tungsten atomic states are of primary interest for absorption measurements, since they are expected to be the most populated in the plasma: $^5D_0$ (ground state); metastable $^5D_1$ (0.207 eV), $^7S_3$ (0.363 eV), $^5D_2$ (0.412 eV), $^5D_3$ (0.599 eV), $^5D_4$ (0.771 eV). There are a number of spectral lines in the UV region between 240 and 260 nm, connected with transitions involving these states, which have considerable strengths. However, higher loses in the UV, compared with the visible region are observed. This can be seen in figure 3, where continuum spectrum of a deuterium lamp has been recorded. The poorer transmittance is connected with losses in the monochromator and the multi-pass cell. The losses in the White cell accumulate with increasing the pass count, due to the multiple reflections at mirror surfaces and passes through the optical windows. Enhancement of the cell transmission will be obtained by multi-layer coating of the optical surfaces and in-tube arrangement of the multi-pass system.

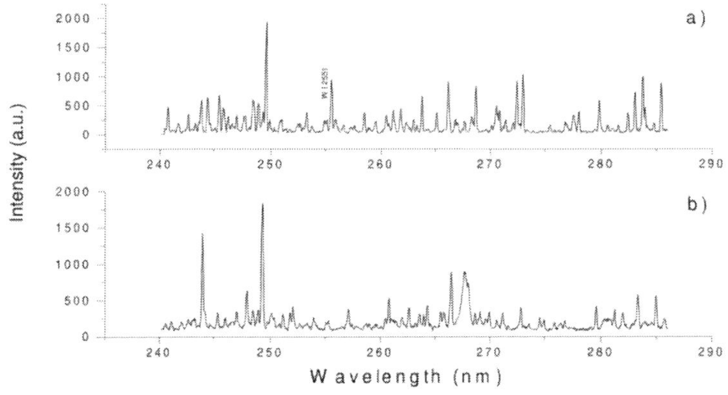

Figure 2. Spectra in the UV region emitted by the tungsten hollow cathode lamp (a) and the absorption tube in the active phase of a pulsed discharge in krypton (b).

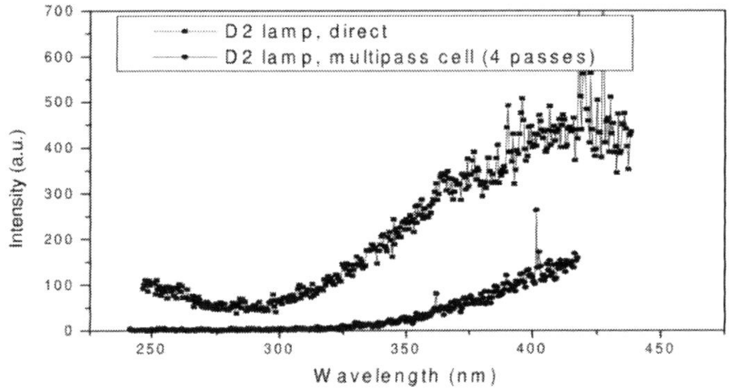

Figure 3. Deuterium lamp spectrum recorded in two arrangements for assessment of the multi-pass system throughput.

Illustration of the work of the system in two-pass arrangement can be seen in figure 4. Time resolved measurements of the density of ground-state tungsten atoms, sputtered in pulsed discharge in argon are shown. The absorption of the resonance line W I 2551 is used for the purpose. The curve is obtained by conversion of the recorded time dependency of the absorption signal (maximum value between 5 and 10 %), based on literature data for the line strength (0.52) [5] and modeled Doppler broadened line profiles. The line profiles are assessed by a model, considering the isotopic abundance of tungsten [6] and isotopic splitting of the $^5D_0$ level, using available data from [7]. Further considerations or profile measurements might be necessary.

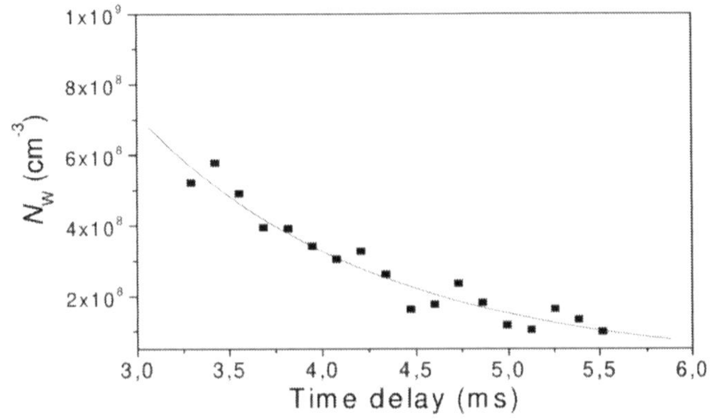

Figure 4. Decay of ground-state ($^5D_0$) sputtered tungsten atoms' density in the afterglow of pulsed discharge in argon. Two-pass arrangement of the optical cell. Gas pressure: 77 Pa; gas temperature: 350 K discharge period: 29 ms; discharge duration: 3 ms; pulse current: 400 mA.

The decay curve from figure 4 exhibits mono-exponential behaviour. It is consistent with a mechanism of diffusion controlled losses in the afterglow period, where the decay rate is dependent on two parameters: the diffusion coefficient and the wall reflection coefficient [8]. Both parameters cannot be derived from a single curve. If however, the reflection of the tungsten ground-state atoms at the cathode wall is negligible (< 30 %), then their diffusion coefficient can be estimated to $D_0 = 210 \pm 30$ cm^{2}s^{-1} at 133 Pa argon pressure, 350 K temperature. Further measurements at various gas pressures are needed to derive more definite values.

3. Conclusion

A multi-pass setup for absorption measurements of sputtered tungsten atoms' density has been put in operation and tested. The relative transmission of the optical system has been measured with impact on future enhancement of the optical system. Based on measured densities, preliminary assessment of the diffusion coefficient of ground-state tungsten atoms in argon has been made.

Acknowledgement

This work is supported by a BG EURATOM Association task and the Sofia University Science Fund.

References

[1] Born M, Lilie P and Uhlenbusch J 2000 *J. Phys. D: Appl. Phys.* **33** 1576–1580
[2] Naujoks D et al 1996 *Nucl. Fusion* **36** 671–687
[3] Maier H et al 2007 *Nucl. Fusion* **47** 222–227
[4] Rusinov I M and Blagoev A B 2006 *J. Phys.: Conf. Series* **44** 210–214
[5] Ballard J, Strong K, Remedios J J, Page M, Johnson W 1994 *J. Quant. Spectrosc. Radiat. Transfer* **52** 677–691
[6] National Institute of Standards and Technology 2008, NIST Atomic Spectra Database, http://physics.nist.gov/
[7] Radcig A A and Smirnov B M 1986 *Parameters of Atoms and Atomic Ions* (Moscow: Energy-Atom) (in Russian)
[8] Kiyoshi Murakawa 1954 *J. Phys. Soc. of Japan* **8** 215–218
[9] Rusinov I M, Paewa G W and Blagoev A B and 1997 *J. Phys D: Applied Phys.* **30** 1878–1874

Journal of Physics: Conference Series **207** (2010) 012018 doi:10.1088/1742-6596/207/1/012018

Ion energy distribution analysis of the TVA plasma ignited in carbon vapours using RFA

C. C. Surdu-Bob, M. Badulescu, C. Iacob, C. Porosnicu, C. P. Lungu

National Institute for Lasers, Plasma and Radiation Physics, Atomistilor No.409, POBox MG-36, Magurele-Bucharest, Romania

cristina.surdubob@plasmacoatings.ro

Abstract. In order to understand plasma processes and to obtain technological control in thin film deposition, the study of surface-plasma interactions is essential. Apart from the type and flux of the impinging ions/neutral atoms on the surface, the ion energy distribution (IED) is an important parameter in understanding surface modification due to the plasma. In this paper, results of ion energy analysis of the Thermionic Vacuum Arc (TVA) plasma ignited in carbon vapours are presented. An in-house, computer-controlled retarding field analyzer was used for determining experimentally ion energy distributions of the carbon ions arriving at the substrate. The correlation of the carbon IED with the applied arc voltage in the TVA plasma was put in evidence for the first time.

1. Introduction

Ion energy distribution is an important parameter in understanding surface modification due to plasmas. Ion energy distribution functions (IEDF) can be measured by energy selective mass spectrometry [1], by time of flight spectrometers [2], by retarding field ion energy analysis (RFA) [3] or can be determined theoretically by Particle-in-cell/Monte Carlo simulations [4].

One of the simplest devices capable of obtaining the IEDF is the RFA. This is an electrical probe which operates as an energy filter. The basic principle of an RFA is the measurement of an electrical current given by electrically charged plasma species against a retarding voltage. The ion energy distribution function is obtained from the first derivative of the current-voltage characteristic I-V. The system can work in either ion mode or in electron mode, depending on the sign of the constant bias applied on the grid in front of the collector. The device consists on a metal collector and a set of grids placed before the collector, which are used to eliminate electrons and to suppress secondary electrons. More details can be found in [5]. The fact that in Thermoionic Vacuum Arc (TVA) the plasma does not fill the vacuum chamber, and thus it does not enter into the analyzer, a simpler design was chosen in this case, with only one electron-repelling grid.

We report here for the first time on the correlation of the plasma operating parameters with the ion energy distribution for carbon ions generated by the TVA plasma.

2. Experimental Setup

The TVA is a localized plasma in a vacuum environment and is ignited using a metal or graphite anode and an electron gun. The ignition is initialized by thermo-electrons generated by the externally heated tungsten filament – the cathode. The experimental arrangement is presented in figure 1 and further details can be found in [6].

The ions created in this localized plasma are accelerated towards the chamber walls (and subsequently towards the substrate) due to the potential difference between the plasma and the grounded walls. Due to a low background pressure (9×10^{-4} Pa) the energy of ions is directly proportional to this potential difference. This is an important feature of the method, as the ion energy can be directly controlled by the operating plasma parameters.

The TVA plasma source was placed in a vacuum chamber pumped down to a base pressure of 9×10^{-4} Pa. The anode consisted of a pure carbon rod of 12.5 mm diameter. As the anode is continuously eroded during plasma running, a computer controlled in-house system which allows a fine mechanical adjustment of the interelectrodic distance was used. The in-situ acquisition of the arc and analyzer I-V characteristics were also performed using current and voltage transducers and LabView software.

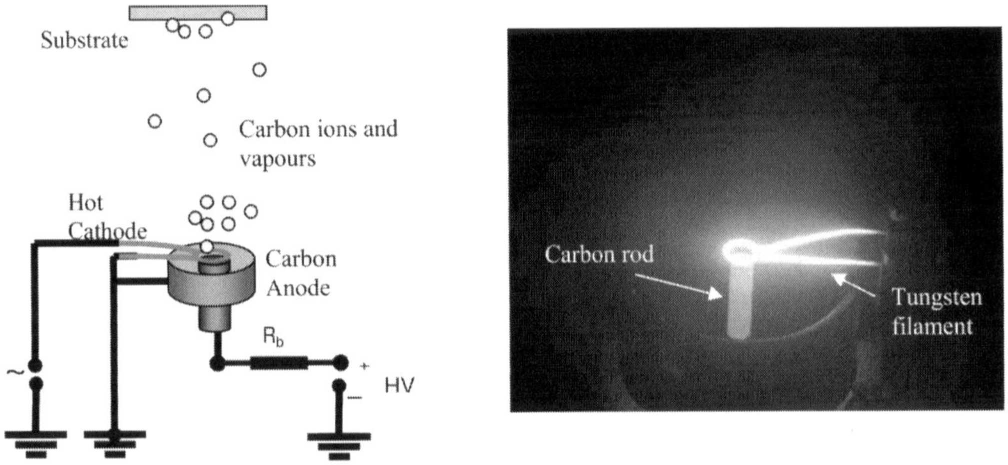

Figure 1. The experimental setup of the TVA plasma

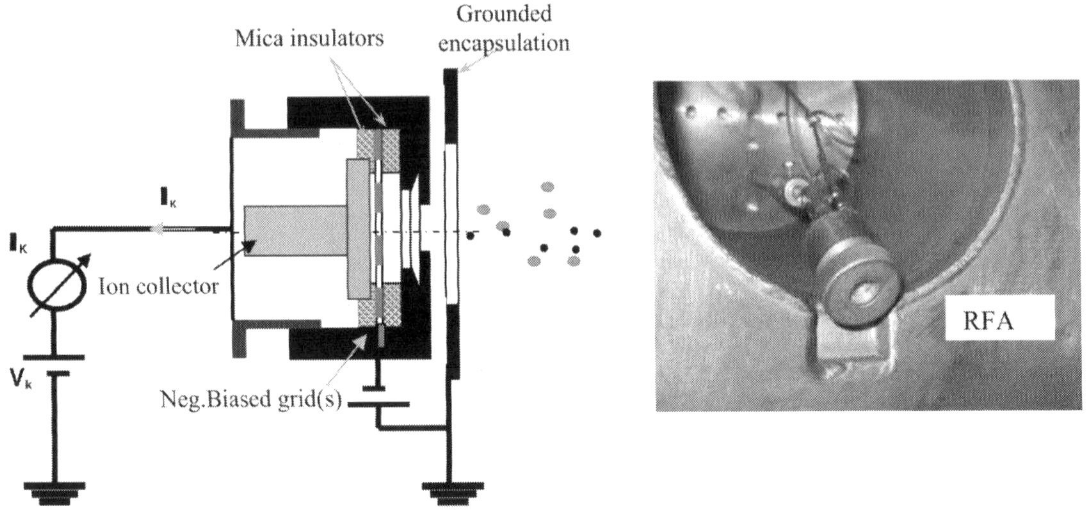

Figure 2. The retarding field ion analyzer.

As the TVA plasma does not fill the vacuum chamber, the energy of the carbon ions formed in the plasma is given by: $E = e(V_0 - V_p)$, where e is the charge of the electron, V_0 is the potential of the arriving point of the ion, and V_p is the plasma potential. Thus, for a given V_0 (which can be ground or any other voltage), the ion energy is directly proportional to the plasma potential which, in turn, is directly correlated to the arc voltage and current.

This paper presents the ion energy distribution at 35 cm distance from the anode.

In order to determine the ion energy distribution, an in-house RFA was used to decelerate and filter ions, according to their energy. A planar RFA consisting of one stainless steel grid (mesh having 30 μm wire and 50 μm gap) and a molybdenum collector, encapsulated in a stainless steel cylinder was used. The grid was polarized with a constant potential. The collector also serves as the retarding electrode. The main components of the RFA module can be observed in figure 2.

3. Results and Discussion

A typical I-V characteristic of the carbon TVA plasma containing the ignition stages of the plasma can be found in [7].

The RFA data presented in figure 3 were obtained for the plasma operating conditions shown in the figure legend.

The voltage applied on the negatively biased grid, which has the role to repel the electrons, did not influence the shape of the I-V characteristic at the collector. The only difference observed was the relative position of this characteristic to the abscissa. The explanation for this behavior is that the negatively biased grid is partially successful in repelling the secondary electrons resulted from the collector due to ion impact. As Bohm [5] pointed out also, in order to obtain the real value of the collector current the electronic contribution to the current must be taken out. Thus, the I-V characteristic of the collector should be shifted with the value given by the saturation of the current signal. For simplicity, we did not shift the characteristics, as at this stage we are interested on the energy distribution function only. In this respect, further studies will be undertaken for the estimation of the ion flux.

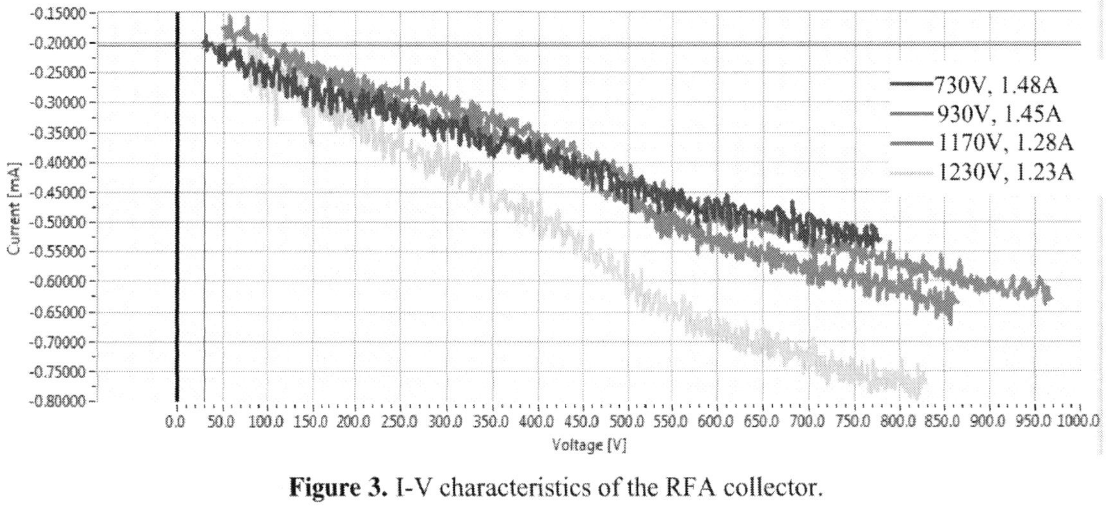

Figure 3. I-V characteristics of the RFA collector.

Figure 4 shows the first derivative of the I-V characteristics presented above and represents the ion energy distribution. From this figure it can be observed that the ion energies have values in the range 300-600 eV and do not depend significantly on the applied voltage at the anode. These are very high values of the carbon ion energy and are not found in other plasma sources. During thin film deposition, high energy ions and neutrals arrive at the substrate. This fact gives them superiority over films obtained by other plasma methods, in terms of higher compactness and smoothness.

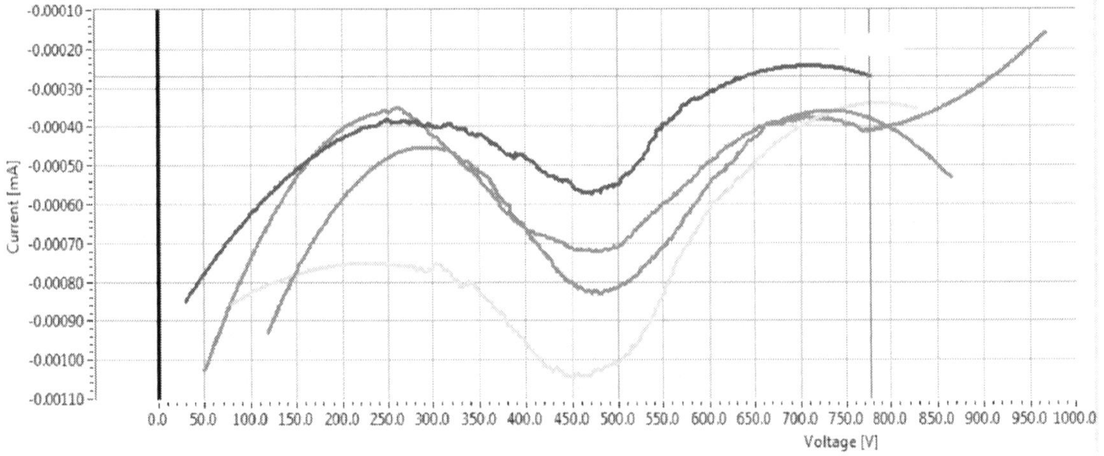

Figure 4. The IEDF of the carbon plasma

4. Conclusions

A relatively high ion energy was found for the carbon ions obtained in the TVA plasma.

It was observed experimentally that the values of the ion energy of carbon ions vary within a relatively small range around 450 eV for arc voltages in the 700–1250 V range. This high energy is not encountered in the ions formed in other types of plasma. This fact gives them superiority over films obtained by other plasma methods, with higher compactness and smoothness.

Further investigations for the measurement of the ion flux for different plasma conditions are envisaged.

References

[1] Olthoff J K, Van Brunt R J, Radovanov S B, Rees J A 1994 *Science, Measurement and Technology* IEE Proceedings **141**/2, 105–10

[2] Richter F 1996 *Physikalische Blätter* **52** 355 (in German)

[3] Ivanov et al *Surf.ace Coat. Technol.* 1997 **92** 1X-156

[4] Lee J K et al 2005 *Plasma Sources Sci. Technol.* **14** 89–97

[5] Bohm C and Perrin J 1993 *Rev. Sci. Instrum.* **64/**1

[6] Surdu-Bob C C, Mustata I and Iacob C 2007 *J. Optoelectron. Adv. Mater.* **9**/9 2932–4

[7] Surdu-Bob C C, Vladoiu R, Badulescu M and Musa G 2008 *Diamond Relat. Mater.* **17**/7–10 1625–8

AC Plasma Induced Modifications in Sb_2S_3 Thin Films

M. Calixto-Rodriguez[1], F. Castillo[2,*], H. Martínez[1], Y. Peña[3], and A. Sanchez-Juarez[4]

[1] Instituto de Ciencias Físicas, Universidad Nacional Autónoma de México, Apartado Postal 48-3, 62210, Cuernavaca, Morelos, México.
[2] Instituto de Ciencias Nucleares, Universidad Nacional Autónoma de México, Apartado Postal 70-543, 04510, México D. F., México.
[3]Universidad Autónoma de Nuevo León, Facultad de Ciencias Químicas, Pedro de Alba s/n, Cd. Universitaria, San Nicolás de los Garza, N.L.
[4]Centro de Investigación en Energía, Universidad Nacional Autónoma de México, Privada Xochicalco s/n Col. Centro, Temixco, Morelos, México; C.P. 62580.

*E-mail: ciro@nucleares.unam.mx

Abstract. Sb_2S_3 thin films, deposited by the chemical bath deposition method, were treated with N_2 plasma at 3.0 Torr during several minutes. The as-prepared Sb_2S_3 thin films and films treated with N_2 plasma have been characterized using several techniques. X-ray diffraction studies have shown that plasma treatment induced recrystallization on the as-prepared Sb_2S_3 thin films. The band gap values decreased from 2.37 to 1.82 eV after plasma treatment, and the electrical conductivity increased from 10^{-9} to 10^{-7} $(\Omega\ cm)^{-1}$ due to the annealing effect.

1. Introduction
Antimony sulfide (Sb_2S_3) has technological applications as target material of television cameras [1], microwave [2], switching [3], and optoelectronic devices [4]. Recently, considerable attention has been given to the preparation and characterization of Sb_2S_3 thin films for use in photovoltaic devices [5-8]. Chemical bath deposition is a very important method used to grow metal chalcogenide materials in thin film form for solar cell applications, due to the simplicity and non-expensive of the technique.

Plasma treatment is an interesting technological method to modify the structural [9] and surface [10] properties of the materials. This paper deals with N_2 plasma treatment used to modify the structural, optical, and electrical properties of the as-prepared samples of Sb_2S_3 thin films obtained by the chemical bath deposition method.

2. Experimental details
Antimony sulfide thin films were prepared using the chemical bath deposition technique, similarly to that already reported in [11]. The bath was prepared using antimony trichloride, $SbCl_3$ and sodium thiosulfate, $Na_2S_2O_3$, and deionized cold (~10°C) water. Microscope glass slides were used as substrates. The substrates were placed vertically in the solution. The deposition was made at ~10 °C for 4 hours without stirring. Thickness of the films was measured with an Alpha Step model 100

* To whom any correspondence should be addressed.

profilometer from Tencor Instruments, the mean value of the film thickness was 240 nm. The as-prepared samples of Sb_2S_3 were treated with N_2 plasma at 3.0 Torr during 15, 30, and 70 minutes.

The experimental apparatus and technique to generate the pulsed plasma was recently reported [12]. The discharge power supply was maintained at an output of 300 Volts and a current of 0.36 A. The glow discharges were monitored by plasma emission spectroscopy using an Ocean Optics Inc. Spectrometer Model HR2000CG-UV-NIR. X-ray diffraction (XRD) patterns were recorded on a Rigaku D-Max X-ray diffractometer using Cu-K_α radiation (λ = 1.5406 Å). The optical transmittance and specular reflectance spectra of the samples were measured with a spectrophotometer Shimadzu model UV-1601PC. For the electrical measurements, current vs. time data were recorded on an automated system using a Keithley 619 electrometer and a Keithley 230 programmable voltage source.

3. Results and discussion

3.1 Structural analysis

The XRD patterns for as-prepared Sb_2S_3 thin films as well as films treated with N_2 plasma are shown in figure 1. It can be seen that no diffraction peaks can be identified on the pattern of the as-prepared film. It is possible that the material is growing with nanometric grains. The diffraction patterns corresponding to films treated with N_2 plasmas during 15, 30, and 70 minutes match well the standard for Sb_2S_3 (JCPDS 42-1393) which has an orthorhombic structure. Films treated with N_2 plasma showed a preferential orientation along the (310) direction. The mean value of the crystallite size for Sb_2S_3 films treated with N_2 plasma was calculated for the (310) diffraction peak, using the Scherrer formula $D = (0.9\lambda)/(\beta \cos\theta)$, where D is the diameter of crystallites, λ is the wavelength of Cu-K_α line, β is full width at half maximum (FWHM) in radians and θ is Bragg's angle. It was found that the value increased from 15.3 nm to 17.8 nm with the increasing in time to the N_2 plasma treatment. This may be due to the recrystallization of the films during the plasma process.

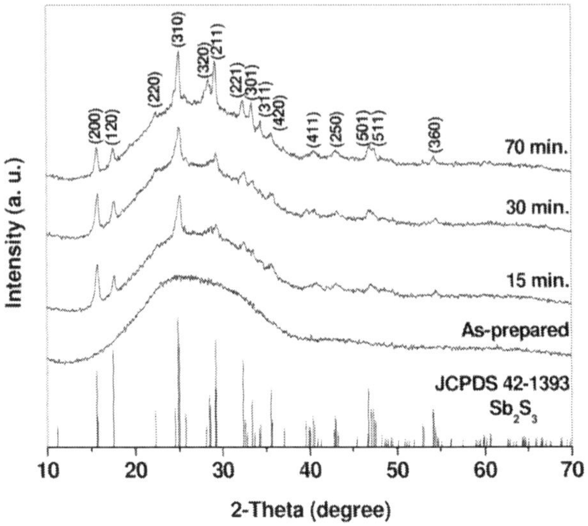

Figure 1. XRD patterns for as-prepared Sb_2S_3 thin films as well as films treated with N_2 plasma.

3.2 Optical emission spectroscopy analysis

Optical emission spectroscopy (OES) measurements were obtained for N_2 glow discharge plasma (figure 2). This fact allowed the analysis of the most luminous area that corresponds to the negative glow near the cathode dark space (it has been subtracted the intensities of the bands coming from the N_2 plasma). The identified species removed from the thin films into the plasma were S_2, NS, Sb and SbS. The majority of the species are vibrationally excited at a fundamental level. There is a flux of material into the plasma that can be deposited somewhere in the chamber and some back on the surface from which it came (in a modified form). By using OES and observing both S_2, NS, SbS and Sb neutrals, it is possible to assume that there is a possible mechanism by which Sb_2S_3 thin film ionization occurs in plasma treatment: electron impact ionization.

$$Sb_2S_3 + e \Rightarrow S_2 + Sb_2S + e \quad (1)$$
$$Sb_2S_3 + N \Rightarrow NS + Sb_2S_2 + e \quad (2)$$
$$Sb_2S_3 + e \Rightarrow Sb + SbS_3 + e \quad (3)$$
$$Sb_2S_3 + e \Rightarrow SbS + SbS_2 + e \quad (4)$$

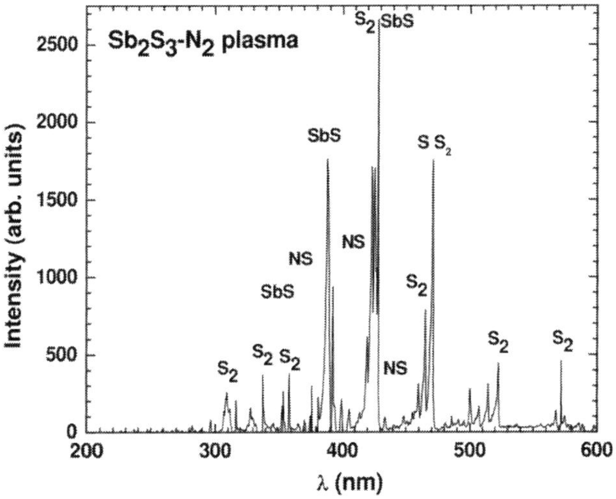

Figure 2. Typical OES spectrum for Sb_2S_3 thin film-N_2 plasma interaction.

3.3 Optical and electrical measurements

The behaviour of the transmittance, T(λ), for as-prepared Sb_2S_3 thin films as well as films treated with N_2 plasma at 3.0 Torr during 15, 30, and 70 minutes are presented in figure 3a. Samples after plasma treatments show a decrease in the optical transmission due to the structural changes. Band gap, E_g, of the samples was determined using the $(\alpha h\nu)^2$ versus $h\nu$ graph by extrapolating the linear portion of the graph to the $h\nu$ axis (figure 3b). The as-prepared Sb_2S_3 thin film had a band gap of 2.33 eV. As the exposition time to the N_2 plasma increased, the band gap decreased. The E_g values for films treated with N_2 plasma during 15, 30, and 70 minutes were: 1.85 eV, 1.80 eV, and 1.73 eV, respectively. This decrease might be due to the increasing in the particle size as observed in the XRD studies.

The electrical conductivity value for the as-prepared thin films was 1.0×10^{-9} $(\Omega\ cm)^{-1}$. The values were ~ 2×10^{-9}, 3×10^{-8}, and 1×10^{-7} $(\Omega\ cm)^{-1}$ for samples treated with N_2 plasma during 15, 30, and 70 minutes, respectively. These results are in agreement with those reported in references [11,13] for Sb_2S_3 thin films deposited by chemical bath deposition and thermal annealed in N_2.

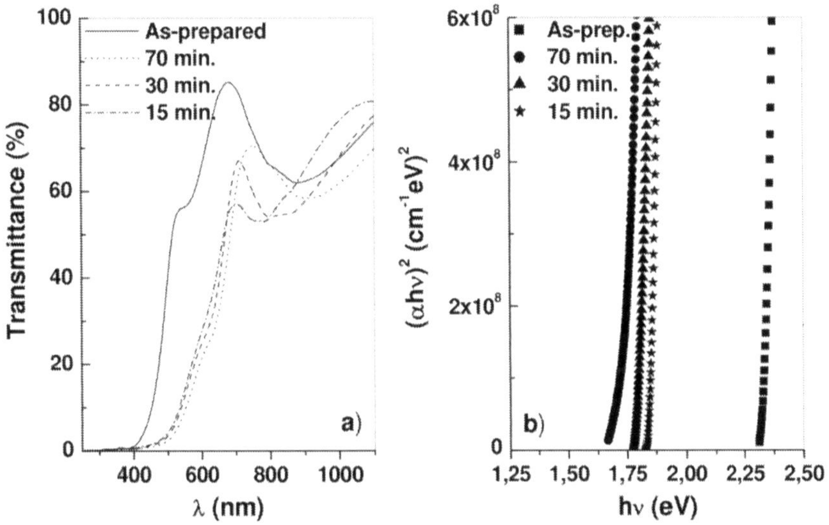

Figure 3. T(λ) and $(\alpha h\nu)^2$ vs $h\nu$ plots for as-prepared Sb_2S_3 film and films treated with N_2 plasma.

4. Conclusions

N_2 Plasma treatments induced recrystallization was observed in Sb_2S_3 thin films due to annealing effect of plasma process. Films treated in N_2 plasma showed optical and electrical properties similar to that obtained through a conventional thermal treatment.

Acknowledgments

The authors are thankful to Maria Luisa Ramón for the XRD measurements, José Ortega (CIE-UNAM), A. Bustos, A. González, R. Bustos (ICF-UNAM), and José Rangel (ICN-UNAM) for technical assistance. One of the authors (MCR) acknowledges the financial support received from DGAPA-UNAM. This research was partially sponsored by DGAPA IN-105707-3, and CONACyT 41072-F.

References

[1] Cope, US Patent 1959 2875359
[2] Grigas J, Meshkanskas J and Orlimas 1976 *Phys. Stat. Solidi (A)* **37** 10
[3] Ablowa M S, Andreev A A, Deb Akaev T T, Melekh B T, Peutsow A B, Sheridel N S and Shivilona L N 1976 *Sov. Phys. Semicond.* **10** 29
[4] Montrimass E and Pazera A 1976 *Thin Solid Films* **34** 65
[5] Rajpure K Y and Bhosale C H 2000 *Mater. Chem. Phys.* **64** 70–4
[6] Rodriguez-Lazcano Y, Nair M T S, Nair and P K 2005 *J. Electrochem. Soc.* **152** G635
[7] Bindu K, Campos J, Nair M T S, Sanchez A and Nair P K 2005 *Semicond. Sci. Technol.* **20** 496
[8] Messina S, Nair M T S and Nair P K 2007 *Thin Solid Films* **515** 5777
[9] Sagar R and Sristava M P 1993 *Phys. Lett. A* **183** 2-3 209
[10] Zhong Z Y and Jiang Y D 2006 *J. Colloid. Interf. Sci.* 302
[11] Nair M T S, Peña Y, Campos J, García V M and Nair P K 1998 *J. Electrochem. Soc.* **145** 6 2113–20
[12] Rosales I, Martinez H, Ponce D and Ruiz J A 2007 *Int. J. Refract. Met. Hard Mater.* **25** 250–5
[13] Grozdanov I 1994 *Semicond. Sci. Technol.* **9** 1234

Measurements of electron temperature, and ion density, in a mixture of O_2 and Ar plasma

P. G. Reyes[1], E. Mendez[1], D. Osorio-González[1], F. Castillo[2], H. Martínez[3].
[1]Facultad de Ciencias, Universidad Autónoma del Estado de México, Toluca, Estado de México, México.
[2]Instituto de Ciencias Nucleares, Universidad Nacional Autónoma de México, Apartado Postal 70543, C. P. 04510, México D. F., México.
[3]Instituto de Ciencias Físicas, Universidad Nacional Autónoma de México, Apartado Postal 48-3, C. P. 62251, Cuernavaca, Morelos, México.

E-mail: pgrr@uamex.mx

Abstract. Optical emission spectroscopy was applied for plasma characterization of a gas mixture glow discharge of O_2 and Ar at a total pressure of 1.0 Torr, power of 150 W, and with 40 l/min flow rate of gases. The main diagnostic techniques applied to determine the plasma parameters were optical and probe diagnostics. We study the dependence of the electron density and temperature as a function of the relative O_2 concentration. The emission bands and lines were measured using a monochromator in the wavelength range from 200 to 1000 nm. We also study the dependence of the principal band intensities observed as the relative O_2 concentration was increasing from 0 to 100 % of the mixture.

1. Introduction

The electron temperature and the electron density are important parameters that describe the operating conditions of low-pressure discharges. Techniques based on optical emission are commonly used to measure those parameters [1–3]. These techniques are usually effective, convenient to use, and non-intrusive. The knowledge of the plasma characteristics is required to understand fully the effects of the different physical processes taking place in the plasma and to deduce from them its properties. These diagnostics includes electrical probes and optical spectroscopy. Usually, only probes are used for measuring the density and temperature of electrons and ions [4,5].

On the other hand, the glow discharge in pure oxygen and its mixture with different rare or molecular gases in pressure range of several Torr can be utilized in miscellaneous applications (e.g. in sterilization, deposition of thin layers, lasers, etc.). The study of the oxygen properties in this pressure range is therefore a very important topic.

With respect to contemporary trends in the application of the plasmas of oxygen and its mixtures a complex study for medium pressures has been started. It is known that the material of the tube wall influences oxygen discharges, namely by the occurrence of surface re-association processes leading to the loss of oxygen atoms, which is important in a lot of applications [6,7].

However, since, there are no systematic studies undertaken to monitor by plasma emission spectroscopy the active species generated in the glow discharge, therefore, it is the aim of the present work to perform analysis of glow discharge by electrical and optical characterization employing pulsed plasma of a mixture of O_2 and Ar gases as a function of the relative O_2 percentage.

2. Experimental details

A DC glow discharge plasma in a mixture of Ar and O_2 gases was generated in a plasma apparatus of the Laboratorio de Física Avanzada, Facultad de Ciencias, UAEM. This experiment is described in detail in [8,9,10]. The reactor consisted of two copper circular plane electrodes, 3 mm thick and 30 mm in diameter. The electrodes are positioned at the centre of the reaction chamber with 20 mm gap spacing. The gas was introduced into the reaction chamber through the lateral flange. A lateral flange was used for the pressure sensor (MKS, model 622A). The plasma chamber was initially evacuated to a base pressure of 1×10^{-6} Torr with a turbo-molecular pump (Alcatel model ATP80) backed by a mechanical pump (Varian D5302, model 949-9325) and purged with the working gas at a pressure of 4.0 Torr. Ultrahigh pure gases (Praxair 99.999 %) were used in the course of the measurements. In another flange, a quartz window was used to monitor the active species generated in the glow discharge by plasma emission spectroscopy; this was performed using a ACTON (SpectraPro 2150i) monochromator equipped with a 1200 lines mm^{-1} holographic grating. The spectrum (200-900 nm) of the emission cell was recorded with a photo multiplier (Hamamatsu model PD471). The wavelength scan interval was 0.1 nm and the dwell time was 500 ms. The data was obtained in a single accumulation with a 0.5 s integration time. The DC plasma was produced in a mixture of Ar and O_2 gases at a pressure of 1.0 Torr. The discharge power supply (Bertan Serie 105) was maintained at an output of 375 V and a current of 0.4 A (150 W), which was measured using a digital Fluke multimeter model 8864. Langmuir probes are commonly used as a diagnostic tool for the determination of plasma parameters, such as electron temperature and ion density. The double Langmuir probe employed in this work consists of two 33 cm length and 0.2 mm radius tungsten wires, which were located inside a 31 cm glass capillary. The wire tips were extended 2 mm beyond the glass capillary.

3. RESULTS AND DISCUSSION

3.1. Electrical measurements

The evaluation algorithm for obtaining the electron temperature was run in the following manner. First the probe characteristic curve is differentiated two times and then smoothed. Smoothing was performed following the procedure described by Savitzky and Golay [11]. For each of the obtained current-voltage characteristic curve at each percentage of O_2, the electron temperature, T_e, is calculated using the equation given in [12]. Following that, a theoretical curve is fitted to the entire measured experimental current-voltage characteristic data curve as given by [13,14] by adjusting only the ion density, n_i.

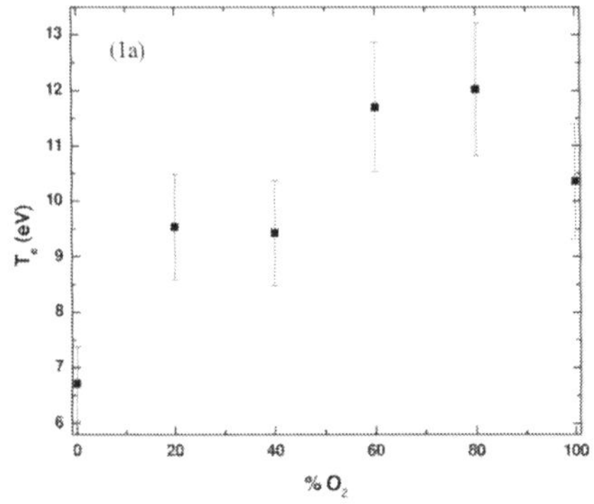

Figure 1. Electron temperature as a function of O_2 percentage.

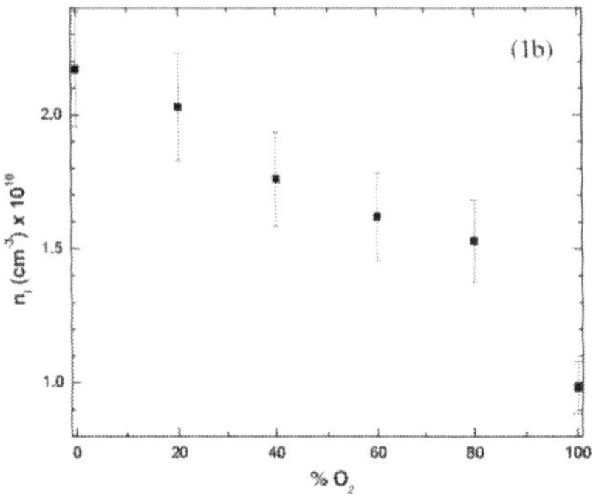

Figure 2. Ion density as a function of O_2 percentage.

Journal of Physics: Conference Series **207** (2010) 012020 doi:10.1088/1742-6596/207/1/012020

Figure 1 shows the electron temperature measured at different percentage values of O_2 in the gas mixture. It shows an increasing pattern. This increasing pattern is rather slow, starting at a 6.8 eV electron temperature for 0 % O_2 and reaching 12.0 eV for 80 % of O_2 in the mixture. Then the electron temperature decreases reaching a value of 10.4 eV at 100 % O_2.

The correlation between the variation of the electron temperature and the relative number of oxygen atoms should be pointed out. Electron impact dissociation is the main source channel for ground state oxygen atoms for the present conditions. Decreasing the Ar amount results in an increase of T_e, thus in the high-energy part of the electron distribution function and, consequently, in the rate of molecular dissociation. Similarly, the measured ion density as a function of the O_2 percentage is shown in figure 2. Note that, compared with electron temperature, the ion density clearly decreased.

3.2. Optical Emission Spectroscopy Analysis

A typical spectroscopic emission measurement of the mixture O_2-Ar glow discharge plasma in a pressure range between 0.0 and 1.0 Torr in percentage of O_2 is displayed in figure 3. This allowed analysis of the most luminous area, which corresponds to the negative glow near the cathode dark space [15]. Considering the population of O_3, O_2, O_2^+, O, Ar and Ar^+ emission bands and lines to be the most abundant, this would imply that the main process for the production of these species is the electron impact dissociation or ionization excitation. The intensities of six main emission bands from the mixture O_2-Ar glow-discharge plasma at 0.4 A discharge current is plotted in figure 4 as function of the percentage of oxygen.

The interplay between the formation and the destruction processes is rather complicated, it depends on many parameters, and a full kinetic model is required in order to get a deeper insight in understanding the mechanisms resulting in the formation of excited O_3, O_2, O_2^+, O, Ar, and Ar^+. Within the experimental parameters of the plasma studied in this work, the plasma generated under each parameters of power and pressure reach an equilibrium at which the above-mentioned radicals are the dominant species.

From figure 4, the intensity of 777.34 nm O displays an increasing behavior as % O_2 increases, while the emission bands and lines (421.2 nm O_2, 439.8 nm O_2^+, and 473.7 nm Ar^+) show a decreasing behavior as % O_2 increases. The emission band (316.4 nm O_3) and line (810.9 nm Ar) display a strong decreasing of the intensity from 0.2 to 0.4 % O_2, and then display an almost constant behavior as a function of % O_2.

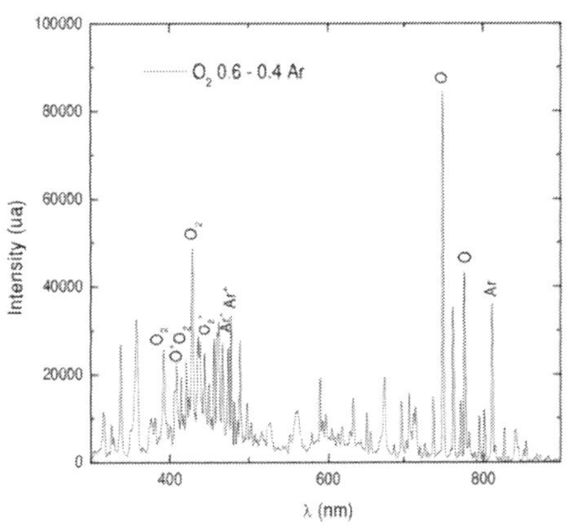

Figure 3. Spectrum emission measurement of the mixture O_2-Ar.

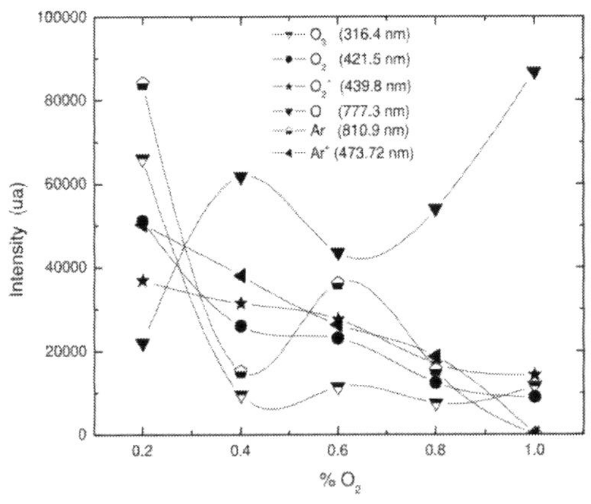

Figure 4. Abundance of elements as function of O_2.

The intensities observed in this work for these species could indicate that although they were produced in the plasma, they are initially in their excited states and decay radiatively or via collisions in reactions with other species.

The behavior of the emission intensities may be correlated to the changes in the ion density and electron temperature. If the pressure is increased, there are more products atoms than gas-phase molecules (fewer reactions, more products), which produced an increment in the ion density (see figure 2) and that increment in the ion density requires a lower electron temperatures (see figure 1) to provide the required ionization rate.

4. Conclusions

The experimental study mainly focused on measurements of electron temperature, ion density and optical characterization of a gas mixture glow discharge of O_2 and Ar at total pressure of 1.0 Torr, a power of 150 W, and a flow of 16.5 l/min. The emission from the DC plasma was scanned in the range of 200–900 nm. It was found to consist of lines and bands corresponding to O_3, O_2, O_2^+, O, Ar and Ar^+. The electron temperature and ion density were determined by a double Langmuir probe. The electron temperature was found in a range from 6.8 eV to 12.0 eV, and the electron concentration in the order of 10^{10} cm^{-3}. The intensity of 777.34 nm O displays an increasing behavior as % O_2 increases, while the emission bands and lines (421.2 nm O_2, 439.8 nm O_2^+, and 473.7 nm Ar^+) show a decreasing behavior as % O_2 increases. The emission band (316.4 nm O_3) and the line (810.9 nm Ar) display a strong decreasing of the intensity from 0.2 to 0.4 % O_2, and then display an almost constant behavior as a function of % O_2.

Acknowledgments

The authors are thankful to A. Bustos, A. González, R. Bustos (ICF-UNAM), and José Rangel (ICN-UNAM) for technical assistance. This research was supported by UAEM 2648/2008U y 2362/2006U.

References

[1] Donnelly V M 2004 *J. Phys. D: Appl. Phys.* **37** R217
[2] Zhu X M and Pu V K 2007 *J. Phys. D: Appl. Phys.* **40** 2533
[3] Garamoon A A, Samir A, Elakshar F F, Nosair A and Kutp E F 2007 *IEEE T. Plasma Sci.* **35** 1
[4] Gudmundsson J T, Takashi Kimura, Lieberman M A 1999 *Plasma Sources Sci. Technol.* **8** 22–30
[5] Xi-Ming Zhu and Yi-Kang Pu 2007 *J. Phys. D: Appl. Phys.* **40** 5202–5
[6] Schmiedt L, Kaňka A and Hrachová V 2006 *Czech. J. Phys.* **56** B1040–4
[7] Aflori M, Dimitriu D G and Dorohoi D 2004 *31th EPS Conf. on Plasma Phys.* (London, UK) ECA **28G** P–4.057
[8] Martinez H and Yousif F B 2008 *Eur. Phys. J. D* **46** 493
[9] Martínez H and Castillo F *Phys. Stat. Sol. C* **5** (4) 897–900
[10] Reyes P G, Mendez E F, Osorio-Gonzalez D, Castillo F and Martínez H 2008 *Phys. Stat. Sol. C* **5** (4) 907
[11] Savitzky A and Golay M J E 1964 *Anal. Chem.* **36** 1627
[12] Amemiya H 1988 *Japan J. Appl. Phys.* **27** 694
[13] Swift J D and Schwar J R 1969 *Electric Probes for Plasma Diagnostics* (New York: Iliffe Books Ltd)
[14] Brockhaus A, Brochhardt C and Engemenn J 1994 *Plasma Sources Sci. Technol.* **3** 539
[15] Pearse R W B and Gaydon A G 1976 in *The identification of molecular spectra* (Cambridge: University Printing House)

Journal of Physics: Conference Series **207** (2010) 012021 doi:10.1088/1742-6596/207/1/012021

Pulsed probe diagnostics of a high-current low-pressure discharge in crossed E×H fields

I.K. Fetisov, I.V. Ivanov, G.V. Khodachenko, D.V. Mozgrin, D.B. Petrov, A.M. Snegirev, T.V. Shukshina*,

Moscow Engineering and Physics Institute (State University), 115409 Kashirskoe shosse 31, Moscow, Russia

E-mail: shukshina_tanya@mail.ru

Abstract. The measurement results for electron temperature, plasma density and emission ability of a high-current low-pressure discharge in crossed **E**×**H** fields by pulsed probe diagnostics method are given in this article.

1. Introduction

The intensive investigations of low-pressure discharges in crossed **E**×**H** fields are at present witnessed thanks to the later wide use in technological magnetron devices to deposit nanostructured coatings onto various materials including thermo-unstable such as *lavsan* (polyethylene terephthalate), fur, paper, plastics and others by employing both high-current pulsed magnetron sputtering systems and high rate ion stimulated etching aimed at polymers surface development [1].

Experimental studies of quasi-steady high-current discharges in crossed **E**×**H** fields in various gases (Ar, N_2, H_2, SF_6) and gas mixtures (Ar/SF_6, Ar/O_2) atmospheres at pressures from 10^{-2} up to 5 Torr under different configurations and magnitudes of the magnetic field have shown the existence of a stable discharge mode without the later being converted into an arc under high current densities. This is the so-called high-current diffuse discharge [2] operating in the current range of 10–2000 A at a cathode current density of up to 90 A/cm^2 and a constant discharge voltage of 75–140 V, the discharge duration being longer than 1 ms; no current contraction being observed in both the discharge plasma and the cathode sheath. Typical current-voltage characteristics of a low-pressure discharge in a magnetic field are shown in figure 1, where a high-current diffuse discharge (3) occupies an intermediate position between the high-current pulsed magnetron discharge (2) and the arc discharge (4).

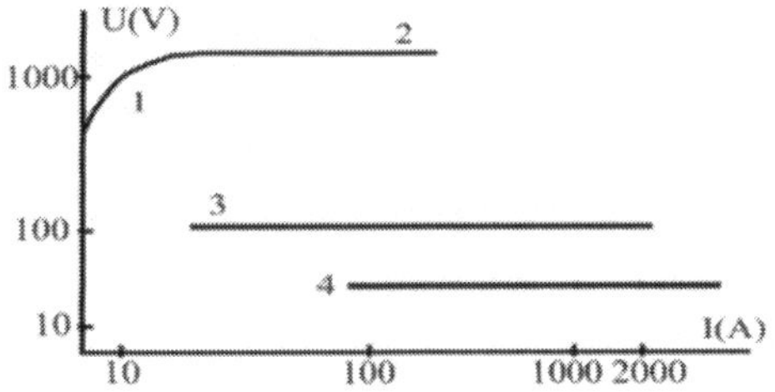

Figure 1. Typical current-voltage characteristics of low-pressure discharge in a magnetic field: 1 – DC magnetron discharge, 2 – high-current pulsed magnetron discharge, 3 – high-current diffuse discharge, 4 – arc discharge.

Since a general theory of the mechanisms governing this type of discharge is still lacking, it is very important to study the parameters (electron temperature, plasma density and emission ability) of the discharge plasma and the cathode sheath. The knowledge of these parameters will allow us to control the processes of coating deposition and surfaces etching.

2. Experimental setup and measurement system

The study the high-current low-pressure discharge in crossed **E**×**H** field has been carried out by using two types of devices: a planar magnetron (figure 2) and a system with profiled electrodes (figure 3).

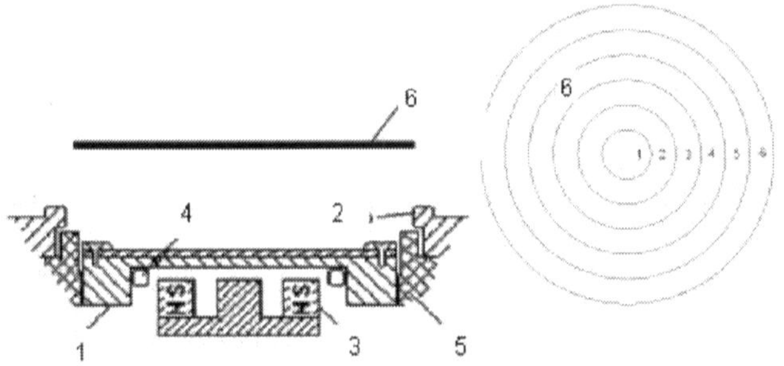

Figure 2. The discharge device "planar magnetron": 1 – cathode, 2 – anode, 3 – magnetic system, 4 – water cooling, 5 – insulator, 6 – probe-collector.

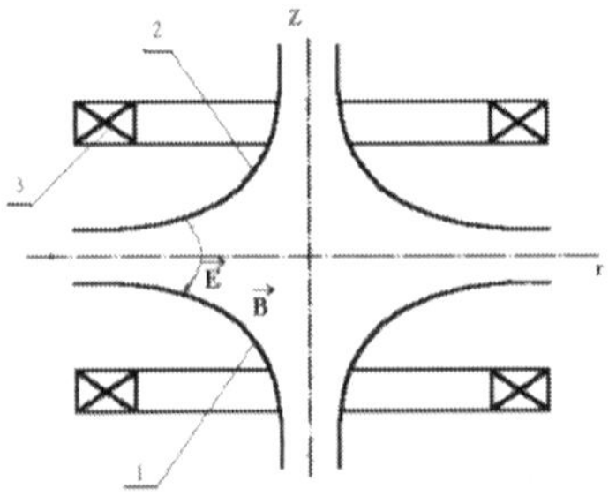

Figure 3. The discharge device with profiled electrodes: 1 – cathode, 2 – anode, 3 – magnetic system.

The planar magnetron consists of a plane cathode 120 mm in diameter and ring-shaped anode 160 mm in diameter. The electrodes are immersed in a magnetic field of annular permanent magnets. The discharge has an annular shape and is closely adjacent to the cathode. The maximum of the magnetic field radial component B at the cathode surface is 1200 G. The cathode is made of Cu or Al placed on a cooled surface. The anode is made of stainless steel.

The discharge device with profiled electrodes consists of two hollow axis-symmetric electrodes and a magnetic system. These electrodes 120 mm in diameter are placed at a distance of 10 mm from one another in a cusp magnetic field produced by two oppositely directed coils. The electrode profile coincides with a magnetic field line to secure the perpendicularity of electric and magnetic fields the entire electrode surface. The ratio of the maximum magnetic field at the symmetry axis, $B_{max}(z, 0)$, to the maximum magnetic field in the symmetry plane, $B_{max}(0, r)$, is $B_{max}(z, 0)/B_{max}(0, r) \cong 3$. The magnetic induction varies from 0 to 1000 G by changing the current in the coils.

To ignite a high-current diffuse discharge, a high-voltage pulse from a forming line, consisting of 19 *LC* section with a total stored energy of 6.3 kJ, is applied to the discharge gap, which is preionized by a steady-state DC magnetron discharge.

The voltage and current of a high-current diffuse discharge are measured by using an ohmic divider and a Rogowski coil (operating as a current transformer), respectively.

The pulsed probe diagnostics scheme of dense gas-discharge plasma includes a probe, a sweep generator, and a measurement system [3]. The probe, in case of the "profiled electrodes", is a conventional cylindrical probe consisting of a stainless steel rod partially covered by a ceramic insulator: the probe diameter is $a = 0.6$ mm, and the area of the working surface is $S = 16$ mm^2. The probe-collector is used for plasma emission ability measurements (in the planar magnetron). The probe-collector is a plane disk 110 mm in diameter consisting of *textolite* covered by an Cu foil. The choice of the probe-collector diameter is linked with the necessity to have the probe covering all the discharge area. The probe surface is divided into 6 annular zones with 1 cm pitch. The probes are located in the center of the electrodes system, on the distances where the magnetic field is of minimum. This allows to process and interpret the results obtained while neglecting the effect of the magnetic field.

3. Experimental results

First, the pulsed plasma parameters measurements of low-pressure high-current diffuse discharge in crossed **E**×**H** fields were carried out in the device with profiled electrodes. The discharge parameters were: magnetic field B = 0–1000 G, pressure $p = 10^{-2} - 1$ Torr, and the discharge current of up to 1500 A. The working gas was argon. Pre-ionization was produced by the DC magnetron discharge with a current of up to 200 mA.

Based on the probe measurements the plasma parameters were calculated. The obtained dependences of the ion density in the center of the system on the current of a quasi steady high-current diffuse discharge for different values of the gas pressure of up to 10^{-2} Torr and magnetic field showed that the plasma density increases almost linearly as the discharge current (figure 4).

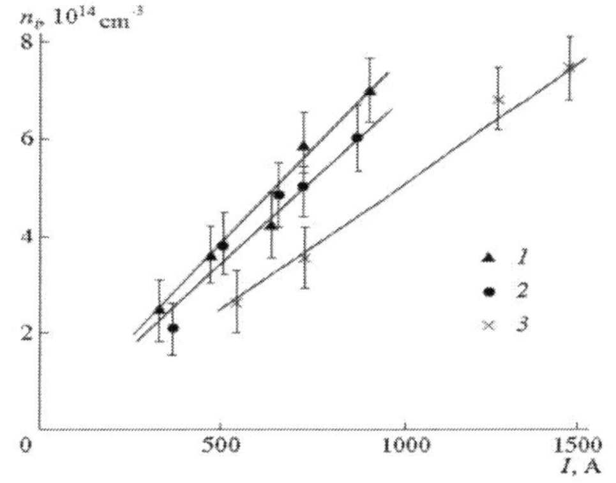

Figure 4. Dependence of plasma density on a discharge current in the discharge device with profiled electrodes at conditions: B = 800 G, Ar: 1 – p = 0.5 Torr, 2 – p = 0.3 Torr, 3 – p = 0.1 Torr.

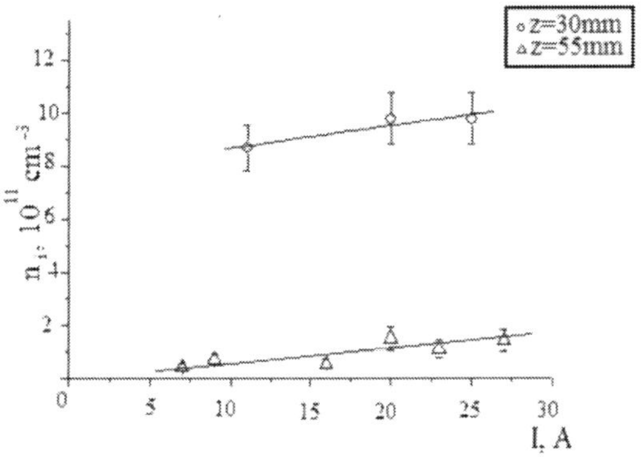

Figure 5. Dependence of plasma density on a discharge current in the planar magnetron at the conditions: $p = 8\times10^{-3}$ Torr, Ar/Air in proportions 7/1.

The electron temperature was 3–8 eV. In this case the density of the probe ion saturation current was up to 15 A/cm^2. As the gas pressure in the discharge gap grows the ion density increases and the electron temperature decreases.

Second, the pulsed plasma parameters of high-current diffuse discharge were measured in the planar magnetron. The discharge parameters were magnetic field B = 0–1200 G and the discharge current of up to 30 A. The working gas was gas mixture Ar/Air in proportions 7/1. Using the Al cathode in the oxygen-containing gas mixture allowed us to lower ion current on the cathode, which was necessary for discharge maintenance and, hence, to reduce high-current diffuse discharge pressure down to 8×10^{-3} Torr. Pre-ionization was produced by the DC magnetron discharge with a current of up to 100 mA. Measurements were carried out on distances of 30 mm and 55 mm from cathode surface. The voltage pulse was applied to the probe with a time delay of 100–1000 μs with respect to the beginning of the pulsed discharge. The electron temperature and the plasma density were measured after having got the quasi steady high-current discharge. The obtained dependences of the plasma density on a discharge current are presented on figure 5. It was seen that the plasma density also increases almost linearly as the discharge current and the distance from cathode surface. At the same time the electron temperature obtained was 3–8 eV. The radial distribution of the plasma density depending on the distance from an axis of the system is presented on figure 6.

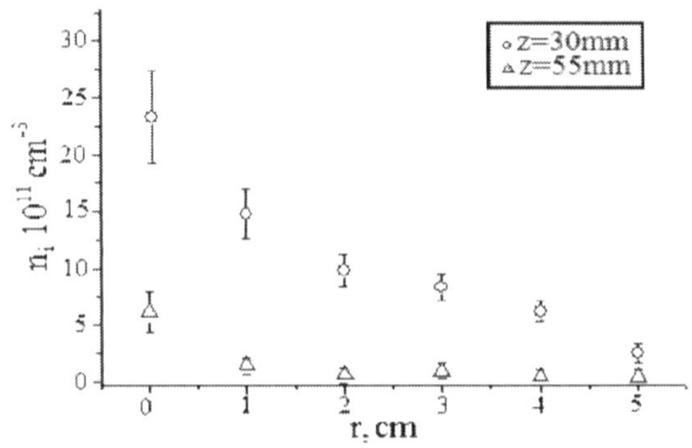

Figure 6. Dependence of the plasma density on the distance from a system axis in the planar magnetron at the conditions: $p = 8\times10^{-3}$ Torr, Ar/Air in proportions 7/1, $I = 30$ A.

The measurements of the probe ion saturations current showed that the emittance of the high-current diffuse discharge with the Al cathode in the oxygen-containing gas mixture can reach 0.003–0.5 A/cm^2 depending on the probe-collector zone. The maximum ion saturation current was observed in the probe centre and fell down to its borders.

On the basis of the plasma obtained and discharge parameters, we estimated: Debye length $r_d = 8\times10^{-4}$ cm, thickness of the cathode layers $d_l = 0.09$ cm, mean length of electron free path in the electron – atom collisions $\lambda = 10$ cm, and height of the electron trajectory above the cathode surface $h = 0.3$ cm. The results reached have shown that the cathode layer of the given discharge is non-collisional and the absence of visible cathode spot traces on the cathode surface.

4. Conclusion

High-current low-pressure diffuse discharge inside crossed **E**×**H** fields with Al cathode placed in oxygen-containing mixture was investigated. The established operations modes are of practical outcome in developing new technologies in particular for forming Al structured surfaces.

Pulsed probe measurements of the plasma parameters for low-pressure high-current diffuse discharge in crossed **E**×**H** fields have shown the presence of a dense plasma (plasma density up to 10^{15} cm^{-3}) providing realization of a non-collisional cathode layer mode that corresponds to the concept of thermal ionization mechanism.

References

[1] Fetisov I K, Filippov A A, Khodachenko G V, Mozgrin DV and Pisarev A A 1999 *Vacuum* **53** 133–6
[2] Mozgrin D V, Fetisov I K and Khodachenko G V 1995 *Plasma Phys. Rep.* **21** 5 400–9
[3] Mozgrin D V, Fetisov I K and Khodachenko G V 1999 *Plasma Phys. Rep.* **25** 2 255

Electrical properties of PMMA ion-implanted with low-energy Si^+ beam

G B Hadjichristov[1,4], V K Gueorguiev[1], Tz E Ivanov[1], Y G Marinov[1], V G Ivanov[2] and E Faulques[3]

[1] Georgi Nadjakov Institute of Solid State Physics, Bulgarian Academy of Sciences, 72 Tzarigradsko Chaussee Blvd., 1784 Sofia, Bulgaria
[2] Sofia University, Faculty of Physics, 5 James Bourchier Blvd., 1164 Sofia, Bulgaria
[3] Institut des Matériaux Jean Rouxel, UMR6502 CNRS, Nantes Atlantic Universities, 2 rue de la Houssiniere, 44 322 Nantes, France

Abstract. The electrical properties of polymethylmethacrylate (PMMA) after implantation with silicon ions accelerated to an energy of 50 keV are studied under DC electric bias field. The electrical response of the formed material is examined as a function of Si^+ fluence in the range 10^{14} – 10^{17} cm^{-2}. The carbonaceous subsurface region of the Si^+-implanted PMMA displays a significant DC conductivity and a sizable field effect that can be used for electronic applications.

1. Introduction

In the last years there has been a growing interest concerning the ion-implanted polymers and their application in electronics, biotechnology and medicine [1–3]. Especially, the controllable modification of the electrical conductivity and electrical properties of the ion-implanted polymers are of great importance for their electronic applications.

As known, the ion implantation of hydro-carbon polymers leads to a formation of conjugated double C=C bonds organized in a nanoclustered carbonaceous conductive material [1–3]. The electrical conductivity of the ion-implanted polymers originates in the π-electrons in the conjugated double C=C bonds and is controlled by the electron hopping mechanism between the conducting C-clusters [1–3].

Poly-(methyl methacrylate) (PMMA) is known to be highly sensitive to accelerated ions [4, 5] and is thereby used in ion-beam lithography and in UV photolithography. For example, masking media (UV-opaque PMMA films) have been produced by implantation of PMMA with 150 – 200 keV silicon ions [6]. Also, very low energy (250 – 1250 eV) Ar^+ irradiation has shown an enhancement of the electrical conductivity of PMMA [7].

Here we report results on the electrical properties of PMMA ion-implanted with a 50 keV silicon ion beam at a fluence in the range from 10^{14} to 10^{17} ions/cm^2. The electrical response of this material, as well as of the corresponding filed-effect structure, is studied under a direct current (DC) electric bias field and is related to the structure formed in the polymer PMMA upon Si^+ implantation.

[4] To whom any correspondence should be addressed. E-mail: georgibh@issp.bas.bg

According to data from [4, 8], a significant contribution of nuclear interactions and of collision damage at the end of the ion range and direct displacement of polymer atoms can be expected under these implantation conditions.

2. Experimental

Six identical PMMA samples of 1×1 cm^2-size with a thickness of 5 mm were implanted with Si^+ ions at an energy of 50 keV and for various fluences ranging from 3.2×10^{14} cm^{-2} to 10^{17} cm^{-2}. The experimental details have been described elsewhere [9]. The thickness of the ion-implanted subsurface region was roughly estimated using Monte Carlo calculations [8]. Usually related to the mean range of ions in polymers, a value of ~100 nm can be considered as the thickness of the electrically active layer in Si^+-implanted PMMA. This value was used for the conductivity calculations.

Narrow, 1 cm-long electric contacts from silver paste were deposited at two opposite edges of the implanted surface of the PMMA samples, as schematically shown in figure 1. The electrical properties of Si^+-implanted PMMA were studied at room temperature. Direct current (DC) measurements were conducted by Keithley 617 Programmable Electrometer controlled by computer using IEEE-488 data interface.

In order to probe the field-effect transistor, (FET)-like properties of the Si^+-implanted PMMA structure, a third (gate) electrode (copper, 3 mm width, 1 cm length) was placed on the implanted PMMA surface between both side-contacts (the drain and the source) (figure 1).

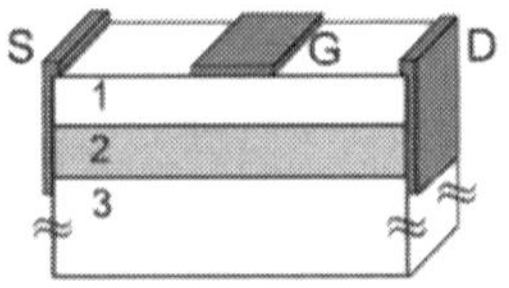

Figure 1. Schematic cross-section of the spatial structure of Si^+-implanted PMMA with source (S), drain (D) and gate (G) electrodes: (1) porous low-conductive ion-modified layer; (2) conductive ion-implanted layer; (3) pristine PMMA.

3. Results and discussion

Si^+-implanted PMMA studied herein possesses electro-transport properties characteristic of the ion-implanted hydro-carbon polymers [1–3]. Electron hopping mechanism of conductivity has been established for PMMA subjected to low-energy ion implantation [7]. The linear current-voltage plots (figure 2) for resistive Si^+-implanted PMMA samples (no gate electrode) reveal ohmic conductance and are consistent with the hopping model [10].

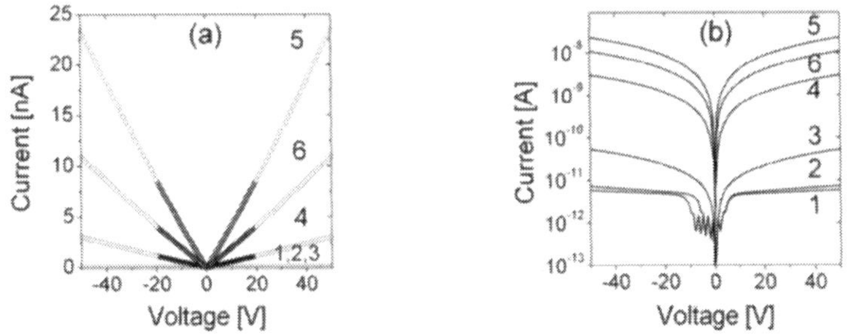

Figure 2. (a) DC current-voltage characteristics of Si^+-implanted PMMA structure shown in figure 1 (without gate); (b) the same, but in a logarithmic scale. The labels from 1 to 6 refer to the fluences as: 3.2×10^{14}, 1×10^{15}, 3.2×10^{15}, 1×10^{16}, 3.2×10^{16} and 1×10^{17} [Si^+/cm^2].

The DC conductivity, σ, of the samples as a function of the Si^+ fluence is presented in figure 3. The ion-induced improvement of σ is consistent with room-temperature data reported for other C-H containing polymers implanted with ions at similar ion energy and fluence levels [11,12]. A two-decade increase in fluence results in a six-decade increase of σ and at 3.2×10^{16} Si^+/cm^2 σ attains values almost eleven orders of magnitude higher than that of unimplanted PMMA (σ_{PMMA} ~ 10^{-15} S/cm, typical for the good electrical insulators). Nevertheless, the DC conductivity of Si^+-implanted PMMA studied herein is still smaller than that (up to 10^2 S/cm) measured in amorphous carbon [13]. Also, the implantation effect is less compared to the conductivity enhancement in polymers irradiated with MeV accelerated ions [14,15].

As evidenced in our previous study [9], a nanoclustered hydrogenated amorphous carbon structure is formed in Si^+-implanted PMMA samples studied herein. The size of the π-bonded C-clusters of fused sixfold rings is closely related to the energy band gap which controls the electrical conductivity of this material – the band gap varies roughly inversely with the cluster size [16]. The altering of the spatial distribution (depth profile) of the layer created within PMMA by Si^+ ion implantation beneath the polymer surface plays also a significant role. Most likely, with increasing ion fluence the depth profile of the implantation-modified layer is moved toward the surface. Thus, by Si^+-implantation one can adjust the spatial characteristics and electrical properties of the ion-modified subsurface region in the polymer PMMA, similarly to the well known tailoring of the refractive index of ion-modified PMMA [17].

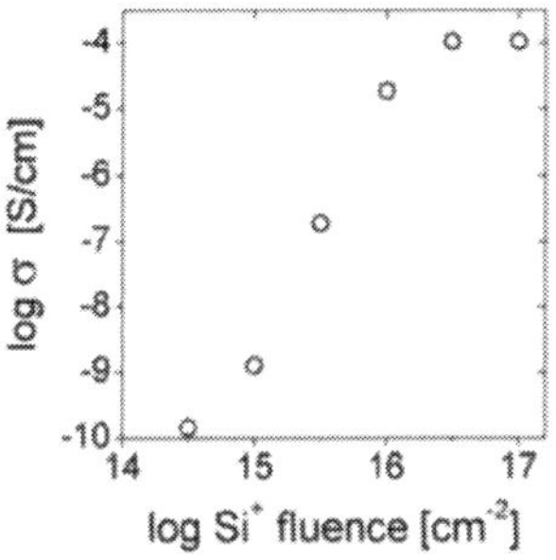

Figure 3. The DC conductivity of Si^+-implanted PMMA samples versus the ion fluence.

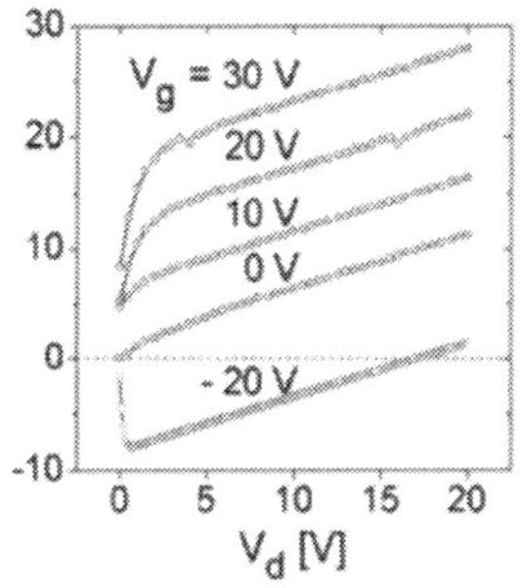

Figure 4. Output characteristics of Si^+-implanted PMMA structure (fluence 3.2×10^{15} Si^+/cm^2).

The possibility to form by ion-beam implantation technique a thin semiconductive buried layer with a controlled degree of modification should allow one to choose implantation conditions that produce a subsurface layer with properties tuned to match a field effect in the formed structure. In addition to the significant enhancement of the DC conductivity with the implantation fluence, we have found a sizable field effect in Si^+-implanted PMMA structure. Figure 4 represents the output characteristics (the drain current, I_d, versus the drain voltage, V_d, for several values of the gate voltage, V_g) of Si^+-implanted PMMA sample with the electrode configuration shown in figure 1. The results imply a typical *N*-channel enhancement transistor behaviour and an electron accumulation mode achieved with a positive V_g. The lack of saturation in the $I_d(V_d)$ plot can be ascribed to the electrons associated with the intrinsic bulk current through the gate and takes also place by negative biasing (figure 4).

Generally, I_d could be enhanced by the optimization of the device architecture. However, the most important is that both the spatial characteristics and the electrical properties of the sandwich structure of the type dielectric/semiconductor/insulator created within the polymer allow the field-effect transistor (FET)-like operation. Thus, the Si^+-implanted PMMA structure with an appropriate electrode configuration enables the control of the channel current by a potential difference and a resistance upon applying an external electric field (gate voltage). Since the implantation conditions (ion energy and fluence) determine the spatial structure, as well as the nano-scale structure, and thereby the electronic and electrical properties of the formed material, they are very important for the formation of a field-effect structure and define the operation in a FET-like regime upon appropriate conditions.

4. Conclusions

A significant enhancement of the conductivity with the implantation fluence, as well as a field effect depending on the ion-implantation conditions, were observed in Si^+-implanted PMMA. The implantation fluence provides the tunability of the implantation-induced electrical conductivity of the formed organic material. These properties show that the Si^+-implanted PMMA appears to be a promising material for highly integrated organic electronic applications, e.g. transistor-like active elements. The bio-compatibility and bio-sensitivity of PMMA subjected to ion implantation make such devices very attractive for electrical bio-sensing and bio-medical systems.

Acknowledgements

This work was supported by the Swiss National Science Foundation (Project No. IB7420-110981/1 "Southern NanoEngineering Network", SONNET), National Science Fund of Bulgaria (contract IRNI 21/2007) and Sofia University (contract UFNI 075/2008).

References

[1] Fink D 2004 *Fundamentals of Ion Irradiated Polymers* (Berlin: Springer Verlag)
[2] Fink D 2004 *Transport Processes in Ion-Irradiated Polymers* (Berlin: Springer Verlag)
[3] Sviridov D V, Odzhaev V B and Kozlov I P 1998 *Electrical and Optical Polymer Systems – Fundamentals, Methods and Applications*, ed D L Wise, G E Wnek, D J Trantolo, T M Cooper and J D Gresser (New York: Marcel Dekker) chapter 11 pp 387–422
[4] Biersack J P and Kallweit R 1990 *Nucl. Instrum. Meth.* B **46** 309
[5] Venkatesan T, Calcagno L, Elman B S and Foti G 1987 *Ion Beam Modification of Insulators*, ed P. Mazzoldi and G.W. Arnold (Amsterdam: Elsevier) pp 301–79
[6] Kenty J L, Puzio L C and Schauerte F J 1983 *J. Vacuum Sci. Techn. B: Microelectron. Nanometer Structures* **1** 1211
[7] Koval Y, Fistul M V and Müller P 2005 *J. Vac. Sci. Technol.* A **23** 1375
[8] Ziegler J F, Biersack J P and Littmark U 1985 *The Stopping and Range of Ions in Solids* (New York: Pergamon Press)
[9] Hadjichristov G B, Ivanov V and Faulques E 2008 *Appl. Surf. Sci.* **254** 4820
[10] Hesto P 1986 *Instabilities in Silicon Devices* vol 1, ed G. Barbottin and A. Vapaille (Amsterdam: North Holland) chapter 5 p 263
[11] Bedell C J, Sofield C J, Bridwell L B and Brown I M 1990 *J. Appl. Phys.* **67** 1736
[12] Wang Y Q, Giedd R E, Moss M G and Kaufmann J 1997 *Nucl. Instrum. Meth.* B **127–128** 710
[13] Hauser J J 1975 *Solid State Commun.* **17** 1577
[14] Venkatesan T, Forrest S R, Kaplan M L, Murray C A, Schmidt P H and Wilkens B J 1983 *J. Appl. Phys.* **54** 3150
[15] Kaplan M L, Forrest S R, Schmidt P H and Venkatesan T 1984 *J. Appl. Phys.* **55** 732
[16] Robertson J and O'Reilly E P 1987 *Phys. Rev.* B **35** 2946
[17] Biersack J P and Kallweit R 1990 *Nucl. Instrum. Meth.* B **46** 309

Journal of Physics: Conference Series **207** (2010) 012023 doi:10.1088/1742-6596/207/1/012023

The main properties of microwave argon plasma at atmospheric pressure

E Benova, M Pencheva

Department for Language Teaching and International Students, University of Sofia, 27 Kosta Loulchev Street, BG-1111 Sofia, Bulgaria

E-mail: benova_phys@deo.uni-sofia.bg

Abstract. Plasma torch sustained by surface wave at atmospheric pressure is theoretically studied by means of 1D model. A steady-state Boltzmann equation in an effective field approximation coupled with a collisional-radiative model for high-pressure argon discharge is numerically solved together with Maxwell's equations for an azimuthally symmetric TM surface wave. The axial dependences of the electrons, excited atoms, atomic and molecular ions densities as well as the electron temperature, the mean power per electron and the effective electron-neutral collision frequency are determined. A strong dependence of the plasma properties on the discharge conditions and the gas temperature is obtained.

1. Introduction

Plasma production at atmospheric pressure is a field in plasma physics which develops headlong because of the easy way for operating with these type of plasmas and thus for numerous practical applications. Among the atmospheric pressure plasmas the ones produced and sustained by means of microwaves draw a particular interest. Because the electrodes or excitation structures, serving to impose the electric field that maintains the plasma, do not need to be located within the plasma vessel, this eliminates electrode corrosion problems, reduces gas contamination, and substantially lowers the production costs. The specified advantages results in wide opportunities for applications in industry [1], materials treatment [2,3], environmental protection [4–6], sterilization [7], biomedical treatment [8] and etc.

This paper deals with the numerical modeling of surface wave sustained plasmas at atmospheric pressure which are a particular class of microwave discharges. They are distinguished for independence of the plasma properties on the type of the wave launcher which allows studying of various configurations. The devices generating the plasma are simple, efficient and easy to operate with. The surface wave technique allows for an exceptionally flexible control of the plasma parameters and of its dimensions, in a wide range of wave frequencies and gas pressures. The created plasma is over-dense plasma ($n > n_{\text{cutoff}} = m\omega^2/4\pi e^2$), which characterizes with both axial and radial inhomogeneity.

The organization of the paper is as follows: section 2 depicts the physical pattern accounted in the model. In section 3 a description of the self-consistent model is given. Section 3.1 presents the main electrodynamic equations as well as the boundary conditions used for obtaining the wave propagation characteristics. In section 3.2 the processes occurring in the plasma, the atomic and molecular energy levels, and the equations considered in order to build a collisional-radiative model of high pressure argon discharge are described. Section 3.3 looks into the self-consistent connection between the former two parts of the modelling. Finally, section 4 shows the results obtained by means of applying the complete model.

2. Physical pattern

The discharge in a surface wave plasma source is sustained by electromagnetic wave. Propagating along the interface between the plasma and the dielectric tube the field of the wave penetrates in the plasma, therefore the name of the wave comes up. The propagating surface electromagnetic wave creates the discharge and at the same time the discharge is a medium for the wave propagation. The wave gradually transfers its power to the plasma, more precise to the electrons. The energy the electrons receive from the wave is expended on different elementary processes which ensure the creation and sustaining the plasma. When the absorbed power is no longer enough for compensation the losses of radiation, diffusion and recombination toward the walls of the tube the discharge quenches.

3. Self-consistent model

The model, as it was mentioned above, consists of two parts: electrodynamic and kinetic ones, and a link which allows deriving self-consistently all discharge and wave characteristics.

3.1. *Electrodynamic part*

The model is created under the following assumptions:

- the plasma is in a steady state
- the electromagnetic wave maintaining the plasma is an azimuthally symmetric (m = 0) HF TM surface wave with components $\mathbf{E}_{\mathrm{hf}} = (E_r, 0, E_z)$ and $\mathbf{B}_{\mathrm{hf}} = (0, B_\varphi, 0)$
- the plasma density n, the wave number k and the wave amplitude are slowly varying functions of the axial coordinate z
- this is numerical one-dimensional axial model where the radially averaged plasma density is used, $n \equiv \bar{n} = \frac{2}{R^2}\int_0^R dr\, r\, n(r)$

The electrodynamic part of the model is based on Maxwell's equation for an azimuthally symmetric TM surface wave with angular frequency ω propagating along a plasma cylinder with radius R surrounded by vacuum. Maxwell's equations give the wave equations for the radial, azimuthal and axial components of the wave. The one for the E_z component of the wave electric field in cylindrical coordinates is:

$$\frac{1}{r}\frac{\partial}{\partial r} r \frac{\partial}{\partial r} E_z + \frac{\partial^2}{\partial z^2} E_z + \left(\frac{\partial}{\partial z} \ln \varepsilon\right) \frac{\partial}{\partial z} E_z + \frac{\omega^2}{c^2} \varepsilon E_z = 0 \tag{1}$$

Here ε is the plasma permittivity of the medium which for vacuum is $\varepsilon_v = 1$ and for plasma:

$$\varepsilon_{\mathrm{p}} = 1 - \frac{\omega_{\mathrm{p}}^2}{\omega(\omega + i\nu_{\mathrm{eff}})} = 1 - \frac{\omega_{\mathrm{p}}^2}{\omega^2}\left(1 + \frac{\nu_{\mathrm{eff}}^2}{\omega^2}\right)^{-1} + i\frac{\nu_{\mathrm{eff}}}{\omega}\frac{\omega_{\mathrm{p}}^2}{\omega^2}\left(1 + \frac{\nu_{\mathrm{eff}}^2}{\omega^2}\right)^{-1} \tag{2}$$

The wave equations are solved for the case $\nu_{\mathrm{eff}} \neq 0$ without any assumptions although in our case $\nu_{\mathrm{eff}} \geq \omega$, ω_{p} is the usual electron plasma angular frequency and ν_{eff} is the effective electron–neutral collision frequency for momentum transfer.

Similar expressions are obtained for the radial electric field component E_r and for the azimuthal magnetic field component B_ϕ. With assumption for week axial density inhomogeneity we seek the solutions of the wave equations in the form:

$$(E_r, E_z, B_\varphi)(r,z,t) = \mathrm{Re}\left[(F_r, F_z, F_\varphi)(r,z)E(z)\exp\left(-i\omega t + i\int_0^z dz' k(z')\right)\right] \tag{3}$$

where $E(z) = B(z) \equiv E(r = R)$ and F_z, F_r, F_φ are cylindrical functions. The solutions we are looking for are combinations of the modified Bessel's functions zeroth and first order.

Using the boundary condition $F_\varphi^{\mathrm{p}} = F_\varphi^{\mathrm{d}}$ at $r = R$ we obtain the local wave dispersion relation:

$$\frac{\varepsilon_{\mathrm{p}}}{a_{\mathrm{p}}}\frac{I_1(a_{\mathrm{p}})}{I_0(a_{\mathrm{p}})} + \frac{1}{a_{\mathrm{v}}}\frac{K_1(a_{\mathrm{v}})}{K_0(a_{\mathrm{v}})} = 0 \tag{4}$$

with $a_{\mathrm{p}}^2 = \tilde{k}^2 - \sigma^2\varepsilon_{\mathrm{p}}$ and $a_{\mathrm{v}}^2 = \tilde{k}^2 - \sigma^2$, where $\tilde{k} = k_{\mathrm{r}}R + ik_{\mathrm{i}}R$ is the dimensionless complex wave number with k_{r} the propagation coeficient and k_{i} the attenuation coefficient, $\sigma = \omega \mathrm{R}/c$ and I_0, I_1, K_0, K_1 are the modified Bessel functions of zeroth and first order.

The second main equation of the electrodynamic part of the model is the wave energy balance equation given in the form:

$$\frac{d}{dz}S = -Q \tag{5}$$

where S is the energy flux along the plasma column and Q is the power absorbed by the electrons at unit column length. The energy flux is the sum of the axial components of Poynting's vectors averaged over the wave period and integrated over the plane normal to the plasma column:

$$S(z) = 2\pi\int_0^1 d\rho\rho S_z^{\mathrm{p}} + 2\pi\int_1^\infty d\rho\rho S_z^{\mathrm{v}} \tag{6}$$

which in our case reads:

$$S = \frac{1}{4}\omega R^4 E^2 \mathrm{Re}\left[\frac{\tilde{k}\varepsilon_{\mathrm{p}}}{a_{\mathrm{p}}^{*2} - a_{\mathrm{p}}^2}\left(\frac{1}{a_{\mathrm{p}}}\frac{I_1(a_{\mathrm{p}})}{I_0(a_{\mathrm{p}})} - \frac{1}{a_{\mathrm{p}}^*}\frac{I_1(a_{\mathrm{p}})}{I_0(a_{\mathrm{p}}^*)}\right)\right]$$

$$+\frac{1}{4}\omega R^4 E^2 \mathrm{Re}\left[\frac{\tilde{k}}{a_{\mathrm{v}}^{*2} - a_{\mathrm{v}}^2}\left(\frac{1}{a_{\mathrm{v}}^*}\frac{K_1(a_{\mathrm{v}}^*)}{K_0(a_{\mathrm{v}}^*)} - \frac{1}{a_{\mathrm{v}}}\frac{K_1(a_{\mathrm{v}})}{K_0(a_{\mathrm{v}})}\right)\right]$$

The wave power dissipated per unit length is given by the relation:

$$Q(z) = 2\pi\int_0^R drr\langle j.E\rangle = \frac{\omega}{4}\mathrm{Im}(\varepsilon_{\mathrm{p}})R^2E^2\int_0^1 d\rho\rho\left(\left|F_r^{\mathrm{p}}\right|^2 + \left|F_z^{\mathrm{p}}\right|^2\right) \tag{7}$$

and is equal to:

$$Q = \frac{\omega}{4}\mathrm{Im}(\varepsilon_{\mathrm{p}})R^2E^2\frac{1}{a_{\mathrm{p}}^{*2} - a_{\mathrm{p}}^2}\left[\tilde{k}\,\tilde{k}^*R^2\left(\frac{1}{a_{\mathrm{p}}}\frac{I_1(a_{\mathrm{p}})}{I_0(a_{\mathrm{p}})} - \frac{1}{a_{\mathrm{p}}^*}\frac{I_1(a_{\mathrm{p}}^*)}{I_0(a_{\mathrm{p}}^*)}\right) + a_{\mathrm{p}}^*\frac{I_1(a_{\mathrm{p}}^*)}{I_0(a_{\mathrm{p}}^*)} - a_{\mathrm{p}}\frac{I_1(a_{\mathrm{p}})}{I_0(a_{\mathrm{p}})}\right] \tag{8}$$

3.2. Kinetic part

The detailed modelling of any kind of plasma requires the knowledge of the discharge processes. In this study we consider argon plasma. In general the kinetic description consists of:

- electron kinetics

Journal of Physics: Conference Series **207** (2010) 012023 doi:10.1088/1742-6596/207/1/012023

- heavy particles kinetics
- gas thermal balance equation

The electron kinetics includes calculation of the electron energy distribution function (EEDF), which determines the electron transport parameters, the rates of elementary processes, and several other important quantities as $\nu_{\rm eff}$ – the effective electron–neutral collision frequency for momentum transfer, a key parameter for the modeling of MW discharge maintained of surface wave. The EEDF is derived from the electron Boltzmann equation, solved in our case using the two-term Legendre polynomials expansion. The electron energy balance equation giving the energy for sustaining an electron-ion pair in the discharge θ, which is another key parameter:

$$\theta \equiv \theta^{\rm elast} + \theta^{\rm exc} + \theta_0^{\rm ion} + \theta^{\rm ion} + \theta^{\rm diff} + \theta^{\rm rec} - \theta^{\rm Pen} \tag{9}$$

as well as the electron particle balance equation are considered here. The superscripts denote the energy expanded on different processes, respectively elastic, inelastic collisions, direct and stepwise ionization, diffusion, recombination and Penning ionization. In additions of these processes the associative ionization, dissociative and three-body recombination, conversion to molecular ions and radiative transitions are taken into account.

The heavy particles kinetics involves calculation of the particle balance equations for the excited atoms as well as for the ions. The species considered in our model are Ar, Ar(4s), Ar(4p), Ar(3d), Ar(5s), Ar(5p), Ar(4d), Ar(6s), Ar_2^*, Ar^+, Ar_2^+ and Ar_3^+. The calculations are done assuming all excited states as blocks of levels with effective energy corresponding to the energy of the center. The heavy particle kinetics coupled with the electron kinetics allows us to determine the atomic and molecular ion number densities, the populations of the excited states as well as θ and $\nu_{\rm eff}$ as a function of the electron number density.

3.3. *Self-consistent link*

The self-consistent link between the electrodynamic and kinetic parts of the model is the balance of the wave and electron energy. On one hand Q is the wave power absorbed per unit length in the discharge. On the other hand, Q is the power expended by electrons in elastic and inelastic collisions:

$$Q = \pi R^2 n_e \theta \tag{10}$$

So the values of Q derived from the both parts of the model must be equal.

The relation between S and Q is given by the wave energy balance equation

$$\frac{d}{dz} S = -Q \tag{11}$$

The local dispersion relation yields the electron density for a given wave number, and the solution of the wave energy balance equation provides the dependence of the electron density on the axial position $n_e(z)$. Having obtained the axial dependence of the electron density, the kinetic part is recalculated again for appropriate axial position. This provides the spatial distribution of the discharge characteristics.

4. Results

Numerical calculations have been done for an argon plasma column sustained by electromagnetic wave with frequency $\omega/2\pi$ = 2.45 GHz, wave power: 500 W, and gas

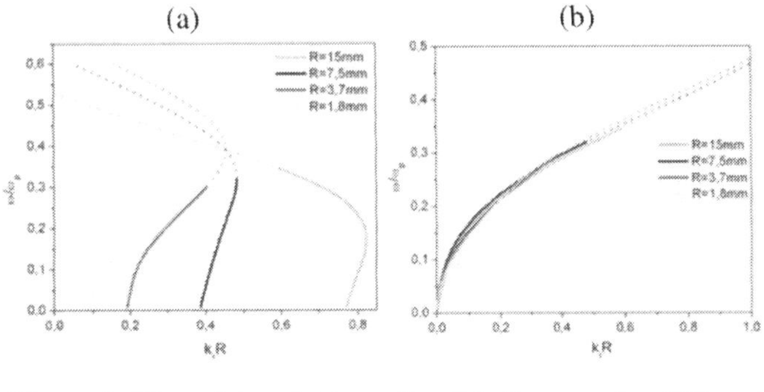

Figure 1. Phase (a) and attenuation (b) diagrams for different radii. The parts displayed with solid lines correspond to the condition $k_r \geq k_i$ and the dotted lines to $k_r < k_i$

pressure: 1 atm, varying the gas temperature for fixed plasma radius R = 3.7 mm and varying the tube radius for fixed gas temperature T_g = 3500K.

In Figure 1 the phase (a) and the attenuation (b) diagrams for the four plasma radii are shown. The parts of the diagrams displayed with solid lines satisfied the requirement the phase coefficient to be higher or equal to the attenuation coefficient $k_r \geq k_i$ and the parts of them presented with dashed lines correspond to $k_r < k_i$. With the decrease of the radius the value of the wave number, for which $k^* \equiv k_r = k_i$ is satisfied, also decreases but the plasma frequency ω_p (and the plasma density n_e) corresponding to this point gets higher.

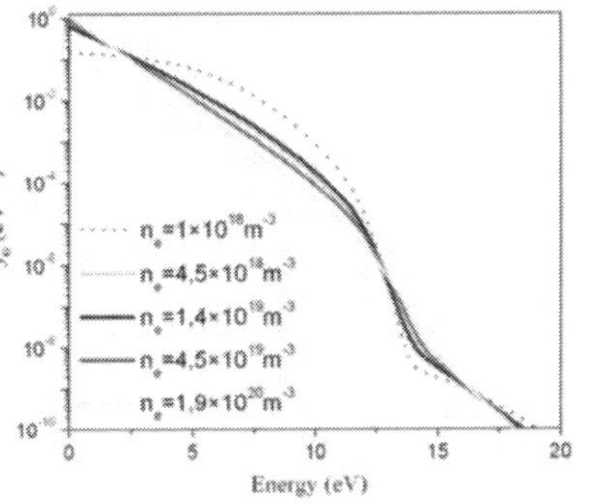

Figure 2. Electron energy distribution function at different plasma densities corresponding to discharges with different radii.

The numerical calculations in the kinetic part of the model allow us to obtain the dependence of the electron energy distribution function on the plasma radius. In Figure 2 the EEDFs calculated for four different discharge radii (R = 15, 7.5, 3.7, and 1.8 mm) are plotted. The solid lines in colours denote the distribution functions for the electron densities close to the launcher corresponding to the fourth radii (n_e = 4.5 x 10^{18}, 1.4 x 10^{19}, 4.5 x 10^{19}, 1.9 x 10^{20} m^{-3} respectively) and the black dotted line denotes the electron density close to the end of the discharge, which is of the same order 1 x 10^{18} m^{-3} for the fourth cases. It can be seen that with the reduction of the radius the electron density increases and the EEDF gets closer to Maxwellian one.

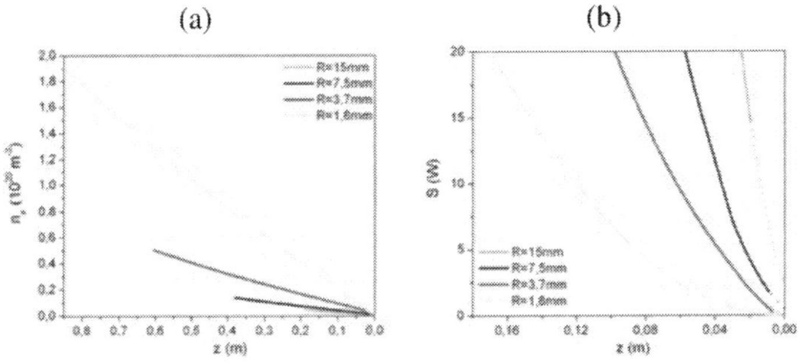

Figure 3. Axial distribution of the electron density (a) and wave energy flux (b) for different radii of the discharge

Using the complete self-consistent model the axial distributions of the wave and discharge parameters are obtained. Figure 3 presents the axial distributions of the plasma density (a) and the wave power flux (b) as functions of the discharge dimensions. One can see that the increase of the plasma radius results in a shorter plasma column and lower electron number density (figure 1(a)). For smaller plasma radii the electron density which corresponds to $k^* \equiv k_r = k_i$ is close to the value where the wave power flux S = 0 W. With the increase of the radius the attenuation coefficient becomes higher than the propagation one before this point, e.g. before the power flux to be dissipated completely (figure 1(b)). Bearing in mind that the regions of the phase and attenuation diagrams corresponding to $k_r < k_i$ (figure 1, dotted lines) cannot be associated with wave propagation so we can conclude that in this case the end of the discharge should be set by the requirement $k_r = k_i$.

Journal of Physics: Conference Series **207** (2010) 012023 doi:10.1088/1742-6596/207/1/012023

Figure 4 shows the axial dependences of the argon excited states population as well as for the argon dimmer and atomic and molecular ions densities for discharges with radii 7.5 (figure 4 (a) and (b)) and 3.7 mm (figure 4 (c) and (d).

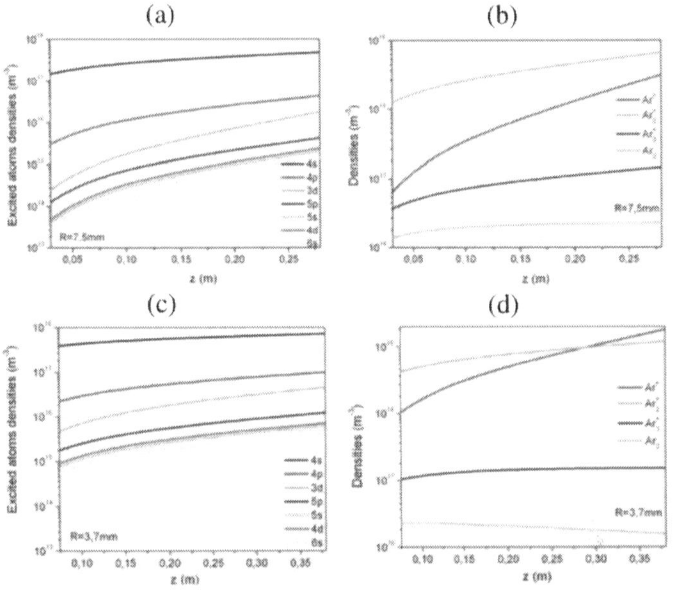

Figure 4. Axial distributions of the excited atoms (a, c) and the ions (b, d) densities for discharge radii 7,5 and 3,7 mm

5. Conclusions

The wave and discharge characteristics are obtained for different discharge conditions. The calculations show that: (i) the stationary discharge in pure Argon at atmospheric pressure can be sustained only in small radius tubes; (ii) the EEDF is non-Maxwellian one in most of the discharge conditions and gets closer to Maxwellian with increasing the plasma dencity (which is also a result of plasma radius reduction); the electron temperature is always higher then the gas temperature; the plasma density (even at the column end) is much higher than the cut-off limit, $n_e >> n_{cut\text{-}off}$.

6. Acknowledgements

This work was supported by the Bulgarian National Fund for Scientific Research under Grant F 1401/04 and by the Fund for Scientific Research of the University of Sofia under Grant No 157/2008.

References

[1] Roth J R 1995 *Industrial Plasma Engineering* (IOP Publishing, Bristol) Vol. 1
[2] Sekiguchi H, Orimo T 2004 *Thin Solid Films* **457** 44–47
[3] Ganashev I, Sugai H 2003 *Surface and Coatings Technology* **174–175** 15–20
[4] Gibalov V G and Pietsch G J, 2000 *J. Phys. D* **33** 2618
[5] Kobouzi Y, Moisan M, Rostaing J C, Trassy C, Guérin D, Kéroack D, Zakrsewski Z 2003 *J. Appl. Phys.* **93** 12 9483–9496
[6] Sekiguchi H, Mori Y 2003 *Thin Solid Films* **435** 44
[7] Boudam M K, Moisan M, Saoudi B, Popovici C, Gherardi N and Massines F 2006 *J. Phys. D: Appl. Phys* **39** 3494–3507
[8] Stoffels E, Kieft IE and Sladek REJ 2003 *J. Phys. D: Appl. Phys.* **36** 2908–2913
[9] Pencheva M, Petrova Ts, Benova E and Zhelyazkov I 2005 *J. Phys.: Conf. Series* **44**

Journal of Physics: Conference Series **207** (2010) 012024 doi:10.1088/1742-6596/207/1/012024

Similarities in Calculated Stark Broadening Parameters of Argon Spectral Lines

M Christova[1], M S Dimitrijević[2], A Kovačević[2]

[1]Department of Applied Physics, Technical University-Sofia, BG-1000 Sofia, Bulgaria
[2]Astronomical Observatory, Volgina 7, 11060 Belgrade 38, Serbia

E-mail: mchristo@tu-sofia.bg, mdimitrijevic@aob.bg.ac.yu, andjelka@matf.bg.ac.yu

Abstract. The similarities of the calculated Stark broadening parameters of four argon spectral lines within a spectral series $3p^5nd \rightarrow 3p^54p$ have been analyzed. The Stark broadening parameters have been calculated using Sahal-Bréchot theory within the semi-classical perturbation formalism. The dependeces of the Stark width and shift versus the effective quantum number have been presented.

1. Introduction

The broadening of spectral lines of heavy elements in plasmas is principally determined by two factors, the plasma environment and the atomic structure of the emitting atom or ion. Since atomic structures exhibit many great regularities and similarities, one must expect the same for the width and shift parameters of plasma broadened lines. A comprehensive study and analysis of regularities and similarities of the existing experimental data for the Stark broadening parameters have been presented in [1], based on the *L-S* coupling scheme.

One of the most useful gases for laboratory plasmas, argon, is also interesting in astrophysics since with the development of space-borne spectroscopy, the importance of atomic data, including line broadening parameters for trace elements like argon [2], is increasing. For example argon is found in CVn binary σ^2 Coronae Borealis [3], and "Chandra's" X-ray spectra of young supernovas 1998S and 2003bg revealed an argon over-abundance [4]. Recently, argon lines are observed in the optical spectrum of the Be star Hen 2-90 [5], as well as in planetary nebulae and H II regions in the two dwarf irregular galaxies Sextans A and B [6]. Also argon abundance has been determined from spectral lines, e.g. for LSE 78, an extreme helium star [7], for the similar star BD-9°4395 [8], for DY Cen [9] and γ Peg [10], as well as for the Sun [11]. Consequently, Stark line broadening parameters for neutral and ionized argon are of interest for the modeling and investigation of astrophysical plasmas. Often, the modeling of astrophysical objects needs atomic data for thousands and sometimes millions transitions. It is difficult and cumbersome to calculate the Stark broadening parameters for all these lines, so that methods enabling interpolation and extrapolation of the calculated results on the basis of similarities and systematic trends are of interest. For example we can obtain new Stark broadening parameters from regularities within spectral series.

In this work, the systematic trend of Stark broadening parameters of Ar I 522.1, 549.6, 603.2 and 737.2 nm spectral lines within the same spectral series with the corresponding transitions $3p^57d - 3p^54p$, $3p^56d - 3p^54p$, $3p^55d - 3p^54p$ and $3p^54d - 3p^54p$ has been investigated. The behavior of these parameters with the increase of the effective quantum number of the initial energy levels of the corresponding transition within the spectral series has been obtained. The results for Ar I 522.1, 549.6 and 603.2 are given in [12] where the calculations are oriented to laboratory plasmas. The examined perturbers are electrons, argon ions and protons. The presented calculations in this work are oriented for astrophysical purposes and the studied perturbers are electrons, protons and helium ions.

2. Theory

The Stark broadening parameters have been calculated using Sahal-Bréchot theory within the semi-classical perturbation formalism [13, 14], where the full width (W) at half maximum and the shift (d) of an isolated line originating from the transition between the initial level i and the final level f is expressed as:

$$W = 2n_e \int_0^\infty v f(v) dv \left[\sum_{i' \neq i} \sigma_{ii'}(v) + \sum_{f' \neq f} \sigma_{ff'}(v) + \sigma_{el} \right] \tag{1}$$

$$d = \int_0^\infty v f(v) dv \int_{R_3}^{R_d} 2\pi\rho d\rho \sin 2\varphi_p , \tag{2}$$

where i' and f' are perturbing levels, n_e and v are the electron density and the velocity of perturbers, respectively, and $f(v)$ is the Maxwellian distribution of the electron velocities.

The inelastic cross sections $\sigma_{ii'}$ (v) (respectively $\sigma_{ff'}$ (v)) can be expressed by an integration of the transition probability $P_{ii'}$ over the impact parameter ρ:

$$\sum_{i' \neq i} \sigma_{ii'}(v) = \frac{1}{2} \pi R_1^2 + \int_{R_1}^{R_d} 2\pi\rho d\rho \sum_{i' \neq i} P_{ii'}(\rho, v) . \tag{3}$$

The elastic collision contribution to the width is given by:

$$\sigma_{el} = 2\pi R_2^2 + \int_{R_2}^{R_d} 8\pi\rho d\rho \sin^2 \delta \tag{4}$$

$$\delta = \left(\varphi_p^2 + \varphi_q^2\right)^{1/2} . \tag{5}$$

The phase shifts φ_p and φ_q are due to the polarization and to the quadrupole potential, respectively. The cut-off parameters R_1, R_2, R_3, the Debye cut-off R_d, and the symmetrization procedure are described in [13, 14].

The impact approximation is valid when the duration of collisions is much shorter than the separation time between strong collisions (collisions whose resulting phase shift is bigger than 1 rad): $n_e \pi \rho_{typ}^3 << 1$, where ρ_{typ} is the typical impact parameter for strong collisions.

The resulting profiles are Lorentzian. This condition is well verified for electron collisions for a large range of densities. For ion collisions the impact approximation might fail, especially for high densities or low temperatures. Within the impact approximation the ion broadening contribution has a Lorentzian shape and the total full width or shift is simply the sum of the corresponding full widths or shifts for electron and ion broadening. If the impact approximation is not applicable to ion broadening, it is possible to apply the quasistatic approximation given by Griem in [15]. Then the ion broadening parameters obtained within the quasistatic approximation, denoted here with the index iq, are:

$$W_{iq} = 1{,}75.10^{-4} n_e^{1/4} A [1 - 0{,}068 n_e^{1/6} T^{-1/2}] W_e \tag{6}$$

$$d_{iq} = 10^{-4} n_e^{1/4} A [1 - 0{,}068 n_e^{1/6} T^{-1/2}] W_e , \tag{7}$$

where W_e is the impact contribution of electrons to the total width, T is in Kelvin, and n_e in cm^{-3}. A is the quasistatic parameter, defined in [15] as follows:

$$A = \left(\frac{e F_0^2}{\hbar W_e} \left| \alpha_i - \alpha_f \right| \right)^{3/4} , \tag{8}$$

where $F_0 = 2\pi(4/15)^{2/3} e n_e^{2/3}$ is the normal field strength. The polarizability of the initial level α_i (resp. α_f is the polarizability of the final level) is expressed as:

$$\alpha_i = 4a_0^3 \sum_{i' \neq i} f_{ii'} \left(\frac{I_H}{\Delta E_{ii'}} \right)^2 , \tag{9}$$

where a_0 is the Bohr radius, $f_{ii'}$ is the oscillator strength, and I_H is the ionization energy of hydrogen.

3. Results

The calculations were performed for particular lines within multiplets (spin-orbital interaction is included). The most appropriate *j-L* coupling scheme for argon atoms has been used. The values of the energy levels have been taken from the NIST catalogue [16]. The oscillator strengths (*j-L* coupling) have been calculated within the Bates & Damgaard approximation. The calculations have been made for a set of temperatures (2.5 – 5.0)×10^4 K at a perturber (electrons, protons and helium ions) density of 10^{16} cm^{-3}.

In this work we present the calculated Stark broadening parameters for the Ar I 737.2 nm spectral line. In table 1 the wavelengths of the studied argon lines, the corresponding transitions in *j-L* coupling, the perturbing levels *i'* and *f'*, the energy values, and the effective quantum number of the initial level are presented.

Table 1. Basic data on the considered Ar I spectral lines. Here λ denotes the wavelength, *i* and *f* are the initial and final level of the transition (within the frame of *j-L* coupling), *i'* and *f'* are the corresponding perturbing levels, E_i and E_f are the energy values, and n^* is the effective quantum number of the initial level.

λ (nm)	Transition ($i - f$)	i' levels	f' levels	E_i (cm^{-1})	E_f (cm^{-1})	n^*
522.1	$3p^5 7d - 3p^5 4p$ $^2[7/2]^\circ_4 - ^2[5/2]_3$	5f, 6f, 7f, 8f, 9f, 5p, 6p, 7p, 8p, 9p	4s, 5s, 6s, 3d, 4d, 5d, 6d	124610	105463	6.62
549.6	$3p^5 6d - 3p^5 4p$ $^2[7/2]^\circ_3 - ^2[5/2]_3$	4f, 5f, 6f, 7f, 4p, 5p, 6p, 7p	4s, 5s, 6s, 3d, 4d, 5d, 6d	123653	105463	5.63
603.2	$3p^5 5d - 3p^5 4p$ $^2[7/2]^\circ_4 - ^2[5/2]_3$	4f, 5f, 6f, 7f, 4p, 5p, 6p, 7p	4s, 5s, 6s, 3d, 4d, 5d, 6d	122036	105463	4.65
737.2	$3p^5 5d' - 3p^5 4p$ $^2[7/2]^\circ_4 - ^2[5/2]_3$	4f, 5f, 6f, 4p, 5p, 6p	4s, 5s, 6s, 3d, 4d, 5d, 6d	119024	105463	3.68

The results are presented in table 2 for the Ar I 737.2 nm argon line. The calculated Stark widths (full width at half intensity maximum) *W* and shifts *d* for collisions with electrons (W_e, d_e), He^+ ions – (W_{He+}, d_{He+}), and protons – (W_p d_p) are given. The ratio *C/W* gives the value of the maximum density for which the line is isolated for the corresponding perturber.

For the atmospheres of the main branch stars the total Stark width *W* and shift *d* are approximatelly:

$$W = W_e + 0.9W_p + 0.1W_{He+} \tag{10}$$

$$d = d_e + 0.9d_p + 0.1d_{He+} . \tag{11}$$

Table 2. Stark broadening parameters for Ar I 737.2 nm for a perturber (electrons, helium ions, and protons) density of 10^{16} cm^{-3} and temperatures from 2500 up to 50000 K.

Perturbers are:		Electrons		Ionized helium		Protons	
Transition	T (10^3 K)	W_e (0.1 nm)	d_e (0.1 nm)	W_{He+} (0.1 nm)	d_{He+} (0.1 nm)	W_p (0.1 nm)	d_p (0.1 nm)
4d $^2[7/2]_4$	2.5	0.572	0.367	-	-	0.171	0.787×10^{-1}
4p $^2[5/2]_3$	5	0.646	0.425	0.168	0.0756	0.182	0.980×10^{-1}
737.2 nm	10	0.739	0.428	0.176	0.0917	0.193	0.117
	20	0.871	0.371	0.185	0.108	0.205	0.136
$C = 0.65\times$	30	0.964	0.318	0.19	0.117	0.213	0.148
10^{19}	50	1.07	0.275	0.198	0.13	0.224	0.163

In figures 1 and 2 the temperature dependences of the electron impact width and shift (electron density 10^{16} cm^{-3}) of Ar I 737.2 nm are illustrated.

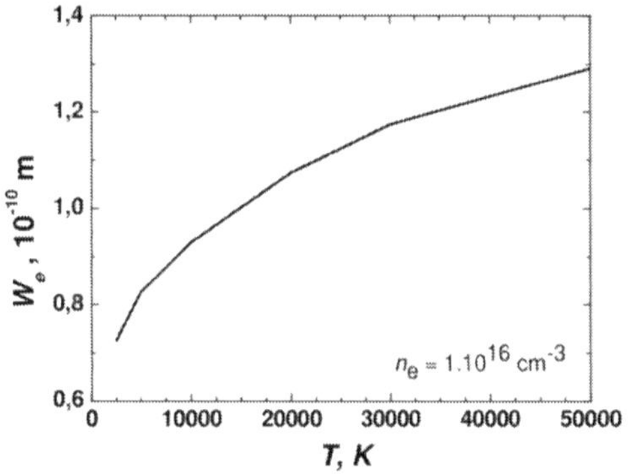

Figure 1. Electron impact width of the Ar I 737.2 nm spectral line versus the temperature for a 10^{16} cm^{-3} electron density.

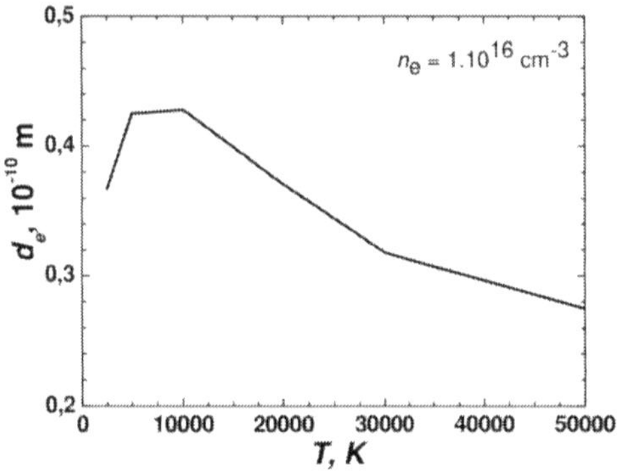

Figure 2. Electron impact shift of the Ar I 737.2 nm spectral line versus the temperature for a 10^{16} cm^{-3} electron density.

The contributions of the different collisions between emitters and perturbers (inelastic and elastic (polarization) collisions) in the Stark width of the examined argon lines from one spectral series show a similar behavior versus the temperature. One example for this behavior is presented in figure 3 for the Ar I 522.1 nm argon line. The broadening coefficient ($\beta = W / n_e$) and their components (β_{in} – contribution of inelastic collisions, β_{el} – contribution of elastic collisions, and β_p – contribution of polarization (the polarization interactions are elastic ones) collisions) have been analyzed. The inelastic component increases with the temperature while the elastic and polarization ones decrease for the argon lines from the studied series. The quadruple component (β_q) depends only by the initial and final states of the emitters and does not depend on the temperature and perturber density, and is not included in the figure.

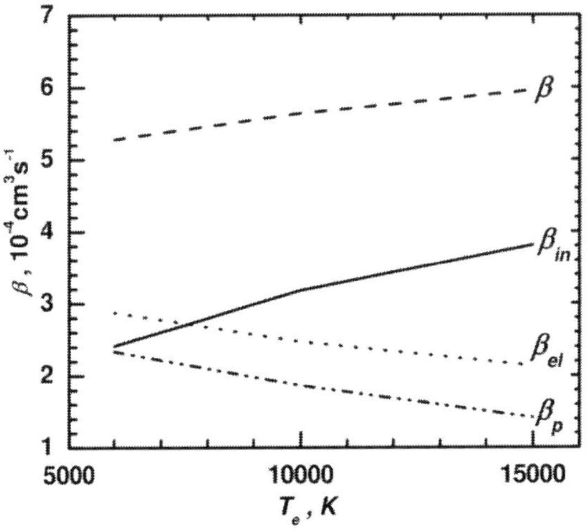

Figure 3. Stark broadening coefficient for Ar I 522.1 nm ($\beta = W / n_e$) and their components: $\beta = \beta_{in} + \beta_{el}$, where β_{in} is the contribution of the inelastic argon emitter-perturber collisions, β_{el} is the contribution of elastic ones, $\beta_{el} = \beta_p + \beta_q$ (β_p gives the contribution of polarisation interactions and β_q – of the quadruple interactions). The quadruple interactions do not depend on the temperature and the perturber density so it is not presented in the figure.

In figures 4 and 5 the electron impact width and shift, respectively, are given. From one side, the argon spectral lines are broadened mainly due to the collisions with electrons, usually. From the other side, it is more correct to study the width and shift trends of the spectral lines for each kind of perturbers separately.

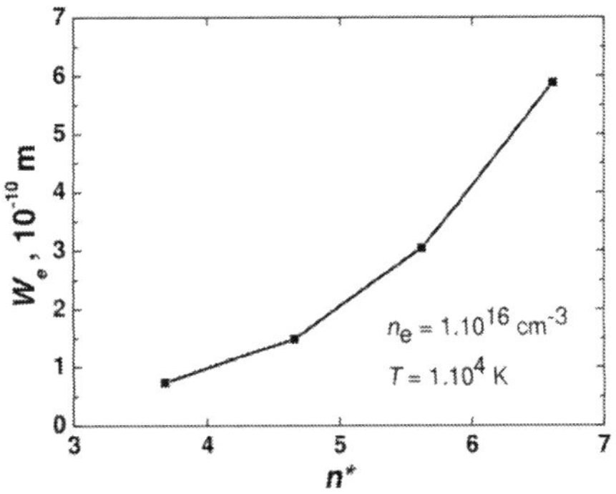

Figure 4. Electron impact widths of Ar I spectral lines within the spectral series versus the effective quantum number of the initial energy level.

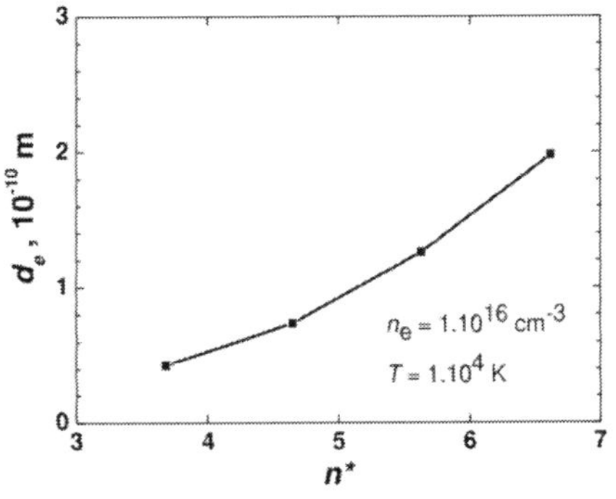

Figure 5. Electron impact shifts of Ar I spectral lines within the spectral series versus the effective quantum number of the initial energy level.

One can see that the behavior of the electron impact widths and shifts within the considered spectral series is so regular, that interpolation and extrapolation of new data is possible. The complete analysis of the results of this paper and the comparison with other theoretical and experimental data available in the literature will be given in [17].

Acknowledgments
This work was partially financed by the Technical University – Sofia and is a part of the project 146001 supported by the Ministry of Science and Technological Development of Serbia.

References

[1] Konjević N, Lessage A, Fuhr J R and Wiese W L 2002 *J. Phys. Chem. Ref. Data* **31** 819
[2] Werner K, Rauch T and Kruk J W 2007 *Astron. Astrophys.* **466** 317
[3] Suh J A, Audard M, Güdel M and Paerels F B S 2005 *Astrophys. J.* **630** 1074
[4] Lewin W 2005 Chandra Proposal ID #07500185
[5] Kraus M, Borges F M, De Araújo F X and Lamers H J G L M 2005 *Astron. Astrophys.* **441** 289
[6] Kniazev A Yu, Grebel E K, Pustilnik S A, Pramskij A G and Zucker D B 2005 *Astron. J.* **130** 1558
[7] Jeffery C S 1993 *Astron. Astrophys.* **279** 188
[8] Jeffery C S and Heber U 1992 *Astron. Astrophys.* **260** 133
[9] Jeffery C S and Heber U 1993 *Astron. Astrophys.* **270** 167
[10] Peters G J 1976 *Astrophys. J. Suppl. Series* **30** 551
[11] Anders E and Grevesse N 1989 *Geochim. Cosmochim. Acta* **53** 197
[12] Dimitrijević M S, Christova M and Sahal-Bréchot S 2007 *Phys. Scripta* **75** 809
[13] Sahal-Bréchot S 1969 *Astron. Astrophys.* **1** 91
[14] Sahal-Bréchot S 1969 *Astron. Astrophys.* **2** 322
[15] Griem H R 1974 *Spectral line broadening by plasmas* (New York: Academic)
[16] http://physics.nist.gov/
[17] Christova M, Dimitrijević M S and Kovačević A (to be published)

Stark Broadening Parameters of Ne I 837.8 nm Spectral Line

M Christova[1], M S Dimitrijević[2], Z Simić[2] and S Sahal-Bréchot[3]

[1]Department of Applied Physics, Technical University-Sofia, BG-1000 Sofia, Bulgaria
[2]Astronomical Observatory, Volgina 7, 11060 Belgrade 38, Serbia
[3]Observatoire de Paris, 92195 Meudon Cedex, France

mchristo@tu-sofia.bg, mdimitrijevic@aob.bg.ac.yu, zsimic@aob.bg.ac.yu, sylvie.sahal-brechot@obspm.fr

Abstract. The Stark broadening parameters of Ne I 837.8 nm spectral line corresponding to the transition $2p^5 3d\ ^2[7/2]^o_4 - 2p^5 3p\ ^2[5/2]_3$ have been calculated using Sahal-Bréchot theory. The temperature dependence of the width and shift has been obtained.

1. Introduction

Stark broadening parameters of neon lines are of importance not only for laboratory plasma diagnostics and for technological plasma investigations and development but also for astrophysics. Neon is the most abundant element in the universe after hydrogen, helium, oxygen, and carbon, and it is, for example, one of the products of hydrogen and helium burning in the orderly evolution of stellar interiors. Consequently, its lines are present in stellar spectra and, for example, neon spectral lines have been used to determine its abundances in late- to mid-B stars [1]. The neon spectral line at 837.8 nm, corresponding to the transition $2p^5 3d\ ^2[7/2]^o_4 - 2p^5 3p\ ^2[5/2]_3$, is one of the most useful for plasma diagnostics [2]. It is intensive, well isolated, and in the visible part of the spectrum. The Stark broadening of this line was theoretically investigated, for example, by Griem in [3] and experimentally by Purić et al. [4] and del Val et al. [5]. In this paper, calculations of the Stark broadening parameters of Ne I 837.8 nm for astrophysical purposes are presented.

2. Theory

In this work the semi-classical theory of Sahal-Bréchot [6,7] for Stark broadening calculations, also presented in [8], has been used.

3. Results

3.1. Atomic data

In table 1 the wavelengths of the studied neon line, along with the corresponding transition in *j-L* coupling scheme, the perturber energy levels, the energy values of the initial and final level of the transition, and the effective quantum number of the initial level are included.

Table 1. Basic data for the considered Ne I spectral line. Here λ denotes wavelength, i and f are initial and final level of the transition (within the frame of j-L coupling), i' and f' are the corresponding perturbing levels, E_i and E_f are the energy values, and n* is the effective quantum number of the initial level.

λ (nm)	Transition ($i-f$)	i' levels	f' levels	E_i (cm^{-1})	E_f (cm^{-1})	N*
837.7	$2p^5 3d' - 2p^5 3p$ $^2[7/2]^{\circ}_4 - {}^2[5/2]_3$	4f, 5f, 3p, 4p, 5p	3s, 4s, 5s, 3d, 4d, 5d	161590.3	149657.0	2.98

3.2. *Calculations*

The calculations have been oriented for the diagnostic of astrophysical objects where the perturbers are electrons, helium ions, and protons, and have been made for a set of temperatures $(2.5 - 5.0)\times10^4$ K at a perturber density of 10^{16} cm^{-3}. The values of the energy levels have been taken from the NIST catalogue [9]. The oscillator strengths (j-L coupling) have been calculated within the Bates & Damgaard approximation.

3.3. *Temperature dependence*

The calculated results for the Stark width and shift of the examined neon spectral line are included in table 2. The contribution of the interactions between: (i) emitters and electrons (W_e, d_e); (ii) emitters and helium ions (W_{He+}, d_{He+}) and (iii) emitters and protons (W_p, d_p) in the Stark width and shift are given. The value of the maximum density for which the line is isolated for the corresponding perturber can be calculated using the ratio C/W, where C is the parameter, included in table 2 and W is the total width. For the ordinary stellar plasma the total Stark width W and shift d, respectively, are approximatelly:

$$W = W_e + 0.9W_p + 0.1W_{He+} \quad (1)$$

$$d = d_e + 0.9d_p + 0.1d_{He+}\,. \quad (2)$$

Table 2. Stark broadening parameters for Ne I 837.7 nm for a perturber (electrons, helium ions and protons) density of 10^{16} cm^{-3} and temperatures from 2500 up to 50000 K.

Perturbers are:		Electrons		Ionized helium		Protons	
Transition	T (10^3 K)	W_e (0.1 nm)	d_e (0.1 nm)	W_{He+} (0.1 nm)	d_{He+} (0.1 nm)	W_p (0.1 nm)	d_p (0.1 nm)
$3d\ ^2[7/2]_4$	2.5	0.306	0.21	0.0788	0.031	0.918×10^{-1}	0.502×10^{-1}
$3p\ ^2[5/2]_3$	5	0.35	0.233	0.0831	0.0386	0.974×10^{-1}	0.601×10^{-1}
837.7 nm	10	0.383	0.239	0.0868	0.046	0.104	0.701×10^{-1}
	20	0.417	0.222	0.0909	0.0535	0.112	0.806×10^{-1}
$C = 0.65\times10^{19}$	30	0.436	0.196	0.0935	0.058	0.117	0.870×10^{-1}
	50	0.465	0.162	0.0973	0.064	0.125	0.956×10^{-1}

3.3.1. *Width.* The values of the three components of the width, corresponding to the different perturbers, and the total Stark width increase with the temperature in the examined interval (perturber density 10^{16} cm^{-3}). The temperature dependence of the electron width is illustrated in figure 1.

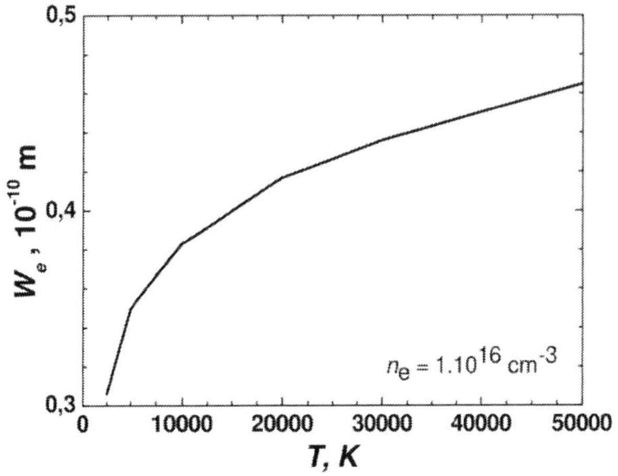

Figure 1. Electron impact width of the Ne I 837.7 nm spectral line versus the temperature for electron density 10^{16} cm^{-3}.

In ordinary stellar plasma for temperatures $(0.25 - 5)\times10^4$ K and at an electron density 10^{16} cm^{-3}, the contribution of the collisions between neon emitters and electrons in the total Stark width is 77 – 79 %, of emitters-helium ions collisions – 1–2 %, and the contribution of the emitters-protons is 19–20 %.

3.3.2. *Shift.* The electronic component of the shift due to the collisions between the emitters and electrons increases with the temperature up to near 10^4 K and after that decreases. This dependency governs the temperature variation of the total Stark shift, as it is shown in figure 2. The values of the shift due to the collisions of the emitters with the helium ions and protons increase with the temperature in the considered temperature interval.

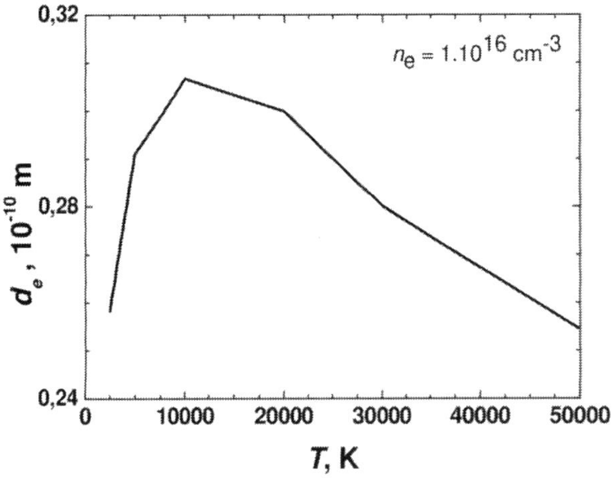

Figure 2. Electron impact shift of Ne I 837.7 nm spectral line versus the temperature for electron density 10^{16} cm^{-3}.

For ordinary stellar plasmas, for example in the temperature interval $(0.25 - 5)\times10^4$ K and at an electron density 10^{16} cm^{-3}, the contribution of the collisions between neon emitters and electrons in the total Stark shift is 64–81 %, this one of the emitters-helium ions collisions – 1–2.5 %, and the contribution of the emitters-protons is 17–33 %.

A complete analysis of the obtained results and a comparison with the theoretical and experimental results available in the literature will be given in [10].

Acknowledgments
This work was partially financed by the Technical University – Sofia and is a part of the project 146001 supported by the Ministry of Science and Technological Development of Serbia.

References

[1] Sigut T A A 1999 *Astrophys. J.* **519** 303
[2] Konjević N, Lessage A, Fuhr J R and Wiese W L 2002 *J. Phys. Chem. Ref. Data* **31** 819
[3] Griem H R 1974 *Spectral line broadening by plasmas* (New York: Academic)
[4] Purić J, Srećković A, Djeniže S, Labat J and Ćirković Lj 1988 *Phys. Lett A* **126** 280
[5] del Val J A, Aparicio J A and Mar S 1999 *Astrophys. J.* **513** 535
[6] Sahal-Bréchot S 1969 *Astron. Astrophys.* **1** 91
[7] Sahal-Bréchot S 1969 *Astron. Astrophys.* **2** 322
[8] Christova M, Dimitrijević M S and Kovačević 2008 Similarities in Calculated Stark broadening Parameters of Argon Spectral Lines *J. Phys.: Conf. Ser.* Third International Workshop and Summer Scool on Plasma Physics 02
[9] http://physics.nist.gov/
[10] Christova M, Dimitrijević M S, Simić Z and Sahal-Bréchot S (to be submitted)

Dynamic effects in dual-frequency capacitively coupled discharges

D Voloshin[1], **O Proshina and T Rakhimova**

Skobeltsyn Nuclear Physics Institute of Lomonosov Moscow State University, 119991, Moscow, Russia

E-mail: dvoloshin@mics.msu.su

Abstract. Investigations of the discharge current shape in a low frequency single frequency discharge and in a dual frequency discharge are performed in this paper. For this purpose a Particle in cell with Monte-Carlo Collisions numerical simulation of the capacitively coupled discharge in argon, operated at frequencies of 1.76 MHz & 81 MHz, is carried out. As a result the non-sinusoidal, close to triangle shape of the discharge current is obtained even for the case of a symmetric discharge. The physical reason of such nonsinusoidal behavior of the discharge current is the dynamic structure of the discharge sheaths. Corresponding to the discharge current, the sheath edge motion is also non-sinusoidal and non-symmetric. These results may be important for the appropriate choice of the current and voltage waveforms in analytic modeling of single and dual frequency discharges.

1. Introduction

Radio-frequency capacitively coupled plasma (CCP) discharges have been intensively studied both experimentally and theoretically due to their wide application in plasma processing and technology.

In modern plasma processing devices the CCP discharge is usually driven by two rf sources. Such dual-frequency excitation allows achieving more flexibility in the control of plasma properties. The typical value of the low-frequency is about 2 MHz and lower. Thereby there is additional interest in the study of the plasma dynamic properties in low-frequency (LF) discharges and the influence of the LF on the plasma in dual-frequency (DF) discharges.

Recent experimental investigations in DF discharges were performed for argon plasmas [1] and hydrogen plasmas [2]. These experiments showed the significant influence of the low-frequency on the discharge behavior.

Single frequency LF discharges have been studied for a long time [3–7], but some physical phenomena are still poorly understood.

In [3] an equivalent circuit model was presented to obtain voltage and current waveforms in the LF CCP discharge. These waveforms were nonsinusoidal for both symmetric and asymmetric discharges, and were explained by the nonlinear properties of the plasma sheaths. However, the physical reasons of this specific current shape were beyond the framework of the equivalent circuit model.

Different analytical and numerical approaches have been developed for analysis of CCP discharges. Such approaches usually use the discharge current or the electrode voltage as an input parameter, and the sinusoidal discharge current is often assumed in single frequency (SF) and DF

[1] To whom any correspondence should be addressed.

discharges [6 – 11]. However, since the DF CCP discharge is a strongly non-linear system, the question about the discharge current waveform is very important.

In view of the great interest in modeling dual frequency discharges and the importance of the appropriate choice of the current and voltage waveforms, the numerical investigation of the discharge current in the LF 1.76 MHz SF discharge and the 1.76 MHz & 81 MHz DF discharge is performed in this paper.

2. Particle in Cell model

A numerical simulation of SF and of DF CCP discharges based on the Particle in cell with Monte-Carlo collision method (PIC MCC) is performed.

The used PIC MCC model of RF discharge was described in detail in [12]. The one dimensional model of CCP discharge with an electrode gap of 2.4 cm and electrode radii of 5 cm is applied. The ion and electron behaviors are kinetically simulated by solving Newton's equations and with a MC treatment of collisions processes. The sets of electron-neutral and ion-neutral collision cross sections for Ar were described in [12]. In addition, in this work we take into account ionization of Ar by Ar^+ ions and the cross section for this process is taken from [13].

Since the large applied voltages are considered, the secondary electron emission from electrodes may play an essential role. The probability for secondary emission induced by the ion impact, γ_i, depends on the electrode material and on the energy of impinging ions. We use the energy dependence of the coefficient γ_i that is suggested in [14]. Besides Ar^+ ions, high energy argon atoms, Ar^{fast}, generated in the sheaths due to charge transfer collisions of Ar^+ ions, may cause secondary electron emission at the electrode surface. As the following calculations have shown, in the conditions under study the relative contribution of the secondary electron emission by Ar^{fast} atoms may be comparable with the secondary emission due to ion impact. That is why we track motion of fast atoms in the present PIC MCC simulation. The initial velocities of the generated Ar^{fast} atoms are equal to the "parent" ions velocities. Then fast atoms move in the sheaths and loss their energy in elastic and ionization collisions with "cold" Ar atoms. The momentum transfer and ionization cross-section of Ar^{fast} were taken from [13]. According to [14] the probability of the secondary electron emission induced by the fast argon atoms, γ_a, depends on the energy of the Ar^{fast}, and this process is not effective for low energies. The used in this work the analytical expression for γ_a from [14] and an energy threshold of 32 eV. Therefore we stop track the Ar^{fast} atom if its energy drops below 30 eV. It is an irreversible process since neutral atoms are not accelerated in the electric field. As a result, Ar^{fast} are distributed mainly in the sheaths and the total number of super particles representing Ar^{fast} atoms is only about 10-15% of the total numbers of super particles for electrons and ions at the same particle weights. In that way the kinetic description of Ar^{fast} atoms does not slow down the PIC MCC simulation.

The anisotropic scattering of electrons in collisions due to high electron velocities is taken into account [15].

2.1. *Single frequency discharge*

The numerical simulation of the SF discharge was performed for the following discharge parameters: rf frequency is f =1.76 MHz, the input power is P = 60 W, the gas pressure is p = 45 mTorr. In figure 1 the discharge current is shown along with the electron, ion and displacement currents to the left electrode. It can be seen that the total current has the non-sinusoidal, close to triangle shape.

The physical reason of such nonsinusoidal behavior of the discharge current is caused by dynamic structure of the discharge sheaths.

The main temporal characteristic of the sheath is the relation between the ion transit time through the sheath, τ_{ion}, and the rf period, τ_{RF}. There are two limit regimes for the SF discharge: the low frequency regime ($\tau_{ion} << \tau_{RF}$) and the high frequency (HF) regime ($\tau_{ion} >> \tau_{RF}$) [16]. In the HF regime ions pass the sheath for many rf periods and respond to the average sheath potential [16]. In the

LF regime ions cross the sheath during the small part of the rf period and trace the instantaneous value of the electric field in the sheath.

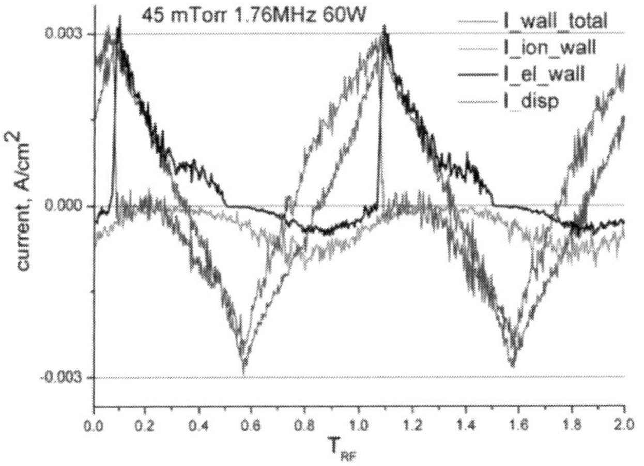

Figure 1. The total discharge current and the electron, ion and displacement currents at the electrode obtained from the PIC MC simulation.

As it follows from [17], in the LF case the ion velocity is close to the Bohm velocity during a considerable part of the rf period. And as the PIC MCC simulation has shown, the time when the ions move with Bohm velocity is about half of the rf period. Therefore considering the LF case of the SF discharge (or the DF discharge, where one frequency is relatively low) the following additional temporal characteristic should be taken into account:

$$\tau_B = \frac{s_m}{\sqrt{T_e / M}}, \tag{1}$$

where τ_B is the Bohm time, s_m is the electrode sheath width, T_e is the electron temperature in the plasma-sheath edge region, and M is the ion mass. The Bohm time describes the time which is necessary for the ion flow from the plasma bulk to maintain the ion concentration profile in the sheath region.

When

$$\omega \tau_B << 1, \tag{2}$$

or $\tau_B << \tau_{RF}$, the quasi-stationary case takes place. In this case the ion flow from plasma maintains the ion profile in the sheath and at each phase of the rf period the plasma is screened by the positive space charge in the sheath. The sheath edge motion corresponds to the sinusoidal electrode voltage.

When inequality (2) fails, or $\tau_B >> \tau_{RF}$, the low-frequency dynamic case takes place [17]. In this case the electrode sheath doesn't correspond to the ordinary Child-Law sheath and the ion flow to the electrode varies during rf period.

In the studied discharge conditions, the sheath temporal characteristic may be evaluated as follows:

$$\tau_{RF} = 1/f = 0.5682 \times 10^{-6}\ \text{s},$$

$$\tau_{ion} \sim \frac{s_m}{\sqrt{eU_{pl} / M}} = 0.26 \times 10^{-6}\ \text{s},$$

$$\tau_B \approx \frac{s_m}{\sqrt{kT_e / M}} = 5.8 \times 10^{-6}\ \text{s},$$

where U_{pl} = 680 V is the plasma potential; the sheath width is s_m = 0.9 cm; the electron temperature is T_e = 1 eV. The used numerical values are taken from the PIC MCC calculation.

It can be seen that in the discharge conditions under study the inequality (2) fails and the low-frequency dynamic case takes place.

In the dynamic case the ion flow from the plasma has no time to restore the ion concentration profile as the electrode potential changes relatively fast. As a result the sheath edge moves away from the electrode during the one half of the rf period to screen the plasma by the positive space charge in the sheath. When the electrode voltage change its sign in the middle of the rf period the sheath edge moves very rapidly towards the electrode. This is illustrated by the PIC MC simulation results in figure 2. In comparison with the quasistationary case, where the sheath edge moves away and towards the electrode with equal velocities, in this dynamic case the sheath edge moves towards the sheath 3 times faster than it moves away from the sheath.

As a result of the fact that sheath edge moves very rapidly towards the electrode, the electron current to the electrode has the triangle shape which can be seen in figures 3, 4. The sheath resistivity becomes higher when the rf frequency decreases. At the present frequency of 1.76 MHz the ion and electron currents are of the same order as the total discharge current, and during a part of the LF period the electron current becomes equal to the total discharge current. Thus in this dynamic case the discharge current takes the non-sinusoidal form, close to the triangle one.

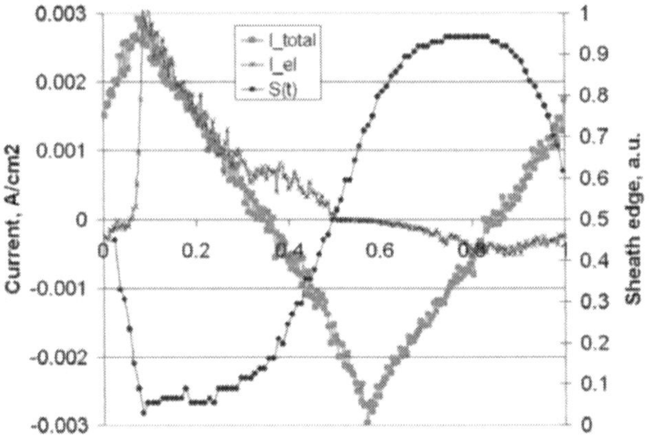

Figure 2. The sheath edge motion during rf period in the LF dynamic case. The discharge parameters are the same as in figure 1.

We should note once again that this phenomenon is caused by the temporal effects of the ion motion in the sheath, because there are also several different cases when the non-sinusoidal discharge current is possible. First, it is the case of an asymmetric discharge with electrodes of different areas. But in this paper we consider the symmetric discharge. Second, the secondary electrons emitted from the electrode may cause the non-sinusoidal shape of the current [18]. But this case takes place at higher gas pressures of several hundreds mTorr [18]. To prove that secondary electrons do not affect much the current shape for the case discussed in this paper, we performed the calculation without secondary electron emission from the electrode at the following discharge parameters: the rf frequency is 1.76 MHz; the pressure is 100 mTorr (since the discharge at 45mTorr does not turn on without secondary electron emission taken into account), and the input power is 60 W. The results are presented in figure 3. It can be seen that the total discharge current is non-sinusoidal, close to triangle. In comparison with figures 1, 2, the electron current to the electrode has positive value everywhere since electrons are not emitted from the electrode.

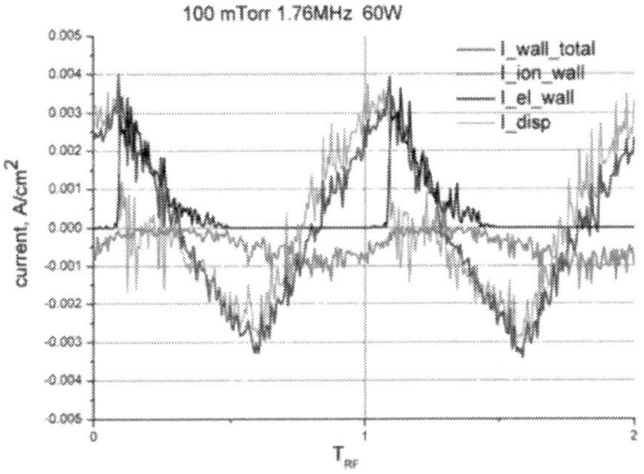

Figure 3. The discharge current from the PIC MCC simulation without secondary electron emission ($\gamma = 0$). 1.76 MHz, 100 mTorr, 60 W.

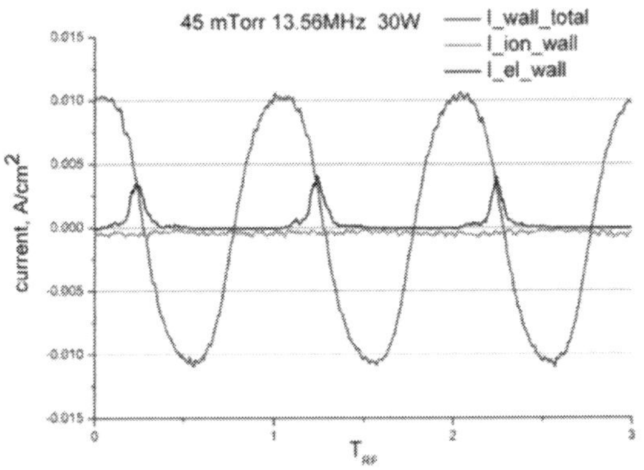

Figure 4. Discharge current from the PIC MCC simulation. 13.56 MHz, 45 mTorr, 30 W

The discharge parameters for the quasi-stationary LF case are far from the typical parameters for the real applications of CCP discharges. Therefore in order to illustrate the case of the sinusoidal discharge current, the discharge in argon at a rf frequency of 13.56 MHz, pressure of 45 mTorr, and input power of 30 W is simulated. Such parameters correspond to the HF regime for the ion motion in the sheath. The sheath edge moves sinusoidal in this case and the electron current has a symmetrical shape. Certainly its value is smaller and the electrons reach the electrode during a smaller part of the rf period in comparison with the 1.76 MHz discharge. The electron current does not disturb the total discharge current and the last one is close to sinusoidal as it can be seen in figure 4.

2.2. *Dual frequency discharge*

The numerical simulation of the DF discharge is performed for the following discharge parameters: $f_l = 1.76\,\text{MHz}$; $f_h = 81\,\text{MHz}$; $p = 45$ mTorr; the input power at low frequency is $P_l = 10$ W, the input power at high frequency is $P_h = 11$ W.

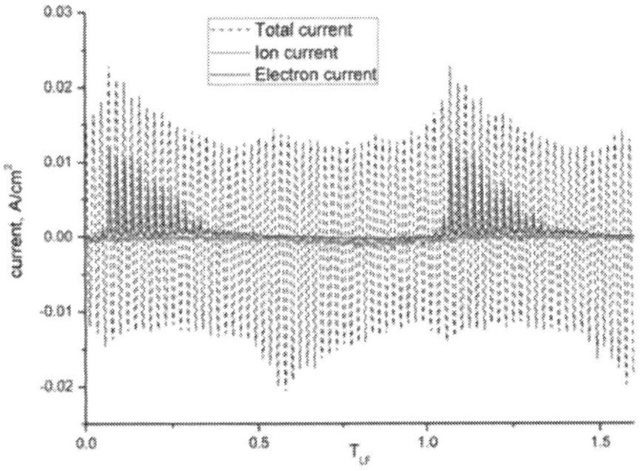

Figure 5. The DF discharge current from the PIC MCC simulation.

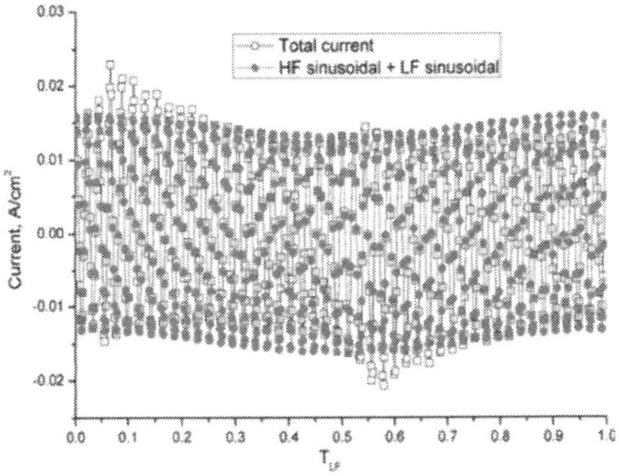

Figure 6. The DF discharge current versus the sinusoidal approximation.

The total discharge current, and electron and ion currents to the electrode are shown in figure 5.

It can be seen that in spite of the high frequency oscillation the envelope curve of the total current (or the current averaged over the high frequency) has the non-sinusoidal shape. The electron current to the electrode is not equal to the total current at its maximum, but nevertheless the electron current is high enough to disturb the shape of the total current.

We can test the error that the sinusoidal approximation of the total current brings in this case. We have taken a Fourier analysis of the total current to find the amplitude of the oscillation at low frequency and at high frequency. In figure 6 the total discharge current is compared with the sum of sinusoidal functions at HF and LF. The maximal total current is 1.5 times higher than sum of LF and HF sinusoidal component. So, the usage of the sinusoidal current shape may lead to errors in the treatment of the discharge processes. The sheath dynamics discussed in this paper may be important also for the treatment of electron heating and resonance phenomena [19] in SF and DF discharges.

3. Conclusion

The low frequency 1.76 MHz discharge and the dual frequency 1.76 MHz & 81 MHz discharge were investigated with the PIC MCC model. The resulting "triangle" form of the discharge current is explained on the base of the ion dynamics in the sheath. It is shown that the sinusoidal approximation of the discharge current is not valid in these regimes.

Acknowledgments

This research was supported by the Russian Foundation of Basic Research (08-02-00465-a) and Key Science School (SS-133.2008.2).

References

[1] Braginsky O V, Kovalev A S, Lopaev D V, Rakhimova T V, Rakhimov A T, Voloshin D G, and Vasilieva A N 2007 *Proc. XXVIII Int. Conf. Phenom. Ionized Gases, Prague, Czech Republic, Jul. 15–20* 5P09-21 2114–7
[2] Lisovskiy V, Booth J-P, Landry K, Douai D, Cassagne V and Yegorenkov V 2008 *Plasma Sources Sci. Technol.* **17** 025002
[3] Metze A, Ernie D W and Oskam H J 1986 *J. Appl. Phys.* **60** 3081
[4] Vallinga P M and de Hoog F J 1989 *J. Phys. D: Appl. Phys.* **22** 925
[5] Deepak Bose, Govindan T R and Meyyappan M 2000 *J. Appl. Phys.* **87** 7176
[6] Panagopoulos T and Economou D J 1999 *J. Appl. Phys.* **85** 3435
[7] Edelberg E A and Aydil E S 1999 *J. Appl. Phys.* **86** 4799
[8] Lieberman M A 1988 *IEEE Trans. Plasma Sci.* **16** 638
[9] Lieberman M A 1989 *IEEE Trans. Plasma Sci.* **17** 338
[10] Robiche J, Boyle P C, Turner M M and Ellingboe A R 2003 *J. Phys. D: Appl. Phys.* **36** 1810
[11] Franklin R N 2003 *J. Phys. D: Appl. Phys.* **36** 2660
[12] Rakhimova T V *et al.* 2007 *IEEE Trans. Plasma Sci.* **35** 1229
[13] Phelps A V 1994 *J. Appl. Phys.* **76** 747
[14] Phelps A V and Petrovich Z Lj 1999 *Plasma Sources Sci. Technol.* **8** R21
[15] Surendra M, Graves D B and Jellum G M 1990 *Phys. Rev. A* **41** 1112
[16] Lieberman M A and Lichtenberg A J 1994 *Principles of Plasma Discharges and Material Processing* (New York: Wiley)
[17] Nikandrov D S and Tsendin L D 2006 *Technical Physics Letters* **32** 719
[18] Surendra M and Graves D B 1991 *IEEE Trans. Plasma Sci.* **19** 144
[19] Czarnetzki U, Mussenbrock T and Brinkmann R P 2006 *Phys. Plasmas* **13** 123503

Journal of Physics: Conference Series **207** (2010) 012027 doi:10.1088/1742-6596/207/1/012027

Electromagnetic modeling of axis-symmetric microwave devices

R Álvarez and L L Alves

Instituto de Plasmas e Fusão Nuclear, IST, Av. Rovisco Pais, 1049-001 Lisboa, Portugal

E-mail: `llalves@ist.utl.pt`

Abstract. This paper presents calculations of electromagnetic field distributions with different microwave-driven devices. Results are obtained by using a two-dimensional numerical simulation tool that solves Maxwell's equations for an axis-symmetric TM mode, by considering a harmonic time description. The tool is applied to a coaxial waveguide, a circular waveguide with a shortening plate at the top, and a plasma reactor operated by an axial injection torch. This reactor consists of a circular cavity powered by a coaxial waveguide assembled to a nozzle, and creates a micro-plasma that affects only very locally the electric field distribution. We compare the solutions obtained for the three devices. We also carry out a simulation benchmarking in terms of accuracy, by using the analytical solution to the problem (in cases where it is available), or in terms of precision, by using the numerical calculation of the average power absorbed (in the presence of plasma).

1. Introduction

The electromagnetic (EM) modeling of microwave plasma sources is of importance for applications, as it can contribute to the optimization of these devices. This paper presents a tutorial on the calculation of EM field distributions and power balance, with different microwave-driven devices operating at fixed $\omega/(2\pi) = 2.45$ GHz frequency. We have developed a two-dimensional (2D) simulation tool [1, 2] that solves Maxwell's equations for an axis-symmetric TM mode, by considering a harmonic time description. Recently, the same kind of simulations was carried out also by other authors [3, 4], with the purpose of modeling (some times self-consistently) different microwave devices with coaxial geometry. Model equations are solved for a given electrical permittivity, which depends on the medium considered (vacuum, neutral gas or plasma). The closure of the problem assumes perfect conductor walls and imposes an absorbing boundary condition upon open-boundaries. Notice that this study can seldom be carried out using commercial tools, especially in the presence of plasma or if a high resolution is required.

The simulation tool is first applied to both a coaxial waveguide and a circular waveguide with a shortening metal plate at the top, for which analytical solutions can be retrieved. These devices are some of the elements constituting a plasma reactor operated by an axial injection torch (AIT). This reactor consists of a circular cavity powered by a coaxial waveguide assembled to a nozzle, which acts as a microwave antenna, and creates a micro-plasma that affects only very locally the electric field distribution. In such cases, the decoupling of the real, complex device into smaller work domains, corresponding to simpler systems for which individual solutions can be easily worked out, is a recommended strategy to map the general solution pattern, before starting a self-consistent, global calculation.

2. Model description

2.1. Maxwell's equations

The model [1, 2] solves Maxwell's equations to calculate the 2D distribution of the EM fields with axis-symmetric microwave devices, by choosing a TM mode solution and by considering a harmonic time description [5]. With these suppositions, Maxwell's equations are written in cylindrical coordinates as

$$\frac{\partial E_z}{\partial r} - \frac{\partial E_r}{\partial z} = \frac{ik_0}{\sqrt{\varepsilon_0/\mu_0}} \frac{rH_\varphi}{r} \tag{1a}$$

$$\frac{1}{r}\frac{\partial (rH_\varphi)}{\partial r} = ik_0\varepsilon\sqrt{\frac{\varepsilon_0}{\mu_0}} E_z \tag{1b}$$

$$\frac{1}{r}\frac{\partial (rH_\varphi)}{\partial z} = -ik_0\varepsilon\sqrt{\frac{\varepsilon_0}{\mu_0}} E_r \quad , \tag{1c}$$

where $E_r(r, z)$ and $E_z(r, z)$ are the complex radial and axial electric field components, respectively; $H_\varphi(r, z)$ is the complex azimuthal magnetic field; ε_0 and μ_0 are the vacuum electrical permittivity and magnetic permeability, respectively; $k_0 \equiv \omega/c$ (c is the vacuum speed of light); $\varepsilon(r, z)$ is the relative permittivity of the background medium [for vacuum or for a neutral gas $\varepsilon = 1$; for plasma $\varepsilon = 1 - i\sigma/(\omega\varepsilon_0) \simeq 1 - \omega_p^2/[\omega(\omega - i\nu)]$, with σ the electron plasma conductivity, $\omega_p \equiv (n_e e^2/\varepsilon_0 m_e)^{1/2}$ the electron plasma frequency (e and m_e are the electron charge and mass, respectively), and ν the electron-neutral collision frequency].

2.2. Boundary conditions

Equations (1*a*-1*c*) are solved subject to the following boundary conditions. At device axis ($r = 0$), the symmetry condition $H_\varphi = 0$ is applied. Perfect-conductor conditions are imposed at the device metal boundaries, by setting $E_r = 0$ at horizontal walls and $E_z = 0$ at vertical walls. In practice, because we are solving a discrete equation system for the magnetic field (see subsection 2.3), these Dirichlet conditions on E_r and E_z are better rewritten as Neumann conditions on rH_φ [see equations (1*a*-1*c*)]:

$$E_{r_{\text{horiz walls}}} = 0 \quad \rightarrow \quad \frac{\partial (rH_\varphi)}{\partial z}_{|\text{horiz walls}} = 0 \tag{2a}$$

$$E_{z_{\text{vert walls}}} = 0 \quad \rightarrow \quad \frac{\partial (rH_\varphi)}{\partial r}_{|\text{vert walls}} = 0 \quad . \tag{2b}$$

However, numerical tests show that the direct imposition of boundary conditions (2*a*-2*b*) doesn't restrict the solution, thus leading to divergence. This problem occurs for one-end-shorted circular devices with radii R above the critical value $R_c \simeq 47$ mm (at 2.45 GHz frequency), for which a wave-mode solution develops [1, 2, 6]. To circumvent this difficulty, we have replaced the Neumann condition at $r = R$ with an equivalent Dirichlet-type boundary condition on rH_φ. The latter is tailored using a Newton-Raphson iterative algorithm, according to which the values of $RH_\varphi(R, z)$ follow a convergence path towards the physical solution $E_z(R, z) = 0$. Details on the numerical algorithm adopted can be found in [2].

At open boundaries (OB) an absorbing boundary condition (ABC) is considered in order to eliminate non-physical reflections, coming from the modeling of the device as a finite system [5, 7]. For a H_φ^{ref} component, associated with a reflected plane-wave, the first-order ABC writes (for a time-harmonic description) [8]

$$\frac{\partial H_\varphi^{\text{ref}}}{\partial z}_{|\text{OB}} - ik_0 H_{\varphi_{\text{OB}}}^{\text{ref}} = 0 \quad . \tag{3}$$

In practice, $H^{\rm ref}_{\varphi_{\rm OB}}$ is generally written as $H^{\rm ref}_{\varphi_{\rm OB}} \equiv H_{\varphi_{\rm OB}} - H^{\rm exc}_{\varphi_{\rm OB}}$, where $H_{\varphi_{\rm OB}}$ is the calculated magnetic field at the open boundary and $H^{\rm exc}_{\varphi_{\rm OB}}$ relates to the magnetic component of the excitation fields, in cases where power is delivered through this boundary.

2.3. Numerics

Maxwell's equations are discretized onto a two-dimensional, non-uniform grid, using a technique based on Yee's method [9]. The number of grid points (typically 10^4) is set as to provide numerical accuracy, while ensuring acceptable run times ($\sim$ 10 min-3 hrs on a 2.5 GHz CPU, depending on the number of grid points). After discretization, Maxwell's equations are merged into a Helmholtz-type equation in the single quantity rH_φ. The discretized equation is solved using a Gauss-Seidel iterative algorithm, with a stabilization improvement that uses direct column and row-wise solutions, and a convergence acceleration that adopts the successive over-relaxation acceleration technique. The convergence test checks for a maximum relative error of H_φ smaller than 10^{-7}.

3. Model solution

3.1. Coaxial waveguide (in vacuum)

Equations ($1a$-$1c$) can be solved analytically for an air-filled coaxial waveguide with inner radius R_i and outer radius R. For a combined forward + backward wave system (which yields a more spectacular representation of the EM fields), the TEM solution writes [6]

$$H_\varphi^{\rm TEM} = C_{\rm TEM}\frac{\exp\left[-ik_0 z\right] + \exp\left[ik_0 z\right]}{r}, \tag{4}$$

with $|C_{\rm TEM}|^2 = P_0\sqrt{\varepsilon_0/\mu_0}/\left[\pi \ln\left(R/R_{\rm i}\right)\right]$, where P_0 represents the time-average incident power on the excitation plane, which is an input parameter to the model.

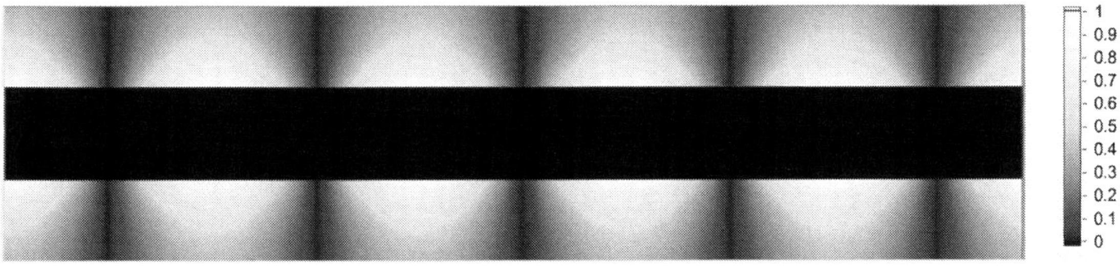

Figure 1. Contour plot of the normalized magnetic field modulus, numerically calculated for an air-filled coaxial waveguide with $R_i = 5.3$ mm and $R = 14.5$ mm.

Figure 1 plots the numerically calculated stationary-wave pattern (represented by the magnetic field modulus, normalized to its maximum value), for a section of the coaxial waveguide with $L = 300$ mm length. We can therefore observe a sequence of about $Lk_0/(2\pi) \simeq 2.5$ wave oscillations, in agreement with the solution given by equation (4).

3.2. Circular waveguide shorted by metal plate (in vacuum)

Equations ($1a$-$1c$) can be solved analytically for an air-filled circular waveguide of radius R, shorted by a metal plate at $z = L$. For a TM$_{0m}$ symmetric mode the solution splits into oscillatory ($H^{\rm TM}_{\varphi,\rm oscil}$) or evanescent ($H^{\rm TM}_{\varphi,\rm evan}$), depending on the values of R at fixed ω, according to [6]

$$H^{\rm TM}_{\varphi,{\rm oscil}}(r,z) = C_{\rm TM}\cos\left[\beta(z-L)\right] J_1\left(\frac{p_{01}}{R}r\right)$$
$$({\rm for}\ \beta^2 > 0\ {\rm or}\ R > R_c \equiv p_{01}c/\omega) \qquad (5a)$$
$$H^{\rm TM}_{\varphi,{\rm evan}}(r,z) = C'_{\rm TM}\exp\left(-\beta' z\right) J_1\left(\frac{p_{01}}{R}r\right)$$
$$({\rm for}\ \beta^2 \equiv -\beta'^2 < 0\ {\rm or}\ R < R_c \equiv p_{01}c/\omega), \qquad (5b)$$

where J_1 is the first-order Bessel function, p_{01} is the 1st zero of the zero-order Bessel function, and $|C_{\rm TM}|^2 = 8P_0\sqrt{\varepsilon_0/\mu_0}\ (k_0/\beta)/[\pi R^2 J_1^2(p_{01})]$.

Figures 2(a) and (b) present contour-plots of the magnetic field modulus (normalized to its maximum value), numerically calculated for an air-filled circular waveguide, shorted by a top metal plate (at $z = L = 150$ mm), with radius $R = 35$ mm $< R_c$ and $R = 55$ mm $> R_c$, respectively. These figures evidence a qualitative agreement between the numerical and the analytical solutions to the problem [cf. (5a-5b)], for both evanescent [figure 2(a)] and oscillatory [figure 2(b)] field patterns.

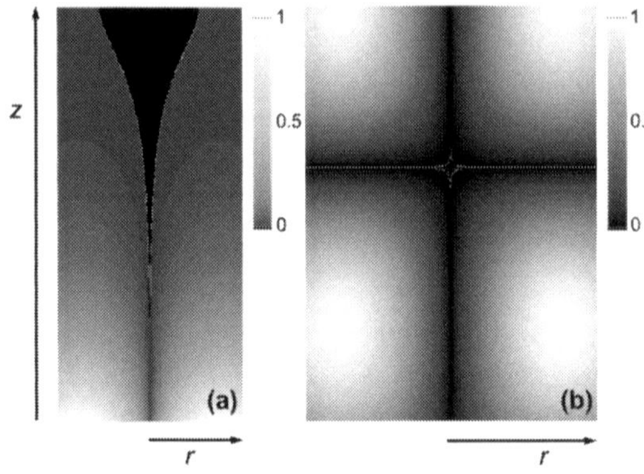

Figure 2. Contour plot of the normalized magnetic field modulus, numerically calculated for an air-filled circular waveguide, shorted by a metal plate (at $z = 150$ mm), with $R = 35$ mm (a) and $R = 55$ mm (b).

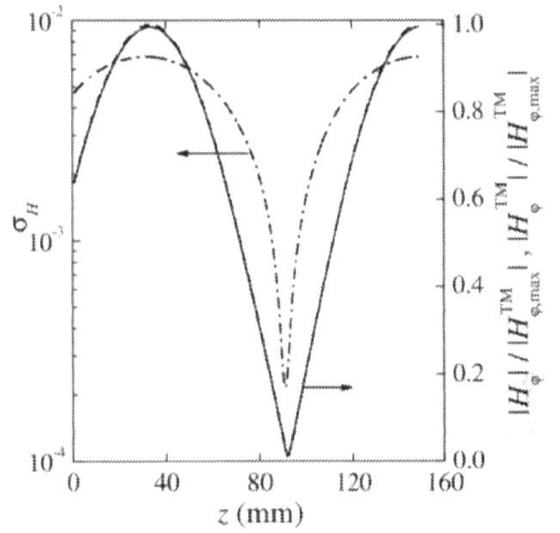

Figure 3. Local relative error σ_H (dashed-dotted curve), and normalized oscillatory solutions $|H_\varphi|/|H^{\rm TM}_{\varphi,{\rm max}}|$ (solid) and $|H^{\rm TM}_\varphi|/|H^{\rm TM}_{\varphi,{\rm max}}|$ (dashed), as a function of z, at radial position where the field is maximum. Results are for the device of figure 2 with $R = 55$ mm.

Quantitatively, the accuracy of the numerical solution is analyzed in figure 3, which plots the axial evolution (at radial position where the field has its maximum value $H^{\rm TM}_{\varphi,{\rm max}}$) of the local relative error $\sigma_H \equiv |H_\varphi - H^{\rm TM}_\varphi|/|H^{\rm TM}_{\varphi,{\rm max}}|$, as an example for the oscillatory solution. This figure presents also the axial profile of the normalized modulus of both the numerical solution $|H_\varphi|/|H^{\rm TM}_{\varphi,{\rm max}}|$ and the analytical solution $|H^{\rm TM}_\varphi|/|H^{\rm TM}_{\varphi,{\rm max}}|$. One observes that the values of σ_H are always below 1%, which confirms the accuracy of numerical results.

3.3. Axial injection torch coupled to a cylindrical reactor (with plasma)

Equations (1a-1c) can be solved numerically for a microwave-driven plasma reactor operated by an AIT. The device is basically a coaxial-type antenna (with inner and outer radii R_i and R_o, respectively), which is placed inside a reactor chamber (with radius R and height L) to provide the coupling of microwave power and the injection of feedstock gases, producing a plasma at atmospheric pressure. Figure 4 represents a schematic diagram of the cylindical AIT-reactor device.

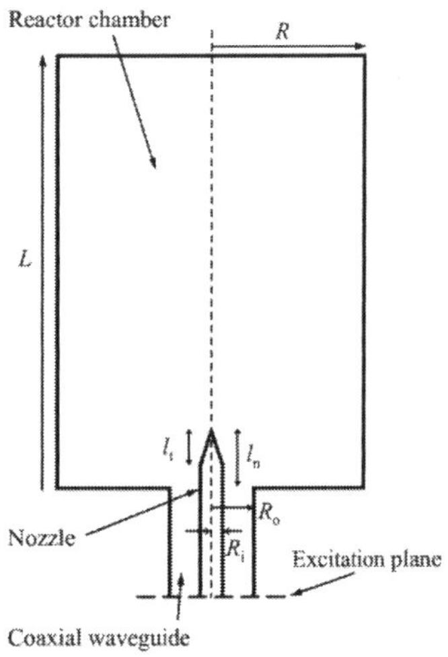

Figure 4. Schematic scale diagram of the cylindrical AIT-reactor device. Dimensions are $R_{\rm i}$ = 5.3 mm, $R_{\rm o}$ = 14.5 mm, $l_{\rm t}$ = 10 mm, $l_{\rm n}$ = 15 mm, R = 55 mm, and L = 150 mm.

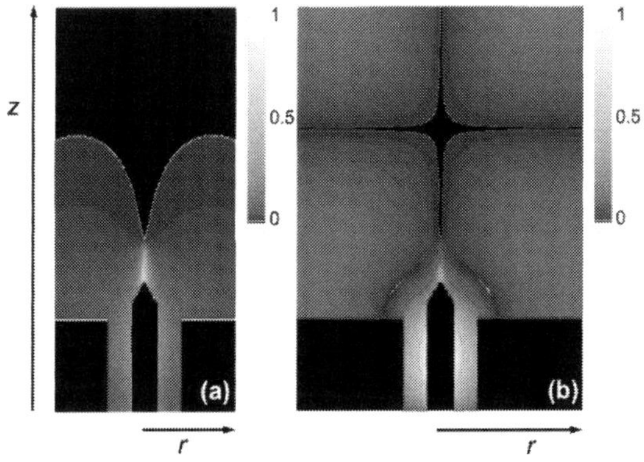

Figure 5. Contour plot of the normalized magnetic field modulus, numerically calculated for an AIT-reactor device with plasma. Results are for R = 35 mm (a) and R = 55 mm (b), at $n_{e,\rm max} = 10^{15}$ cm^{-3}. All other dimensions are as in figure 4.

Model solution is obtained for a helium plasma with 0.5 mm radius and 20 mm height, whose electron density profile has been tailored according to measurements [10, 11]. In axial direction, n_e decreases linearly from a maximum value $n_{e,\rm max}$. In radial direction, we impose a hollow-type profile with four regions of 0.125 mm each, exhibiting the following behaviors (from axis): a constant $0.8n_{e,\rm max}$ value; a linear increase up to $n_{e,\rm max}$; a constant $n_{e,\rm max}$ value; a linear decrease down to zero.

Figures 5(a) and (b) present contour-plots of the magnetic field modulus (normalized to its maximum value), calculated for the AIT-reactor device represented in figure 4, with radius R = 35 mm and R = 55 mm, respectively, at $n_{e,\rm max} = 10^{15}$ cm^{-3}. The self-consistent global solution represented in figure 5(b), for the power-input lower region and the reactor upper region, shows a remarkable similarity with the separate calculation results obtained, respectively, for a coaxial waveguide (see figure 1) and a cylindrical waveguide shorted by a metal plate [see figure 2(b)], with dimensions corresponding to those of the AIT-reactor device. Similar results are also obtained for the reactor region in figure 5(a) and the cylindrical waveguide in figure 2(a). These observations are directly associated to the fact that the plasma size is much smaller than that of the container, and they show that in such cases the general solution pattern can be qualitatively mapped by decoupling the full system into smaller work domains.

The quality of power coupling for the AIT-reactor device is measured by its power transmission coefficient $P_{\rm abs}/P_0$, where $P_{\rm abs}$ is the average power absorbed by the plasma, calculated using either surface of volume integral expressions [6]

$$P^{\rm S}_{\rm abs} \equiv \frac{1}{2}\Re \int_{\rm exc.\,plane} E_r^{\rm TEM}(H_\varphi^{\rm TEM})^\star \, {\rm d}S \qquad (6a)$$

$$P^{\rm V}_{\rm abs} \equiv \frac{1}{2}\int_V \Re[\sigma]\,|E|^2\,{\rm d}V \quad . \qquad (6b)$$

Here, $\Re$ represents *the real part of* and $\star$ refers to the complex conjugate. The input EM fields at the excitation open boundary are taken equal to the TEM mode solution with a coaxial waveguide (see subsection 3.1). Equations (6*a*-6*b*) can be used to test the precision of model results. Figure 6 plots the local relative error $\sigma_P \equiv |P^{\rm S}_{\rm abs} - P^{\rm V}_{\rm abs}|/P_0$ and the power transmission coefficient [using both expressions (6*a*-6*b*)], as a function of R. An observation of this figure shows that the values of σ_P are always below 0.1%, which confirms the precision of numerical results. Notice that power coupling is strongly affected by variations in R due to the development of resonant wave-cavity modes ($P_{\rm abs}$ varies from $\sim P_0$ down to 0, over only a few mm).

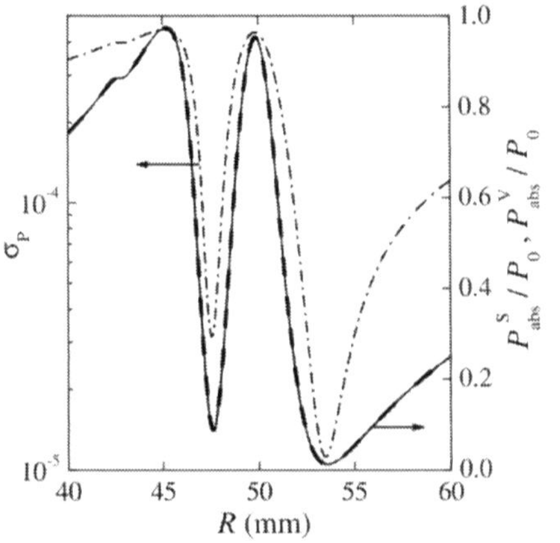

Figure 6. Relative error σ_P (dashed-dotted curve), and power transmission coefficient $P^{\rm S}_{\rm abs}/P_0$ (solid) and $P^{\rm V}_{\rm abs}/P_0$ (dashed), as a function of R, for an AIT-reactor device with plasma at $n_{e,\max} = 10^{15}$ cm^{-3}. All other dimensions are as in figure 4.

4. Concluding remarks

We have developed a 2D numerical simulation tool that solves Maxwell's equations for an axis-symmetric TM mode, by considering a harmonic time description. The tool was successfully used to calculate the EM field distributions with a coaxial waveguide, a circular waveguide with a shortening metal plate at the top, and a cylindrical plasma reactor operated by an AIT. Results were obtained for both evanescent and oscillatory field patterns, the latter leading to extra numerical difficulties in achieving model convergence. Simulations featured a 99% accuracy and precision errors lower than 0.1%, for a 10^4 point grid. Results demonstrated the interest of a multi-domain analysis to qualitatively map the EM solution pattern, and confirmed the important role of device dimensions in changing the distribution of EM fields, hence the quality of power coupling.

Acknowledgments

This work was supported by the Portuguese Foundation for Science and Technology (Grant No. BPD/26420/2005 and Project PTDC/FIS/65924/2006).

References

[1] Álvarez R and Alves L L 2007 *J. Appl. Phys.* **101** 103303
[2] Álvarez R and Alves L L 2008 *J. Phys. D: Appl. Phys.* (accepted for publication)
[3] Lebedev Yu A and Tatarinov A V 2004 *Plasma Sources Sci. Technol.* **13** 1
[4] Gritsinin S I Kossyi I A Kulumbaev É B and Lelevkin V M 2006 *Plasma Physics Report* **32** 872
[5] Bilgic A M, Garloff K and Voges E 1999 *Plasma Sources Sci. Technol.* **8** 325
[6] Johnson C C 1965 *Field and Wave Electrodynamics* (New York: McGraw-Hill)
[7] Mur G 1981 *IEEE Trans. Electromagn. Compat.* **23** 377
[8] Engquist B and Majda A 1977 *Math. Comp.* **31** 629
[9] Yee K S 1966 *IEEE Trans. Antennas Propag.* **14** 302
[10] Rodero A, Quintero M C, Sola A and Gamero A 1996 *Spectrochim. Acta* Part B **51** 467
[11] Álvarez R, Quintero M C and Rodero A 2004 *Spectrochim. Acta* Part B **59** 709

Third International Workshop & Summer School on Plasma Physics 2008 IOP Publishing
Journal of Physics: Conference Series **207** (2010) 012028 doi:10.1088/1742-6596/207/1/012028

Diffuse, constricted-stratified and constricted modes of a DC discharge in argon. Simulation of transitions between these modes

I. A. Shkurenkov, Yu. A. Mankelevich, and T. V. Rakhimova
Skobeltsyn Institute of Nuclear Physics, Lomonosov's Moscow State University, Leninskie gory, Moscow, Russia

E-mail: chkouren@gmail.com

Abstract. This paper presents the results of theoretical studies of high-pressure (~10's and 100s of Torr) dc discharges in argon. The diffuse, constricted-stratified, and constricted discharge modes are studied using a developed model. Two transitions were studied. One of them is transition from the constricted-stratified mode to the constricted mode. This transition occurs smoothly, without any jumps of plasma parameters. The other is a hysteresis transition between the diffuse and the constructed-stratified mode. It occurs by a jump and is accompanied by a several orders of magnitude increase in the electron density at the discharge axis and an appearance of moving striations.

1. Introduction

The phenomenon of the constriction of a direct current discharge in noble gases at the elevated gas pressures (tens and hundreds Torr) is well known in the physics of gas-discharge plasma. At attainment of a critical current the discharge structure and plasma parameters are changed abruptly and the discharge is compressed in a narrow bright filament. The current-voltage characteristic of the discharge is bi-stable. A transition from the diffuse to the constricted mode occurs by a jump and is accompanied by a several orders of magnitude increase in the electron density at the axis of the discharge tube. In the course of this transition, all the discharge parameters (the electric field, the plasma density, etc.) change abruptly. In passing back from the constriction to the diffuse mode, a hysteresis phenomenon is observed; i.e., a transition from the diffuse to the constricted-stratified mode (as the discharge current is increased) and the reverse transition (as the current is decreased) occur at different values of the discharge current. No striations were observed in the diffuse form of the discharge, the positive column is uniform along the tube axis. In wide range of discharge currents (≈20-200 mA) constriction of the discharge is accompanied by the appearance of moving striations, which lead to modulation of plasma glowing regions. The shape of modulation is changed and its amplitude is decreased with current. Pulsation frequency comes down with current and the pulsation shape is changed from sinusoidal to substantially non-sinusoidal. This mode is distinct from purely constricted mode and it is necessary to consider the appearance of the striations (periodic fluctuations of plasma parameters). The constricted-stratified mode differs very much from purely stratified mode. For example, in the stratified mode in the region of high concentration of electrons, their concentration is smoothly distributed over the cross section of the discharge tube, while in the constricted-stratified mode it is concentrated in a narrow region near the discharge axis.

The processes leading to the constriction [1,2,3] and bringing the system out of the constricted [4] mode were studied in theoretical and experimental investigations of constricted discharges in noble gases. But to the best of our knowledge, until now, there were no publications in which the constricted-stratified mode was simulated self-consistently. In [5] there is only an experimental diagram where one can see a transition mode (constricted-sratified) between diffuse and constricted modes.

As mentioned above the constriction of the discharge is accompanied by the appearance of moving striations. The developed theoretical model of a dc discharge in all these modes has allowed us for the first time, to our knowledge, to obtain periodic fluctuations of plasma parameters for the constricted-stratified mode without any artificial assumptions.

In [4] the phenomenon of hysteresis transition was studied. The time dynamics was not studied in this article. The numerical model, that was used, allowed to study the basic processes leading to the discharge constriction and bringing the system out of the constricted mode. In the equation for high-energy part of electron energy distribution function (EEDF) was used the value of the electron temperature, which was found from local approximation instead of the equation for the electron temperature. It also suppresses the time fluctuations. Solving of full equations set (without using any time approximations), leads to appearance of the fluctuations of plasma parameters, which structure resemble striations. The temporal distribution of the distributions of charged particles and excited atoms correlate with the distribution of radiation intensity [2].

In this article, the simulation results for diffuse, constricted-stratified and constricted modes and transitions between these modes are presented.

2. Model

In this work, we simulate the diffuse, constricted-stratified and constricted discharge modes and describe numerically the hysteresis effect at the direct (with increasing discharge current) and reverse (with decreasing current) transitions between diffuse and constricted-stratified modes. A one-dimensional (radial) self-consistent model is developed that involves, as in [4], the continuity equations for electrons, ions (Ar^+ and Ar_2^+), and excited atoms (metastable and resonant), the conservation equation for electron energy; the heat conduction equation for the neutral gas; and Poisson's equation for the radial electric field. The processes included in the model are listed in table 1. At first, similar to [1], we solved a Boltzmann equation with account of electron-electron collisions in the local approximation. However, the determination of the excitation rate constants and the coefficients of direct and step ionization from the local EEDF calculated with account of electron-electron collisions did not allowed us to describe the hysteresis transition from the constricted-stratified to the diffuse mode. Therefore, the model was modified to account for the non-local character of the formation of the EEDF by introducing a two-temperature distribution function [4].

Table 1. List of reactions.

1.Excitation	$Ar + e \rightarrow Ar^* + e$	Calculation k (cm^3/s) using
2.Ionization	$Ar + e \rightarrow Ar^+ + e + e$	EEDF for
3.Step ionization	$Ar^* + e \rightarrow Ar^+ + e + e$	Cross sections [6]
4.Conversion	$Ar^+ + Ar + Ar \rightarrow Ar_2^+ + Ar$	$k=1.8\cdot10^{-31}\cdot(300/T)^{3/4}$ (cm^6/s) [7]
5.Dissociative recombination	$Ar_2^+ + e \rightarrow Ar^* + Ar$	$k=4\cdot10^{-5}\cdot T_e^{-0.67}$ (cm^3/s) [7]
6.Electron impact dissociation	$Ar_2^+ + e \rightarrow Ar^+ + Ar + e$	$k=5.7\cdot10^{-7}$ (cm^3/s) [8]
7.Three-body recombination	$Ar^+ + e + e \rightarrow Ar^* + e$	$k=5.1\cdot10^{-9}/(T[K])^{4.5}$ (cm^6/s) [9]
8.Radiative decay	$Ar^r \rightarrow Ar + h\nu$	$k\approx \theta\cdot A$ (s^{-1}) [4]
9.States mixing	$Ar^* + e \leftrightarrow Ar^r + e$	$k\approx10^{-7}$ (cm^3/s) [4]

As it appears in our model calculations, a transition from the constricted to the diffuse mode can be regarded as a demonstration of the non-local character of the formation of the high energy part of the EEDF: the diffusion of high-energy electrons (capable of producing gas ionization) from the central (constricted) region toward the periphery. The non-local formation of the distribution function is described by a non-local kinetic equation accounting for electron-electron collisions and electron radial transport. The effect of the non-local formation of the EEDF is taken into account by introducing the effective temperature T^* of the high-energy part of the EEDF and solving the equation for the radial profile of $T^*(r)$. It should be noted that the usage only of the non-local equation for mean electron energy $3/2 \cdot T_e$ does not allow to obtain and to describe the constricted-stratified mode and transition to the diffuse mode [4]. The approach with two electron temperatures $T_e(r)$ and $T^*(r)$ allows one to take into account approximately the non-local character of the EEDF without substantial consumption of computer resources. This non-local model makes it possible to numerically simulate the hysteresis transition between the diffuse and constricted-stratified modes, which is impossible in our calculations using the local approximation approach [4].

3. Results

As mentioned above, at wide range of gas pressures direct current discharge can exist in three modes: diffuse, constricted and transition region between them, which is constricted-stratified mode. At first, let's consider diffuse and constricted modes of discharge. We carried out the systematic calculations of these modes. Figure 1 shows the radial profiles of the electron concentration in the diffuse (for a discharge current of 10 mA) and constricted (for a discharge current of 180 mA) modes of a dc discharge in argon at experimental conditions [2]: tube radius R=1.5 cm, a pressure of 40 Torr. In the diffuse mode, the radial profile of the electron density is smooth. In this mode, the diffusive losses prevail and the radial profile of the ionization rate is determined by (E/N profiles) non-uniform gas heating. In the constricted mode, unlike the diffuse mode, the plasma electrons are mainly concentrated in a narrow central region. The discharge constriction manifests itself in a sharp decrease in the diameter of the current channel (which is characterized by the high conductivity and the high electron density n_e) and a sharp increase in the degree of ionization (up to $n_e / N_0 \sim 10^{-5} - 10^{-4}$).

Between these two modes there is a wide region of the constricted-stratified mode. Transition from the diffuse to constricted mode, without passing the constricted-stratified mode at current increase, can exist only under high pressures ($P \cdot R \geq \sim 3$ atm·cm [10]). That's why the hysteresis transitions, which are observed in various experiments [2,3,10], are the transitions between diffuse and constricted-stratified modes. Calculated electric field E in the constricted-stratified mode must be averaged through the regions with high and low electric field to compare with experimental field E, which is usually measured by probe methods. Figure 2 shows the temporal behavior of radial profile of the electron concentration. As it seen, plasma parameters in this discharge mode undergo periodic time fluctuations. The period of such fluctuations is about 0.25 ms. This value is close to the period of striations (0.4ms) which was observed experimentally [2]. In our calculations processes of transport along the discharge tube, as well as longitudinal non-local effects were not taken into account. The correct quantitative description of striations is possible only in 2D (axial-radial) model. The drift of the electrons could increase the period of such fluctuations.

Figure 3 shows the electric field E and electron temperature T_e at the tube axis as functions of time. We can see here a well-defined periodical structure of the fluctuations. We can see that each new fluctuation begins from jump of the electric field. In turn, it causes increase in the electron temperature and in the concentration of metastable atoms.

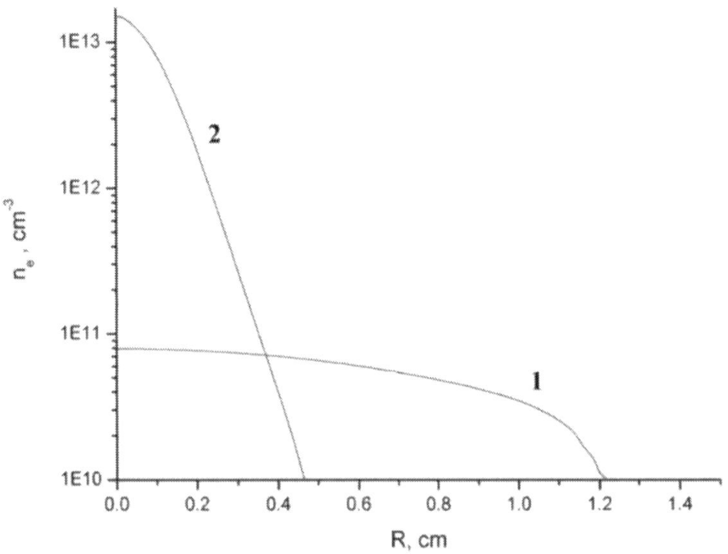

Figure 1. Radial profiles of the electron concentration in the diffuse (10 mA, curve 1) and the constricted (180 mA, curve 2) modes at pressure of 40 Torr.

We carried out systematic calculations of the transition between diffuse and constricted-stratified modes of the discharge. This transition is interesting for two reasons: first, at this transition the phenomenon of a hysteresis in VCC is observed; secondly, it is a serious problem for discharge models to obtain a decay of the constricted mode and a disappearance of the fluctuations at the transition to the diffuse mode. In our case the mechanisms taking out the discharge from the constricted (constricted-stratified) mode, are the effects of the radial non-local formation of the EEDF. It should be noted that this radial non-locality of EEDF tail is of different kind than axial non-locality of EEDF tail, considered in [5,11] as a source of ionization instability (striations). The constricted-stratified discharge by its nature is closer to the constricted mode, than to the stratified.

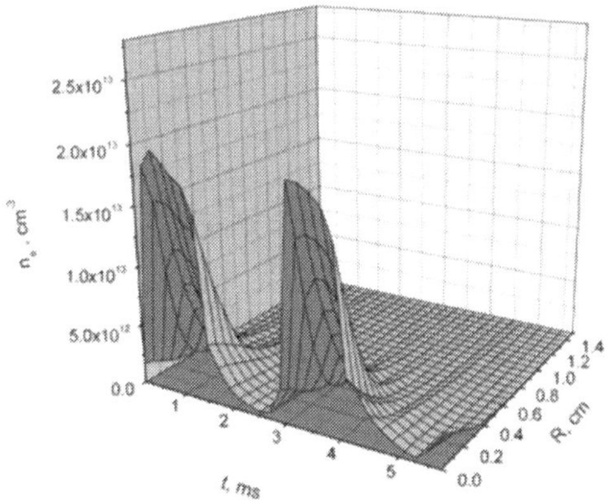

Figure 2. Electron density, electric field and electron temperature at the discharge tube axis as a function of time (current is 30 mA).

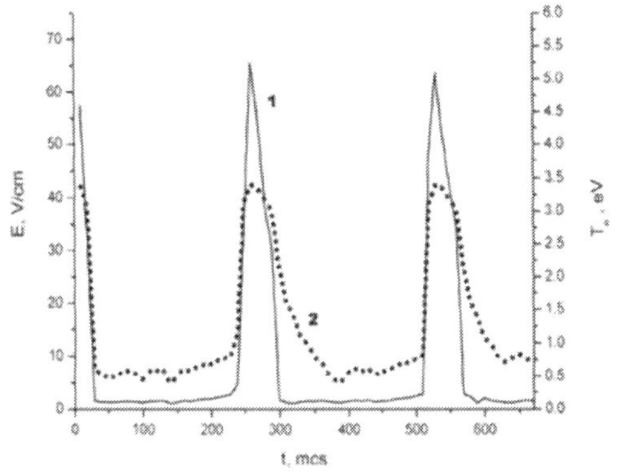

Figure 3. Electric field E (1) and electron temperature T_e (2) at the discharge tube axis as a function of time (current is 30 mA).

Figure 4 shows the calculated current-voltage characteristics of a dc discharge in argon in the conditions [2]. The characteristic demonstrates a well-pronounced hysteresis effect. To the best our knowledge, until now, there were no publications in which the hysteresis transition between the diffuse and constricted-stratified discharge modes was obtained numerically. It is seen that, the calculated current-voltage characteristics well reproduces the measured one and both the direct transition from the diffuse to the constricted-stratified mode and the reverse transition occur at the same values of the discharge current as in the experiment [2]. This indicates that the model takes into account correctly the basic processes leading to the discharge constriction and bringing the system out of the constricted mode.

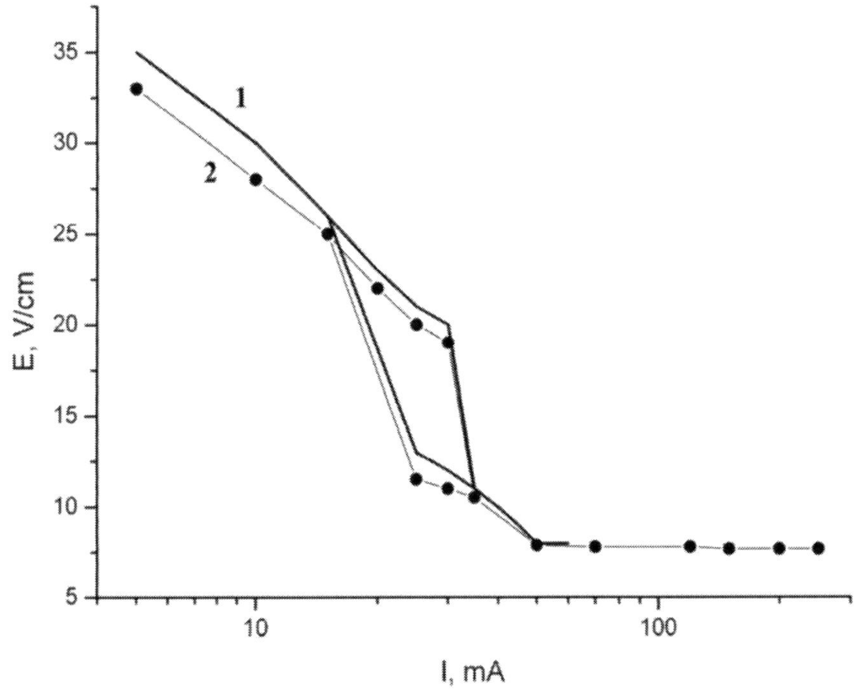

Figure 4. Current-voltage characteristic of a dc discharge in argon in the parameter range corresponding to the transition between the diffuse, constricted-stratified, and constricted modes: 1-experiment [2] and 2-simulations.

At the extremely high currents (for the case under study, 180 mA and higher) the discharge smoothly passes from the constricted-stratified mode into the constricted mode. We traced the transformations of the structure and plasma parameters of the constricted-stratified mode with the discharge current. A transition (at ~180 mA) from the constricted-stratified mode to the constricted mode occurs smoothly (without any jumps). As it is seen from figure 4, the voltage jumps and hysteresis effects are not observed here, contrary to the transition from the diffuse to the constricted-stratified mode. As all the striations in noble gases are moving, we can say a length of this region when we are speaking about duration of the phase of high or low electron density. In the constricted-stratified mode an increase in the discharge current leads to increase in the period of fluctuations. Thus the increase in the length (duration) is specific to the region with the high electron concentration (such as regions at t~(0 ÷ 1ms) and (3 ÷ 4ms) in figure 2) and the length of the region with low electron concentration is decreased. Thus, the ratio of the lengths of the high and low electron concentrations regions (~0.5:0.5 at 30 mA) is steadily increased with discharge current and reaches ~1:0 at I~180mA and the discharge passes to the constricted mode without any fluctuations. So, at value of the discharge current ~180 mA, the adjacent regions (with high electron concentrations) flow together. The nature of the fluctuations in the constricted-stratified mode is in the inconsistency between the ionization rate and the electron energy gain: e.g., low electric field and high ionization rates in the phase of the high electron density. This phase lasts until the energy delivery from the body of the EEDF to the tail of the EEDF by e-e collisions takes place. In our calculations, we observed the quasi steady state T^{*}~0.8 eV against the background of the decrease in T_e from 1.5 down to 1.0 eV.

4. Conclusion

In this work, a one-dimensional self-consistent model for describing high-pressure dc discharges in noble gases has been developed. The model includes the continuity equations for charged and excited particles, the heat conduction equation for the neutral gas and electron energy conservation equation, and Poisson's equation for the radial field, as well as equations for the temperatures of the low- and high-energy parts of the EEDF. Three discharge modes have been considered: the diffuse, constricted-stratified, and constricted ones. The developed theoretical model of a dc discharge in all these modes has allowed for the first time, to our knowledge, to obtain periodic fluctuations of plasma parameters for the constricted-stratified mode without any artificial assumptions. A transition from the diffuse to the constructed-stratified mode occurs by a jump and is accompanied by a hysteresis effect. This effect consists in that the transition from the diffuse to the constricted-stratified mode (as the discharge current is increased) and the reverse transition (as the current is decreased) take place at different discharge currents. A transition from the constricted-stratified mode to the constricted mode occurs smoothly, without any jumps. The developed model allowed to obtain also both of these transitions (hysteresis transition between diffuse and constricted-stratified modes and smooth transition between constricted-stratified and constricted modes).

Acknowledgments

We are pleased to acknowledge support from RF government for Key Science Schools grant № 133.2008.2.

5. References

[1] Petrov M and Ferreira C M 1999 *Phys. Rev. E.* **59** 3571
[2] Dyatko N A, Ionikh Y Z, Kochetov I V, etc 2008 *J. Phys. D: Appl. Phys.* **41** 055204 (14pp)
[3] Golubovskii Yu B, Sonneburg R 1979 *Zh. Tekh. Fiz.* **49** No.2 295
[4] Shkurenkov I A, Mankelevich Yu A, Rakhimova T V 2008 *Plasma Physics Reports* **34** No.9 780
[5] Kolobov V I 2006 *J. Phys. D: Appl. Phys.* **39** 487
[6] Phelps A V 1985 *JILA Information Center Report* **28**
[7] *Physical quantities: reference book*, Ed. by Grigoriev I S, Meilihov E Z 1991 (Moskow: Energoatomizdat) [in Russian]
[8] Marchenko V S 1983 *Plasma Physics Reports* **85** No.8 500
[9] *Encyclopedia of Low-Temperature Plasma*, Ed. by Fortov V E 2000 (Moscow: Nauka Vols. 1, 2) [in Russian].
[10] Golubovskii Yu B, Zinchenko A K, Kagan Yu M 1977 *Zh. Tekh. Fiz.* **47** 1478
[11] Tsendin L D 1995 *Plasma Sources Sci. Technol.* **4** 200

Third International Workshop & Summer School on Plasma Physics 2008 IOP Publishing
Journal of Physics: Conference Series **207** (2010) 012029 doi:10.1088/1742-6596/207/1/012029

Kinetics of Ar-He Atmospheric Pressure Surface Wave Discharges

J Muñoz and M D Calzada[1]
Grupo de Espectroscopía de Plasmas, Edificio Einstein, Campus Rabanales, Universidad de Córdoba, 14071 Córdoba (Spain)

E-mail: fa1cazal@uco.es

Abstract. An atmospheric pressure surface wave discharge generated using several Ar-He mixtures has been studied with emission spectroscopy techniques. Deviations in the excitation (Boltzmann) and ionization (Saha) equilibriums respect to the case of pure Ar discharges have been found for He concentrations as low as 10%. The relevance of these deviations grows as the concentration of He in the discharge does. While no exact equilibrium can be stated without knowledge of the electron temperature, departure of the discharge respect to pLTE is clear and consequently alternative diagnostic methods that do not require LTE must be employed.

1. Introduction

The thermodynamic equilibrium (TE) of plasmas, which is related to the detailed balance principle and so to the proper balance of kinetics processes taking place within a discharge, is a general issue in plasma physics to which special attention must be paid. Although it does not seem relevant for practical applications of discharges, that usually imply far from equilibrium systems except for some cases such as arc discharges and thermonuclear plasmas, it is very important when dealing with plasma diagnosis since some methods, particularly spectroscopic ones, require of the existence of, at least, partial local thermodynamic equilibrium (pLTE).

2. Experimental Setup

Figure 1 shows a schematic block diagram of the plasma source and the optical detection and data acquisition systems for emission spectroscopy measurements.

Microwave power was provided to the plasma by a SAIREM 12kT/t microwave generator of 2000 W maximum power in continuous mode, equipped with a water cooled circulator to avoid power reflection damage. The power was coupled to the plasma by a *surfaguide* device and two impedance matching means of the wave launcher make the power reflected from the applicator back to the generator lower than 5% of the incident power.

The discharge was contained in quartz tubes of several inner (2, 4 and 5 mm) and outer (3, 5 and 6 mm) radii. Since the quartz tubes containing the discharge suffer a great damage for powers over 300 W [1], the discharge tube was coaxially surrounded by another quartz tube of 8.5 mm internal radius in order to avoid this problem. This external tube made up a jacket through which a dielectric liquid (1–Tetradece) or air circulated. In our experiments, the plasma column is extended to both sides of the

---

[1] To whom any correspondence should be addressed.

wave launcher and direct and inverse columns appear. The direct column is considered that one where the gas flux and the propagation of the wave takes place in the same sense contrarily to what happens in the inverse.

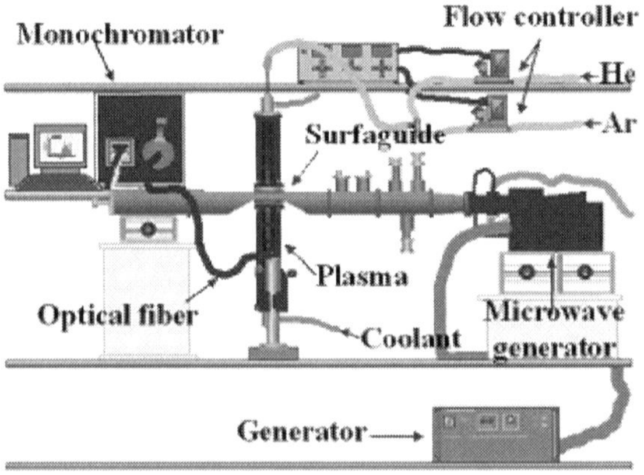

Figure 1. Experimental Setup.

Several gas mixtures of high purity (99.999 %) Ar and He, were used as plasma gases keeping the same total flow equal to 1 slm (standard litre per minute). The flows for Ar, He and mixtures were controlled by HI-TEC flow controllers (IB 31) with different maximum flow limits (0.25 and 5 slm). Several He concentrations where used, ranging 0 – 99 %.

An optical fiber was used to pick up the light emitted from the discharge at z = 2 cm from the plasma end and, also, to drive it to the entrance slit of a Jobin-Yvon-Horiba monochromator previously calibrated and equipped with a 2400 grooves/mm holographic grating. A Hamamatsu R928P photomultiplier and a Symphony CCD were used as radiation detectors. In each measure, spectrums from H_β line were taken for electron density measurement, as well as those of some ArI lines. Hydrogen atoms were present as impurities in the plasma gas.

3. Results

3.1 Excitation temperatures of the Ar atomic system.

Since no atomic lines of He or ionic lines of Ar could be detected, only excitation temperature of the ArI system could be calculated. Two different temperatures were calculated for each experimental data set from the slopes of the Boltzmann-Plots: one from the linear fit connecting the 4p and the 5p states (lower levels) T_1, and another from the linear fit connecting the levels above 5p (upper levels) T_2.

Table 1. Excitation temperatures calculated from the Boltzmann-Plot of the ArI system.

[He] (%)	T_1 (K)	T_2 (K)
0 (r_i = 2 mm)	5400 ± 400	4500 ± 300
30 (r_i = 2 mm)	6100 ± 140	5500 ± 400
50 (r_i = 4 mm)	5800 ± 140	4900 ± 300
70 (r_i = 4 mm)	6100 ± 140	4700 ± 250
90 (r_i = 5 mm)	5900 ± 140	3500 ± 200
99 (r_i = 5 mm)	6100 ± 180	3400 ± 210

Journal of Physics: Conference Series **207** (2010) 012029 doi:10.1088/1742-6596/207/1/012029

In Table 1 it is clear that two very different temperatures are needed to describe the atomic state distribution function for He proportions above 50 % in the discharge, showing variations in the excitation/ionization scheme of the ArI system.

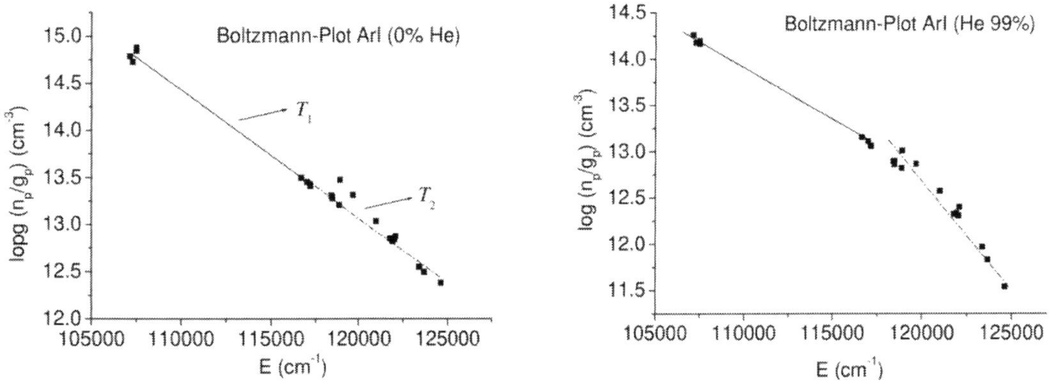

Figure 2. Boltzmann Plot of the ArI system for two different Ar-He discharges containing 0% He (left) and 99% He (right).

A relative depopulation of the higher levels of the ArI system at high He concentrations can be clearly observed in figure 2. This can be attributed to the differences in the frequency for electron-atom collisions for effective momentum transfer between argon and helium [1] and so a larger number of collisions between electrons and He atoms are expected. This induces a lower population on the upper levels of ArI provided that in the stepwise excitation scheme the upper levels of ArI are mainly populated by collisions with electrons of the bulk of the electron energy distribution function.

3.2 Griem's Criterion applied to an Ar-He discharge.

For a given excited atomic state, the effective quantum number p can be derived from the following expression

$$p = z\left(\frac{E_H}{E_{i,n}}\right)^{1/2} \tag{1}$$

where z is the effective charge of the nucleus, E_H is the ionization energy of the hydrogen atom and $E_{i,n}$ is the ionization energy of the n-th excited atomic state.

Griems criterion [3] is the most simple way typically used to identify deviations from local thermodynamic equilibrium. According to this criterion a critical level p_{cr} can be calculated as a function of the electron density (n_e) and temperature (T_e) of the discharge.

$$p_{cr} = 62\cdot\left(\frac{T_e}{n_e^2}\right)^{1/17} \tag{2}$$

According to Griem's criterion, those levels with quantum effective numbers above p_{cr} are considered in saturation, being its population and depopulation processes carried out mainly by electron collisions. Levels whose p falls under p_{cr} are in capture radiative cascade or corona balances, and are populated by electron collisions and depopulated by radiative processes.

In figure 2, p_{cr} values have been calculated for discharges with different Ar/He proportions. Electron density values have been calculated from the Stark broadening of the Balmer series H_β line [4]. Electron temperatures where estimated from the values of T_1, considered as a lower limit to T_e [5] and the behaviour of theoretical available data [6] of similar discharges at reduced pressure. Though this does not provide exact values of the electron temperature, the influence of this parameter on Griem's criterion has been proved to be less important as compared to electron density [3].

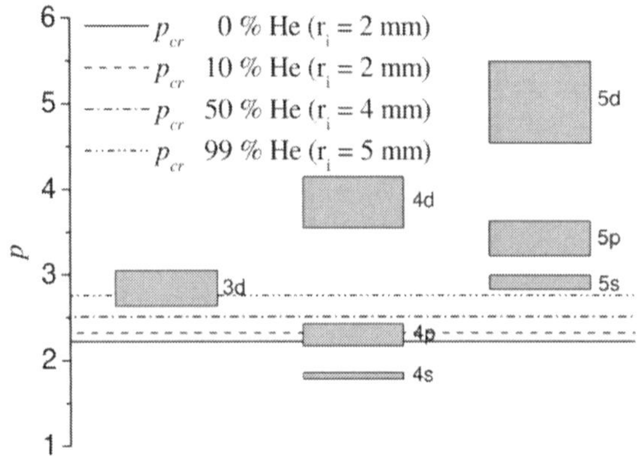

Figure 3. Gotrian diagram of Ar atomic system. The different lines represent critical p values calculated according to Griem's criterion from the experimental data.

As can be seen, even at a low concentration of He in the discharge, 4p levels enter in the capture radiative cascade zone. As the He concentration increases, these levels completely exit the saturation zone and other levels (3d) start being affected.

4. Conclusions.

The need for two excitation temperatures to properly describe the ArI atomic state distribution function as the He concentration grows up over 50%, related to a relative depopulation of the higher levels (above 5d), indicates that the higher levels start to separate from Saha equilibrium.

On the other hand, the departure from equilibrium pointed out by Griem's criterion indicates the lower lying excited levels (4s, 4p, 3d) of the ArI system are populated by electron collisions and depopulated by radiative processes ($p < p_{cr}$) and so the Boltzmann equilibrium is not verified for this levels.

This information, together with an increase of the discharge radius (reduction of the radial contraction) point to a decrease of the influence of processes involving molecular ions such as associative ionization and dissociative recombination, while the influence of molecular dimmers such as He metastable molecules can not be neglected.

Acknowledgements

This work was subsidised by the Ministry of Science and Technology (Spain) in the framework of the project No. ENE2005-00314.

References

[1] Muñoz J and Calzada M D, 2008 *J. Phys. D: Appl. Phys* **41** 135203
[2] Sáinz A, Margot J, García M C and Calzada M D 2005 *J. Appl. Phys.* **97** 113305
[3] Griem H R, 1997 *Principles of Plasma Spectroscopy*, (Cambridge, Cambridge University)
[4] Gigosos M A and Cardeñoso V 1996 *J. Phys. B: At. Mol. Opt. Phys.* **29** 4795–4838
[5] Jonkers J, PhD Thesis 1998 Eindhoven Technical University (Holland)
[6] Berndt J, Makasheva K, Schlüter H and Shivarova A, 2002 *J. Appl. Phys.* **92** 6461

Journal of Physics: Conference Series **207** (2010) 012030 doi:10.1088/1742-6596/207/1/012030

Numerical investigations of sheath structure in presence of magnetic field

E Havlíčková and R Hrach

Department of Surface and Plasma Science, Charles University, V Holešovičkách 2, 180 00 Prague 8, Czech Republic

E-mail: ehavlickova@centrum.cz

Abstract. The contribution is focused on a computational study of the plasma-solid interaction in DC glow discharge in argon plasma with a special interest in the inclusion of external magnetic field and its effects on the sheath structure. In addition, the properties of the sheath and the presheath region have been analyzed in order to estimate the influence of pressure and size of a solid substrate immersed in plasma. The problem is assumed to be two-dimensional and the attention is also given to the efficiency of computer codes in two dimensions.

1. Introduction

The sheath structure and the sheath formation during the interaction of plasma with a solid surface is a fundamental phenomenon of plasma science. During the past decades, the behaviour of the sheath and the presheath region has been a topic of many investigations [1].

Regarding the complexity of studied problems, the computational approach is now being widely used to supplement experimental and theoretical investigations. Moreover, the theoretical description lose its validity in some conditions (e.g. collisional plasma at high pressures) and in this case the computational approach proved to be the best solution.

The contribution presents a computational analysis of the plasma-solid interaction in DC glow discharge in argon plasma in the presence of external magnetic field. The magnetic field significantly influences the sheath and the presheath region, its size and shape. Some of modern plasma technologies employ magnetic fields and the movement of charged particles in these fields is of scientific interest. In addition to the effects of the magnetic field, the influence of pressure and size of a solid surface immersed in plasma on the sheath structure is analyzed.

Besides these physical questions the attention was given to problems of computational physics, concerning the performance of individual algorithms in more dimensions and realistic assumptions about physical processes taking place during the plasma-solid interaction. The problem is considered to be two-dimensional and it is studied using a fluid description of plasma species dynamics. The fluid model of plasma describes an evolution of macroscopic plasma parameters such as plasma density and average plasma velocity and since it is based on solving a set of macroscopic partial differential equations (PDEs), it is not as time-consuming as a particle simulation where the trajectories of all the individual particles are being traced. Therefore, it is particularly suitable for any kind of extended analysis where the particle approach would be too time demanding.

2. Fluid model of plasma-solid interaction

2.1. Basic features of the model

The model uses a square computational domain (x,y) with a regular grid in Cartesian coordinates together with the assumption of translational symmetry along the axis perpendicular to the plane of the model. The model solves the dynamics of plasma species in the vicinity of a cylindrical probe placed in the z direction, perpendicular to the computational plane. Magnetic field $\mathbf{B} = (B_x, 0, 0)$ is set to be parallel with the x axis. These assumptions allow us to work in two dimensions.

The model is solved for each plasma species – electrons and argon ions. Collisions with neutral background are approximated in a simple way as elastic scattering with a given collision frequency. This frequency is constant for each particle species.

Boundary conditions correspond to an assumption of undisturbed plasma at the outer boundaries of the computational domain, while the internal boundary with electrode has to represent an interface with a metal immersed in plasma.

2.2. Mathematical background

Mathematical description is based on a fluid model governed by partial differential equations (PDEs), which can be derived taking velocity moments of the Boltzmann equation. This computational study uses the continuity equation (1) for plasma density n and momentum transfer equation (2) for average velocity of plasma species $\mathbf{u}$. The conservation equations describing a transport of mass and momentum are solved for electrons (k=e) and ions (k=i).

$$\frac{\partial n_\mathrm{k}}{\partial t} + \nabla \cdot (n_\mathrm{k}\mathbf{u}_\mathrm{k}) = 0 \tag{1}$$

$$m_\mathrm{k} n_\mathrm{k}\frac{\partial \mathbf{u}_\mathrm{k}}{\partial t} + m_\mathrm{k} n_\mathrm{k}(\mathbf{u}_\mathrm{k}\cdot\nabla)\mathbf{u}_\mathrm{k} = q_\mathrm{k} n_\mathrm{k}(\mathbf{E} + \mathbf{u}_\mathrm{k}\times\mathbf{B}) - \nabla p_\mathrm{k} - m_\mathrm{k} n_\mathrm{k}\nu_\mathrm{k}\mathbf{u}_\mathrm{k}. \tag{2}$$

Quantities in equations (1) and (2) have their usual meaning – density n, average velocity $\mathbf{u}$, pressure p, mean collision frequency ν, mass m, charge q and electric and magnetic field $\mathbf{E}$ and $\mathbf{B}$. The system of equations is completed by Poisson's equation for electric potential φ

$$\Delta\varphi = \frac{e}{\epsilon_0}(n_\mathrm{e} - n_\mathrm{i}) \tag{3}$$

$$\mathbf{E} = -\nabla\varphi. \tag{4}$$

The model is simplified by an assumption of constant temperature and the energy balance is not solved. Viscosity effect are neglected and only scalar pressure is taken into account $p = nkT$. Collision processes are simplified assuming only elastic scattering with constant collision frequency ν. Considered approximation follows an intent to carry out a qualitative analysis, while more precise model with improved treatment of collisions and the energy equation taken into account is described and solved e.g. in [2]. Compared to other papers, where the momentum equation is strongly simplified, this model works with full momentum equation. It is usual, under certain assumptions being fulfilled, to cancel the inertial and convective terms and to work with a drift-diffusion approximation [3]-[5], which has an algebraic form and describes a flux composed of the drift and the diffusion. In such a case, Scharfetter-Gummel scheme [6] is commonly proposed to solve PDE of the drift-diffusion type.

2.3. Boundary conditions

Boundary conditions of the equations (1)-(3) imposed on density, velocity and electric potential are summarized in Table 1. The problem is fixed by Dirichlet boundary conditions imposed

Journal of Physics: Conference Series **207** (2010) 012030 doi:10.1088/1742-6596/207/1/012030

Table 1. Boundary conditions.

	n_e	n_i	$\mathbf{u}_e$	$\mathbf{u}_i$	φ
exterior (undisturbed plasma)	$n_e = n_0$	$n_i = n_0$	open	open	$\varphi = 0$
interior (solid surface)	open	open	open	open	$\varphi = U_p$

on electric potential, assuming biased electrode with a positive voltage U_p and assuming undisturbed quasineutral plasma on the exterior of the computational domain. Dirichlet boundary condition is used also for plasma density at the outer boundary, setting the density to an experimental value of undisturbed plasma density. All the other conditions are kept open, i.e. we extrapolate values from nearest grid points inside of the computational region to the boundary (see [7]).

2.4. Computational technique

The system of equations (1)-(3) is solved using the finite difference approach. The equations are discretized on a regular rectangular grid of size $L \times L$ with uniform grid spacing in both dimensions $h = \Delta x = \Delta y$.

The fluid equations (1) and (2) can be written in general form as

$$\frac{\partial f}{\partial t} + \frac{\partial}{\partial x}(f v_x) + \frac{\partial}{\partial y}(f v_y) = F \tag{5}$$

with $f = (n, u_x, u_y)$. The same form of the continuity and momentum equations allows us to apply the same method of solution. The equations are discretized in a similar way using the flux-corrected transport (FCT) algorithm [8]-[9] and the solution is being updated to converge to a steady state. The numerical implementation involves three steps: advection, diffusion and anti-diffusion. To update the solution, any traditional high-order numerical scheme can be used, e.g. the two-step Lax-Wendroff discretization is implemented here. Then the diffusive flux is computed using a diffusivity coefficient and the solution is advanced. Final step to update the solution to next time level is to anti-diffuse the solution using a flux limiter. The whole procedure enable to smooth oscillations arising in the numerical solution and steep gradients can be resolved. Detailed description of the implementation is beyond the scope of the paper and it is presented e.g. in [8].

The Poisson's equation (3) is discretized using the traditional five-point stencil

$$\varphi_{i+1,j} + \varphi_{i-1,j} - 4\varphi_{i,j} + \varphi_{i,j+1} + \varphi_{i,j-1} = h^2 \frac{e}{\epsilon_0}(n_e - n_i). \tag{6}$$

The resulting system of linear equations is solved using the UMFPACK library [10]-[11], where the solution of systems of linear equations is based on direct LU decomposition solver.

3. Application of the model and input parameters

The fluid model described above was applied to simulate the interaction between plasma and an immersed solid substrate – a cylindrical probe. The model works with plasma composed of electrons and argon ions on a neutral background of argon atoms. Collisions of charged particles with neutrals are simplified assuming elastic scattering with constant collision frequency ν_e and ν_i. The simulation was performed for various values of external magnetic field B_x, probe radius R_p and pressure p. Pressure corresponds to certain values of undisturbed plasma density

n_0 and mean collision frequency ν_e and ν_i and these values were derived from experimental measurements performed in DC glow discharge in argon plasma.

The input parameters of the model are summarized as follows: size of the computational domain $L = 0.01$ m, probe voltage $U_p = 10$ V, electron temperature $T_e = 23210$ K and ion temperature $T_i = 300$ K. The main subject of interest is to analyze the effect of magnetic field and the influence of pressure and the probe size on the sheath structure. Simulations were performed for various values of B_x, R_p and p and only few simulations were selected to illustrate the effects considered here. Values of analyzed parameters used for results in the paper are listed as follows: magnetic field strength $B_x = 0$ T and $B_x = 0.001T$, probe radius $R_p = 1 \times 10^{-4}$ m and $R_p = 5 \times 10^{-4}$ m and pressure $p = 500$ Pa ($n_0 = 1.9 \times 10^{15}$ m^{-3}, $\nu_e = 3.9 \times 10^9$ s^{-1}, $\nu_i = 2.0 \times 10^7$ s^{-1}), $p = 1000$ Pa ($n_0 = 2.7 \times 10^{15}$ m^{-3}, $\nu_e = 7.8 \times 10^9$ s^{-1}, $\nu_i = 3.9 \times 10^7$ s^{-1}) and $p = 2000$ Pa ($n_0 = 3.9 \times 10^{15}$ m^{-3}, $\nu_e = 1.6 \times 10^{10}$ s^{-1}, $\nu_i = 7.8 \times 10^7$ s^{-1}).

4. Results

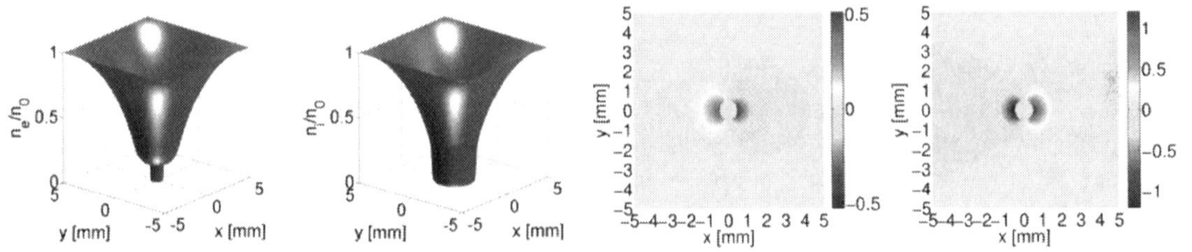

Figure 1. From left to right – electron density n_e/n_0 (1), ion density n_i/n_0 (2), x component of electron velocity u_e^x [10^6 m/s] (3) and x component of ion velocity u_i^x [10^3 m/s] (4). $B_x = 0$ T, $p = 1000$ Pa, $R_p = 5 \times 10^{-4}$ m.

The results of a basic fluid model without magnetic field are shown in Figure 1. The distribution of plasma density and electric potential is symmetric around the probe axis. The radial dependence of plasma density in the presheath region is logarithmic (Figure 2), which is predicted also by a simple theoretical model – analytic solution of the diffusion equation and it corresponds also to experimental investigations.

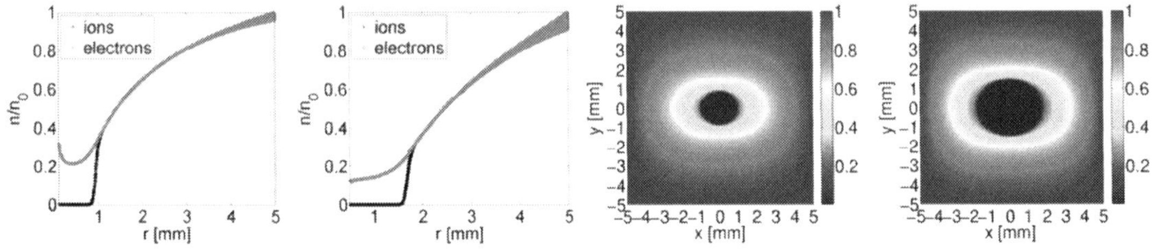

Figure 2. From left to right – radial dependence of plasma density for $R_p = 1 \times 10^{-4}$ m (1) and $R_p = 5 \times 10^{-4}$ m (2) for $B_x = 0$ T and $p = 1000$ Pa. Ion density n_i/n_0 for $R_p = 1 \times 10^{-4}$ m (3) and $R_p = 5 \times 10^{-4}$ m (4) for $B_x = 0.001$ T and $p = 1000$ Pa. The radial coordinate r ranges from $r = R_p$ to $r = 5$ mm.

The structure and the size of the sheath region depends on the probe size, as it is illustrated in Figure 2 together with the effect of the magnetic field. Concerning the simulation without magnetic field, the sheath size becomes larger with increasing probe radius. The radial dependence of plasma density in the presheath region becomes more linear than logarithmic and it is not so steep. The effect of the magnetic field on the sheath structure is clearly seen. The shape of the sheath region is elongated in the x direction, parallel with the magnetic field B_x. Obviously, the extension is larger with increasing magnetic field strength.

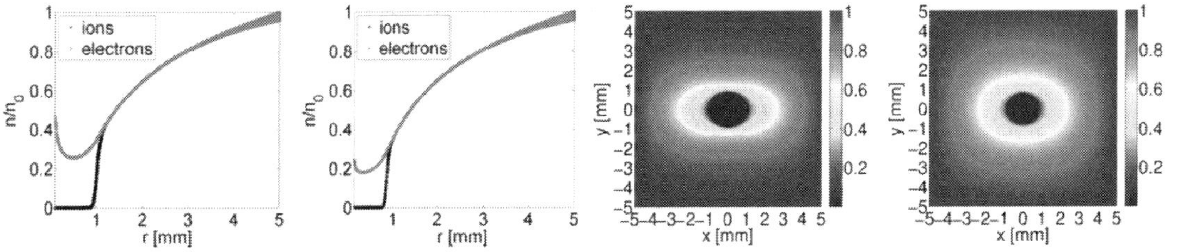

Figure 3. From left to right – radial dependence of plasma density for $p = 500$ Pa (1) and $p = 2000$ Pa (2) for $B_x = 0$ T and $R_\mathrm{p} = 1 \times 10^{-4}$ m. Ion density n_i/n_0 for $p = 500$ Pa (3) and $p = 2000$ Pa (4) for $B_x = 0.001$ T and $R_\mathrm{p} = 1 \times 10^{-4}$ m.

The effect of pressure on the sheath structure can be seen in Figure 3. Simulation without magnetic field shows that pressure influences the width of the sheath region and for lower pressure density of electrons on the probe surface is higher as well as their velocity. The collision processes significantly influence the sheath shape in presence of magnetic field. For lower pressure, when plasma is less collisional, the magnetic field affects the sheath elongation much more strongly in comparison with more collisional plasma.

The spatial resolution of the grid was 400×400 grid points or 200×200 in some simulations. The time resolution was given by Courant-Friedrichs-Lewy (CFL) stability condition and proper value is automatically computed in the code run. Minimum time steps in calculations satisfying the CFL criterion were $\Delta t_\mathrm{e} \sim 1 \times 10^{-11}$ s and $\Delta t_\mathrm{i} \sim 1 \times 10^{-8}$ s. The efficiency of the fluid code depends on considered physical conditions. One iteration of the simulation in Figure 1 with 400×400 grid points takes 0.35 s (Intel compiler, Intel Core2 Duo 2.4 GHz CPU) and sufficient number of time steps to reach a steady state solution was 3×10^4.

5. Discussion and conclusions

The size and the shape of the sheath region surrounding a solid substrate embedded into plasma depends on plasma properties and processes taking place during plasma-solid interaction. In particular, the magnetic field strength and pressure (rate of collision processes) strongly influence the sheath structure. Even in the case of relatively weak magnetic field the effect can be clearly seen.

The presented model was simplified by neglecting non-elastic collisions and assuming constant temperature. These aspects could influence the results quantitatively, but the qualitative effect and the tendency would remain the same. More exact model from this point of view has been also developed and it was described elsewhere [2]. However, the model presented in [2] doesn't take into account the magnetic field, the equations are simplified in different manner and the numerical technique is different. It would be rather straightforward to upgrade the model described here and include the energy balance and better treatment of collision processes. On the other hand, the model solves full momentum equation compared to common approach, when the momentum equation is simplified and an algebraic expression for flux is used [4].

It should be stated, that for collisionless plasma in presence of the magnetic field it is characteristic to observe the perturbation of the sheath and the presheath region relatively far from the solid surface. In such a case of strong extension of the sheath shape by the magnetic field, the ratio of the width of the perturbed region in the direction of magnetic field lines and in the direction perpendicular to magnetic field lines can be one order of magnitude. In these conditions, the validity of the model is limited by an assumption of undisturbed plasma outside of the computational region and then a proper boundary conditions should be used or the size of the grid should be modified to correspond to physical conditions. In this study we keep the magnetic field and pressure in such a range to satisfy the boundary conditions.

Acknowledgments

The authors acknowledge the support of Ministry of Education of the Czech Republic, research plan MSM 0021620834, Grant Agency of Charles University, Project GAUK-220/2006 and GAUK-51808/2008 and Grant Agency of the Czech Republic, Grant No. 202/08/H057.

References

[1] Barnat E V and Hebner G A 2007 *J. Appl. Phys.* **101** 013306
[2] Havlíčková E, Bartoš P and Hrach R 2007 *J. Phys.: Conf. Series* **63** 012019
[3] Passchier J D P and Goedheer W J 1993 *J. Appl. Phys.* **74** 3744
[4] Donkó 2001 *Phys. Rev. E* **64** 026401
[5] Chen G and Raja L L 2004 *J. Appl. Phys.* **96** 6073
[6] Scharfetter D L and Gummel H K 1969 *IEEE Trans. Electron Devices* **16** 64
[7] Baboolal S 2002 *J. Phys. D: Appl. Phys.* **35** 658
[8] Boris J P and Book D L 1973 *J. Comput. Phys.* **11** 38
[9] Zalesak S T 1979 *J. Comput. Phys.* **31** 335
[10] Davis T A 2004 *ACM Trans. Math. Softw.* **30** 165
[11] Davis T A 2004 *ACM Trans. Math. Softw.* **30** 196

The role of a metal screen on the coaxial discharge wave propagation characteristics

T Bogdanov [1], **E Benova**[2]

[1]Faculty of Physics, University of Sofia, 5 James Bourchier Blvd., BG-1164 Sofia, Bulgaria,

[2] Department for Language Teaching and International Students, University of Sofia, 27 Kosta Loulchev Street, BG-1111 Sofia, Bulgaria,

Email: benova_phys@deo.uni-sofia.bg

Abstract. The purpose of this work is to investigate theoretically the propagation characteristics of the electromagnetic wave that can produce and sustain plasma in a coaxial structure, as well as the wave field components. We have investigated the coaxial structure which consists of a metal rod in the centre, a dielectric tube, plasma outside the tube and with or without a metal screen. The plasma is both radially and axially inhomogeneous but we consider a radially averaged electron density in describing the plasma and we are presenting here one-dimensional axial model. The basic relation in our model is the local dispersion relation obtained from Maxwell's equations. Since the plasma is axially inhomogeneous the local dispersion relation gives the dependence between the normalized plasma density and the dimensionless wave number, so called phase diagrams. The radial variations of the normalized wave field components are calculated. The behavior of the phase diagrams and the wave field components is compared in four configurations: metal–vacuum–plasma; metal–vacuum–dielectric–plasma; metal–vacuum–plasma–metal; metal–vacuum–dielectric–plasma–metal.

1. Introduction

The coaxial structure is relatively new type of plasma source, which was proposed recently [1,2]. This new type of plasma source is very attractive because of an increasing range of their possible technological applications as UV lamps [6], surface cleaning, material treatments, plasma sterilization, thin film deposition [7] producing large volume plasma. Typical configuration for such kind of gas discharges is presented in figure 1. The plasma at low pressure in the coaxial structure is produced in a discharge chamber which usually consist of two coaxial dielectric tubes and a metal cylinder of small radius is arranged at their axis. The plasma is both radially and axially inhomogeneous. In some cases the second dielectric tube is replaced by a metal tube. The purpose of this work is to investigate the role of this metal screen on the wave propagation characteristics as well as the influence of the first dielectric tube radius. We studied

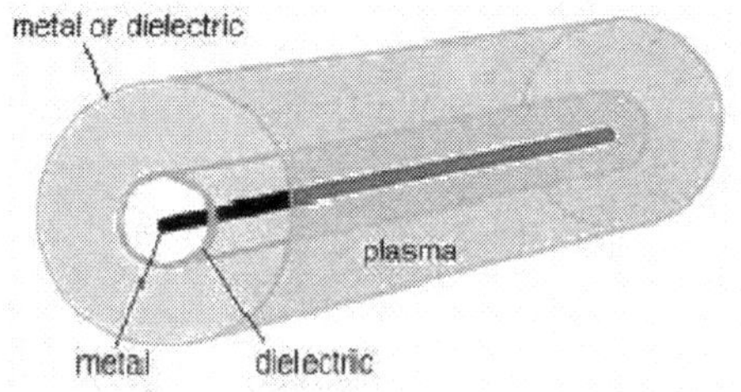

Figure 1. Coaxial discharge configuration

theoretically four configurations: (i) metal–vacuum–plasma (mvp); (ii) metal–vacuum–dielectric–plasma (mvdp); (iii) metal–vacuum–plasma–metal (mvpm); (iv) metal–vacuum–dielectric–plasma–metal (mvdpm).

2. Basic assumptions and equations

At low pressures plasma can be considered as a weakly dissipative medium. At this assumption the ratio of the electron-neutral collisions frequency ν to the wave angular frequency ω is smaller than unity and can be neglected in the plasma permittivity expression, i.e. $\varepsilon_{\rm p} = 1 - \omega_{\rm p}^2/\omega^2$ ($\omega_{\rm p} = \left(4\pi e^2 n/m\right)^{1/2}$ being the plasma frequency). The plasma is produced and sustained by a surface electromagnetic wave ($\omega/2\pi = f = 2.45$ GHz), which propagates along the interface between the plasma and the inner tube. We consider the stationary state of a plasma sustained by azimuthally symmetric ($m = 0$) electromagnetic wave with field components $\mathbf{E} = (E_r, 0, E_z)$ and $\mathbf{B} = (0, B_\varphi, 0)$. We assume that the wave number k and the wave amplitude are slowly varying functions of the axial coordinate and radially averaged plasma density is used:

$$n \equiv \bar{n} = \frac{2}{R^2}\int_0^R dr\, r\, n(r). \tag{1}$$

At these assumptions the coaxial structure is investigated on the base of one-dimensional axial fluid model [8].The fluid model of surface-wave-sustained plasma is applicable at low gas pressures when the main process for electron production is the direct ionization from the ground state and the electron losses are due to the diffusion to the wall. Our model is based on Maxwell's equations from which we obtain the wave equation. Keeping in mind the abovementioned assumptions we consider the solutions in the form:

$$E_z(r,\varphi,z,t) = \mathrm{Re}\left[F_z(r,z)E(z)\exp\left(-i\omega t + i\int_0^z dz' k(z') + im\varphi\right)\right] \tag{2}$$

The solutions of the wave equation give the wave field components amplitude F_z as combinations of Bessel function J_0, modified Bessel functions I_0, K_0, and Hankel function $H_0^{(1)}$ of zero order:

$$\begin{aligned} F_z^{\rm v}(a_{\rm v}\rho) &= C_1 I_0(a_{\rm v}\rho) + C_2 K_0(a_{\rm v}\rho) \\ F_z^{\rm d}(a_{\rm d}\rho) &= C_3 J_0(a_{\rm d}\rho) + C_4 H_0^{(1)}(a_{\rm d}\rho) \\ F_z^{\rm p}(a_{\rm p}\rho) &= C_5 I_0(a_{\rm p}\rho) + C_6 K_0(a_{\rm p}\rho) \end{aligned} \tag{3}$$

where

$$a_{\rm p} = (x^2 - \sigma^2\varepsilon_{\rm p})^{1/2}, \quad a_{\rm d} = (\sigma^2\varepsilon_{\rm d} - x^2)^{1/2}, \quad a_{\rm v} = (x^2 - \sigma^2)^{1/2}, \quad \rho = \frac{r}{R}, \quad x = kR, \quad \sigma = \frac{\omega R}{c}.$$

Similar solutions can be obtained also for the E_r and B_φ wave field components. The boundary conditions are the conditions for continuity of the electric and magnetic field tangential components E_z and B_φ at the vacuum–dielectric and dielectric–plasma interfaces and the condition for annulment of the E_z–component on the metal cylinders. When the outer metal cylinder does not exist the boundary condition is annulment of the E_z–component at infinity. From the boundary conditions we obtain the local dispersion relation:

$$\mathrm{D}(\omega,\ R,\ \varepsilon_{\rm p}, \varepsilon_{\rm d}, \gamma, \eta, \xi, k) = 0, \tag{4}$$

where the geometry of the discharge is included by dimensionless parameters presented in Table 1.

When the plasma is axially homogeneous ($\omega_{\rm p}$ = const) dispersion relation gives the dependence of the wavelength on the wave frequency, plotted usually in the form $\omega/\omega_{\rm p}$ versus kR – dispersion diagrams. In our case the dispersion relation (2) is local because of the plasma axial inhomogeneity. It is solved at fixed wave frequency and gives the dependence of the wave number on the plasma frequency (plasma density) usually also plotted as $\omega/\omega_{\rm p}$ versus kR – the phase diagram.

Table 1.

Geometric factors	Notations	Parameters
Plasma radius	R (outer radius of the dielectric tube)	$\sigma = \omega R/c$ (c – speed of light)
Dielectric tube radius	R_d (inner radius of the dielectric tube)	$\gamma = R_d/R = 1 - d/R$
Tube thickness	$d = R - R_d$	
Radius of the metal rod in the centre	R_{m1}	$\eta = R_{m1}/R$
Radius of the metal screen	R_{m2}	$\xi = R_{m2}/R$

3. Results and discussion

Solving the local dispersion relation at fixed ω we obtain the phase diagrams for the four configurations (figure 2a). The higher plasma density corresponds to the beginning of the plasma

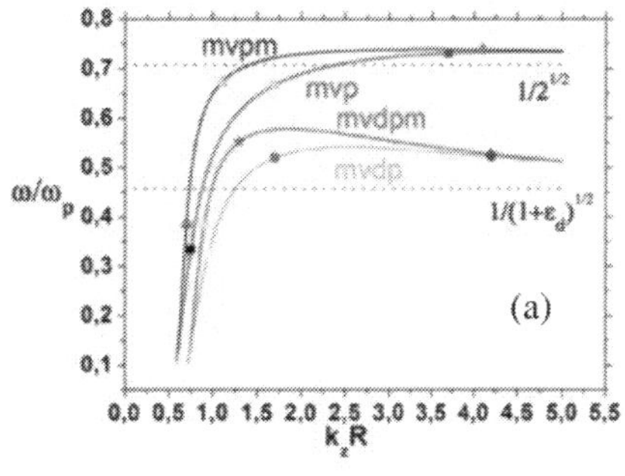

Figure 2a. Phase diagrams for the four configurations under investigation

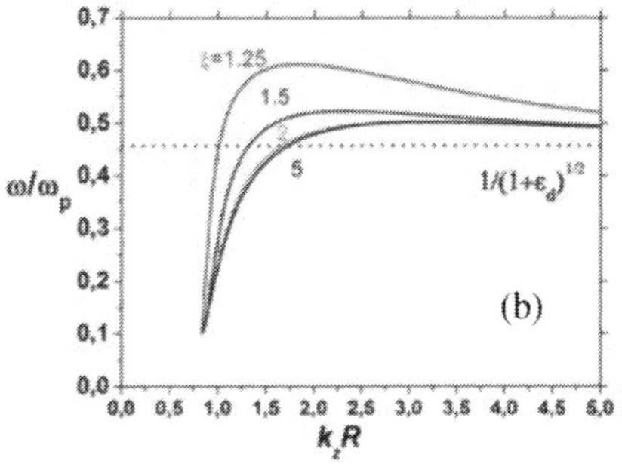

Figure 2b. Phase diagrams at various metal screen radii

column and it is in the region of small wave numbers. With the wave number increasing the plasma density decreases. The cut-off limit for the configurations without dielectric tube corresponds to $(1/2)^{1/2}$ and for configurations with dielectric tube is $1/(1+\varepsilon_d)^{1/2}$. The part of the phase diagram below the ω/ω_p cut-off limit corresponds to overdense plasma and above it – to underdense plasma. One can see from figure 2 that a region of backward wave propagation appears at high wave numbers. In this region the wave cannot sustain plasma and the real end of the plasma column is at the maximum of the phase diagram in the underdense plasma region. When the metal screen is placed the phase diagram moves up, corresponding to a decrease in the plasma density.

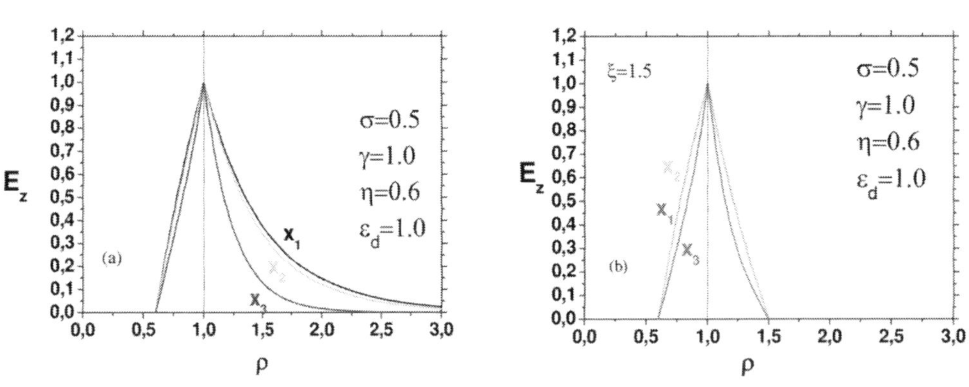

Figure 3. Radial distribution of the E_z-electric field component for the metal–vacuum–plasma configuration without (a) and with (b) metal screen

We have also obtained that increasing the metal screen radius R_{m2} increases the plasma density (figure 2b) and the best position of the metal screen is at $\xi = 2$ ($R_{m2} = 2R$). Further increasing of this radius does not change significantly the plasma density and the phase diagrams at $\xi = 5$ and $\xi = 10$ practically coincide with that at $\xi = 2$.

The radial distributions of the amplitude of the normalized E_z -wave electric field component are presented in figure3 (without dielectric) and figure 4 (with dielectric) at given discharge conditions. For each phase diagram we have calculated the electric field radial distribution at three (x, N)–values presented in figure 2a by different color dots and corresponding to the regions of overdense plasma, underdense plasma and backward wave propagation. For all configurations the electric field amplitude is higher in the overdense plasma region (x_1) than in the underdense plasma region (x_2). In the region of backward wave propagation (x_3) the E_z -electric field amplitude is the lowest.

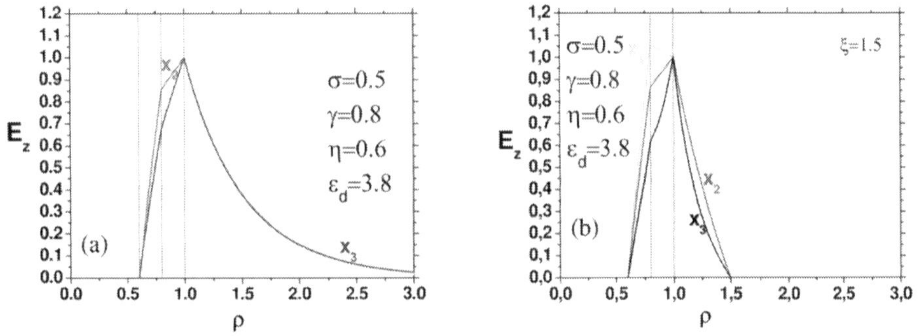

Figure 4. Radial distribution of the E_z-electric field component for the metal–vacuum–dielectric–plasma configuration without (a) and with (b) metal screen

4. Conclusions

From the obtained results one can see that the coaxial plasma can be used for cleaning and modifying the inner surface of metal tubes. The best conditions and plasma characteristics are obtained at metal screen radius $R_{m2} = 2R$. At smaller metal screen radii the plasma density is lower. Taking into account the existence of the dielectric tube we obtain higher plasma density.

Acknowledgments
This work was supported by the Bulgarian National Fund for Scientific Research under Grant F 1401/04 and by the Fund for Scientific Research of the University of Sofia under Grant 157/2008.

References

[1] Räuchle E 1998 *J. Phys. IV France* **8** 99–108
[2] Gritsinin S, Kossyi I, Malykh N, Misakyan M, Temchin S and Bark Y 1999 *Preprint No 1* Russian Academy of Science, General Physics Institute, Moscow
[3] Benova E and Neichev Z 2002 *Czechoslovak J. Phys.* **52** D659–65
[4] Alves L L, Letout S, Boisse-Laporte C and Leprince P 2006 *Microwave Discharges: Fundamentals and Applications* (Yanus-K, Moscow) pp. 177-186
[5] Leroy O, Leprince P and Boisse-Laporte C 2006 *Microwave Discharges: Fundamentals and Applications* (Yanus-K, Moscow) pp. 137-140.
[6] Barkhudarov E M, Misakyan M A, Kossyi I A, Kiselev V A, Kozlov Y N, Taktakishvili I M and Tulupov A V 2006 *Microwave Discharges: Fundamentals and Applications* (Yanus-K, Moscow) pp 279–284
[7] Shullz A, Walker M, Feichtinger J, Räuchler E and Schumacher U 2003 *Microwave Discharges: Fundamentals and Applications* (INP, Greifswald) pp 231–241
[8] Neichev Z, Benova E, Gamero A and Sola A 2007 *Journal of Physics: Conference Series* **63** 012024

Third International Workshop & Summer School on Plasma Physics 2008 IOP Publishing
Journal of Physics: Conference Series **207** (2010) 012032 doi:10.1088/1742-6596/207/1/012032

Pre- and Post-Processing of data for Simulation of Gyrotrons by the GYREOSS Software Package

M Damyanova[1], S Sabchevski[1], I Zhelyazkov[2]

[1]Academician Emil Djakov Institute of Electronics, Bulgarian Academy of Sciences, 72 Tzarigradso Schaussee Blvd., BG-1784 Sofia, Bulgaria
[2]Faculty of Physics, Sofia University, 5 James Bourchier Blvd., BG-1164 Sofia, Bulgaria

E-mail: sabchevski@yahoo.com

Abstract. In this paper, we present the concept of the pre- and post-processing of data for simulation of gyrotrons by the GYREOSS software package, which is being developed now. It is based on the utilization of an appropriate code for description and input of the geometry of the tube in three dimensions, for discretization of the computational region by unstructured tetrahedral mesh and for visualization and manipulation of the results of numerical experiments.

1. Introduction

Modelling and simulation are two indispensable parts of the computer-aided design (CAD) of gyrotrons for fusion research. The current status of the physical models and the simulation tools used for analysis and optimization of the existing tubes as well as for development of novel high-performance devices has been reviewed recently in [1]. Based on the critical examination of both the advantages and the disadvantages of the known and the available computer programs a conceptual design of the next generation software package has been performed. As a result, the main requirements of such codes have been formulated. Among them, the most important characteristics are: (i) the use of adequate self-consistent physical models formulated in a 6D phase space and in three spatial dimensions; (ii) program implementation based on modern programming technologies, environments and numerical libraries that allow easy portability to different platforms, and (iii) extensibility, which would allow more physical factors to be taken into account adding new computational modules. Following the conceptual design, a programming of a novel software package called GYREOSS (which stands for **GYR**otron **E**lectron-**O**ptical **S**ystem **S**imulation) has been initiated and is now in a process of active development. The physical model implemented in the GYREOSS code is embedded in several modules added to the **G**eneral **e**nvironment for the **t**reatment of **D**iscrete **P**roblems (GetDP) [2]. The method used for calculation of the electrostatic potential (boundary value problem for Poisson equation with appropriate Dirichlet and Neumann boundary conditions) by the GetDP solver is the Finite Element Method (FEM), which in 3D requires the computational area to be discretized by unstructured tetrahedral mesh.

It is well known that the most laborious and error prone tasks in performing numerical simulation is the input of the initial data that describe the geometry of the device as well as its subsequent modification. This problem can be solved using the functionality of **Gmsh** (A finite element mesh generator with built-in pre- and post-processing facilities) [3] as geometrical modeller and CAD program. It has both a convenient language and graphical user interface (GUI) for description and manipulation of the geometry. Additionally, **Gmsh** produces optimal tetrahedral mesh in a format that is compatible with the format used by GetDP. Another advantageous feature of this software package is its capability to visualize and manipulate geometrical entities, mesh elements and the results of calculations.

This paper is organized as follows. In section 2 we describe briefly the **Gmsh** package. Its use for input and manipulation of the geometrical information about simulated gyrotrons is presented in section 3. In section 4 we illustrate the capabilities of the package for visualization of the finite element mesh and the results of the calculations. Finally, we present an outlook and discuss some additional programming modules for input of the rest of the information (magnetic system, emission model etc.) necessary for caring out the numerical experiments that are in development now.

2. General description of the Gmsh package

Gmsh is a 3D geometry modeller and finite element mesh generator with built-in pre- and post-processing capabilities including advanced means for manipulation and visualization of the data. Its code is cross-platform and runs on Linux, various flavors of UNIX and Windows operating systems. This is an important feature since, as already mentioned above the GYREOSS software package has to be portable.

To describe *geometries* this mesh generator uses a boundary representation ("b-rep"). Models are created in a bottom-up flow by successively defining points, oriented lines (line segments, circles, ellipses, splines), oriented surfaces (plane surfaces, ruled surfaces, triangulated surfaces) and volumes. Compound groups of geometrical entities (called "physical groups") can also be defined, based on these elementary geometric entities. **Gmsh**'s scripting language allows all geometrical entities to be fully parameterized.

A finite element *mesh* is a tessellation of a given subset of the three-dimensional space by elementary geometrical elements of various shapes (lines, triangles, quadrangles, tetrahedra, prisms, hexahedra and pyramids). The mesh generation is performed in the same bottom-up flow as the geometry creation: lines are discretized first; the mesh of the lines is then used to mesh the surfaces; then the mesh of the surfaces is used to mesh the volumes.

External *solvers* can be interfaced with **Gmsh** through UNIX or TCP/IP sockets, which permit to launch external computations and to collect and process the results directly from within **Gmsh**'s post-processing module. The default solver interfaced with **Gmsh** is GetDP [2]. It is a powerful software environment for numerical solution of integro-differential equations that describe various (including coupled) physical problems by the finite element method. A characteristic feature of GetDP is the closeness between the input data defining discrete problems (written using a convenient language and ASCII format) and the symbolic mathematical expressions (mathematical formulation of the physical model) of these problems. GetDP is extremely suitable especially for solution of electromagnetic problems since it makes use of advanced formulation based on differential forms and finite elements (Whitney edge elements of various orders, Nedelec's elements and so on) tailored specifically for integration of Maxwell's equations. Therefore, it was chosen as the environment for the development of the GYREOSS code.

Gmsh can load and manipulate multiple post-processing scalar, vector or tensor maps along with the geometry and the mesh. Scalar fields are represented by iso-value lines/surfaces or color maps, while vector fields are represented by 3D arrows or displacement maps. *Post-processing* functions include section computation, offset, elevation, boundary and component extraction, color map and range modification, animation, vector graphic output, etc. All the post-processing options can be accessed either interactively or through the input ASCII text files. Scripting permits to automate all

Journal of Physics: Conference Series **207** (2010) 012032 doi:10.1088/1742-6596/207/1/012032

post-processing operations, as for example to create animations. User-defined operations can also be performed on post-processing views through dynamically loadable plugins.
Gmsh's scripting language enables all commands and command arguments to depend on previous calculations. **Gmsh** can display scalar, vector and tensor data sets and can perform various operations on the resulting post-processing views, export plots in many different formats: vector PostScript or encapsulated PostScript, LaTeX, PNG, JPEG, generate complex animations, run on low end machines and/or machines with no graphical interface.

3. Input of the geometrical description of the gyrotron and creation of the computational mesh

The electron optical system (EOS) of a gyrotron consists of several components, namely a magnetron injection gun (MIG), a beam tunnel, a cavity resonator and a collector. The main part is the MIG, which consists of a cathode with annular emitting ring, an anode and a second anode in the case of a triode gun. A 3D-cut view of a diode coaxial MIG of 170-GHz 2-MW gyrotron for ITER is shown in figure 1. The description of the geometry of the 3D physical domain starts from an arbitrary meridional plane (cross-section). The intersections of the electrodes with this plane are lines that we represent as a sequence of segments.

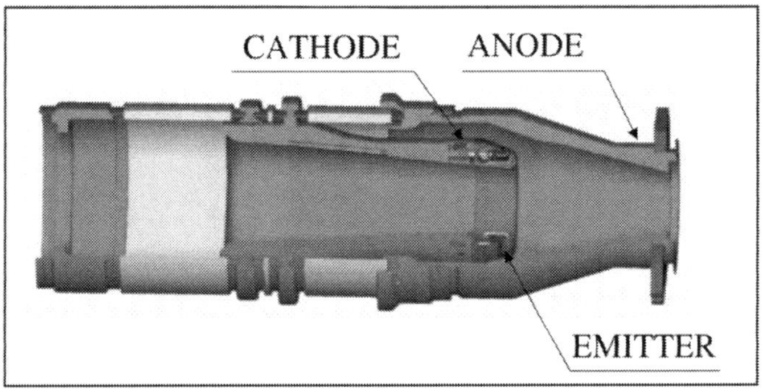

Figure 1. 3D-cut view of a coaxial magnetron injection gun (CMIG). The coaxial insert is not shown.

The segments can be straight lines or arcs of circles. The straight segments are represented by two points, notably the beginning and the end point while the arc segments are represented by three points: the beginning; the center of the circle and the end point. Each point is specified by three Cartesian coordinates (X,Y,Z). Thus, for each electrode contour (line) the points along it are specified [see figure 2(a)]. To each point a fourth parameter, namely the characteristic length lc is associated, which controls the sizes of the finite elements [figure 2(b)]. Then from points we compose segments and from segments we compose the contours. The contours are used to describe closed loops that bound surfaces. Three-dimensional volumes are created either by rotation or by extrusion of surfaces. They can also be specified defining a sequence of surfaces (plane or ruled) that encompass the closed space. Figure 3 shows the 3D geometry of the gun described in such a way. It can easily be manipulated and/or modified using the GUI of the code or editing the ASCII file (*.geo file) that contains the description of the geometrical data. Each of the elementary entities, as well as all the complex elements are identified by unique names and numbers that are used for their description and also for attachment of appropriate boundary conditions to them. Additionally, the GUI allows one to specify the visibility individually to any element or to a combination of elementary entities and in such a way to visualize selectively different parts of the geometry.

The tetrahedral finite element mesh generated by **Gmsh** using the geometrical description presented above is shown in figure 4. Since this figure is produced for illustrative purpose only, the mesh size is increased deliberately in order to make the finite elements more visible.

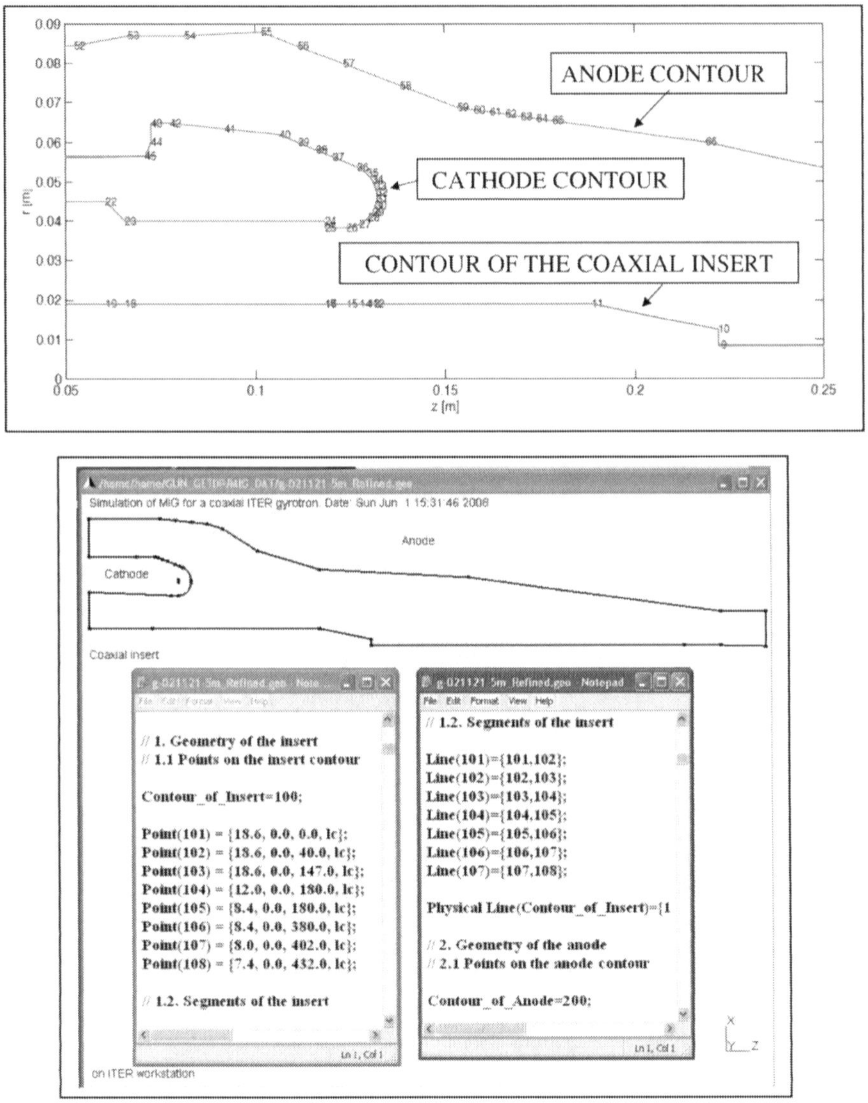

Figure 2. Geometry description: (a) Contours of the electrodes in the meridional cross-section of the gun represented by linear and arc segments. The end points of the segments are numbered consecutively. (b) Input of the coordinates of the points, characteristic lengths and description of the geometrical entities (lines, surfaces, volumes) using the GUI of **Gmsh**.

4. Visualization of the computational mesh and the results of the simulations

In the preceding section we described concisely the capabilities of the **Gmsh** as a pre-processing tool of the GYREOSS code. Here we illustrate its use for post-processing and visualization of the results obtained in the numerical experiments. The simulations provide data that includes scalar fields like the electric potential and space charge distributions, vector fields (electric and magnetic fields), electron orbits and various integral averaged parameters like velocity spread, velocity ratio etc. This data can be presented and visualized as colour maps, schematic diagrams and graphs with the standard tools of the package. As an example, in figure 5 we present the cut view of the electrostatic potential distribution in the electron gun, which geometry and mesh was presented in figures 1–4. Also shown is a part of the mesh used in the calculations.

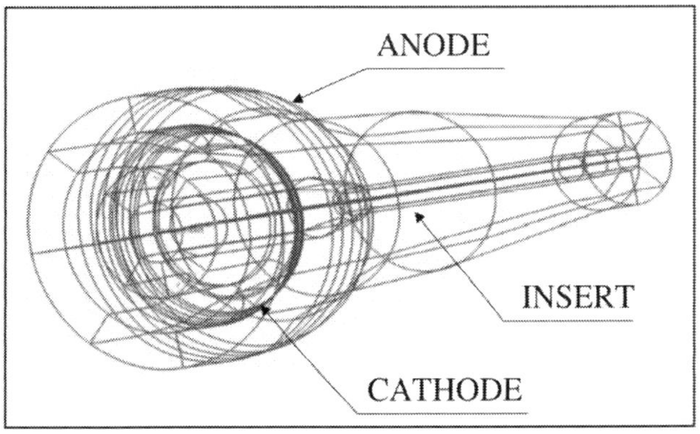

Figure 3. 3D description of the geometry of the gun. Shown are some boundary contours of the electrodes, surfaces of the electrodes and elementary volumes that form the computational space.

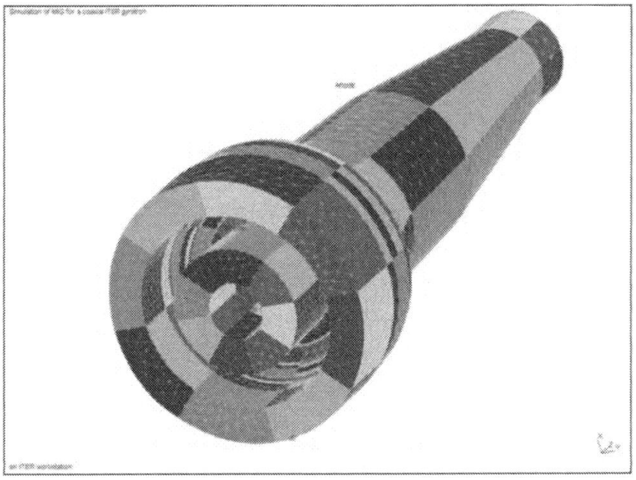

Figure 4. Computational mesh for GYREOSS code generated by the **Gmsh** preprocessing module.

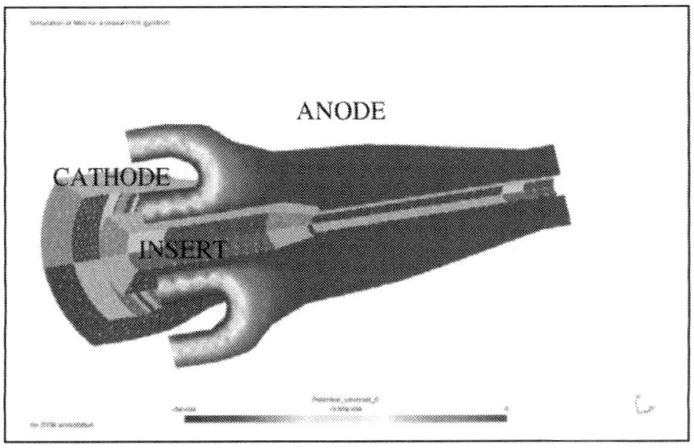

Figure 5. Part of the computational mesh and distribution of the electrostatic potential in a coaxial MIG.

An advantageous feature of the visualization performed by **Gmsh** is the possibility to combine different views and easily change the properties of the image (colours, orientation, selective zooming and so on). Various calculations (for example gradient of the scalar fields, curl of the vector fields, integral quantities) can be performed as well. The plots, as we have already pointed out in section 2, can be exported in many different formats: vector PostScript or encapsulated PostScript, LaTeX, PNG, JPEG. Scripting permits to automate all post-processing operations, as for example to create animations. User-defined operations can also be performed on post-processing views through dynamically loadable plugins. **Gmsh** treats by default all post-processing views as three-dimensional plots, i.e., draws the scalar, vector and tensor primitives (points, lines, triangles, tetrahedra, and so on) in 3D space [3] but can also represent each post-processing view containing scalar points as two-dimensional ("*X–Y*") plots, either space- or time-oriented. This means that in a 2D space plot, the scalar points are taken in the same order as they are defined in the post-processing view: the abscissa of the 2D graph is the curvilinear abscissa of the curve defined by the point series, and only one curve is drawn using the values associated with the points. If several time steps are available, each time step generates a new curve. Similarly, in a "2D time" plot, one curve is drawn for each scalar point in the view and the abscissa is the time step.

5. Conclusions

Presented here concept for pre- and post-processing of the simulation data used/produced by of the GYREOSS software package and its implementation are based on the standard tools provided by the **Gmsh** code. They offer an appropriate and convenient interface for preparation of the input data and for both manipulation and visualization of the results of the numerical experiments carried out aiming optimization, analysis or CAD of electron-optical systems (EOS) of gyrotrons. A significant advantage of this approach is the possibility to use dynamically loadable plugins. The post-processing plugins permit to extend significantly the functionality of the post-processing. While the regular post-processing options can only change the way the data are displayed, the post-processing plugins either create new post-processing views, or modify the data stored in a view. We plan to extend the capabilities of the pre- and post-processing of the GYREOSS data developing novel plugins that are specific for the ray-tracing (trajectory analysis) and particle-in-cell (PIC) numerical simulations and will allow us to present in an informative and well structured manner the main beam quality parameters and the characteristics of the EOS.

Acknowledgements

This work was carried out in the framework of Task P3 of the scientific programme of the Association EURATOM-INRNE.

References

[1] Sabchevski S, Zhelyazkov I, Thumm M, Illy S, Piosczyk B, Tran T-M, Hogge J-Ph, Pagonakis J Gr 2007 Recent Evolution of the Simulation Tools for Computer Aided Design of Electron-optical Systems for Powerful Gyrotrons *Computer Modeling in Engineering and Sciences* **20** 203–220

[2] Dular P and Geuzaine Ch 2006 GetDP: A General Environment for the Treatment of Discrete Problems, Version 1.2 http://www.geuz.org/getdp/

[3] Geuzaine Ch and Remacle J-F 2008 Gmsh: A three-dimensional finite element mesh generator with built-in pre- and post-processing facilities, Version 2.2.0 http://www.geuz.org/gmsh/

Model Calculation of Positron States in Tungsten Containing Hydrogen and Helium

T. Troev[a] , E. Popov, N. Nankov[a] and T. Yoshiie[b]

[a]Institute for Nuclear Research and Nuclear Energy, 72 Tzarigradsko Chaussee Blvd., 1784 Sofia, Bulgaria

[b]Research Reactor Institute, Kyoto University, Kumatori-Cho, Sennan-Gun, Osaka-Fu 590-0494, Japan

E-mail: troev@inrne.bas.bg

Abstract. Tungsten is a candidate material for plasma-facing first wall of a fusion power plant. Understanding of defects, tritium and helium behaviour in plasma facing materials [PFM] is an important issue for fusion reactor from viewpoints of its mechanical properties under neutron irradiation. Experiments with high-Z materials show that erosion of these materials under normal operation condition is considerably lower than the plasma induced erosion of low-Z materials like carbon or beryllium. Quantitative understanding of the experimental results for defects in tungsten needs a comprehensive theory of electron-positron interaction. The properties of defects in tungsten containing hydrogen or helium atoms have been investigated by model positron lifetime quantum-mechanical calculations. The electron wave functions have been obtained in the local density approximation LDA to the density functional theory DFT. On the bases of calculated results, the behaviour of vacancies, empty nano-voids and nano-voids with hydrogen and helium were discussed. It was established that hydrogen and helium in larger three-dimensional vacancy clusters in W change the annihilation characteristics dramatically. The hydrogen and helium atoms are trapped by lattice vacancies. These results provide physical insight for positron interactions in tungsten defects and can be used for prediction of hydrogen-H or helium-He4 and (tritium-H3) generation for the design of fusion reactors.

1. Introduction

The positron is the antiparticle of the electron: it has the same mass as the electron but its charge is positive. Positrons are available *e.g.* from radioactive sources such as the ^{22}Na isotope. Positrons injected in matter quickly lose energy and live in thermal equilibrium for a short before annihilating with electrons at time-scale inversely proportional to the electron density. In a perfect crystal, the positron density is delocalized over the whole crystal where as in presence of vacancies, the repulsion from positive ion cores may cause positrons to be trapped in open volumes with lower electron density thus increasing the lifetime.

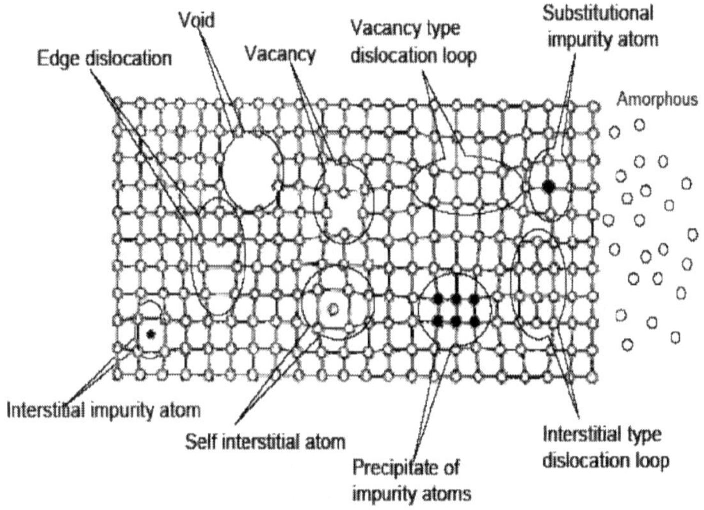

Figure 1. Formation of different defects and impurity atoms in neutron irradiated tungsten.

Positron lifetime spectroscopy is a powerful and sensitive technique for studying of defect concentrations as low as 10^{-6} in metals [1,2]. The method is based on the positron trapping at defects as shown in figure 1. The positron trapping in metals may be used for characterization of monovacancies, divacancies and vacancy-clusters (nano-voids). The annihilation characteristics of a positron trapped at a vacancy or empty nano-voids are different from those of positrons trapped at nanovoids containing hydrogen or helium. High amount of helium (~1000 appm He) can be generated in all materials proposed for fusion reactor first wall applications during the lifetime of a fusion reactor. Hydrogen or helium presences influence the behavior of defects in tungsten samples [3,4]. The interaction of positrons with defects containing hydrogen or helium illustrates interesting physical concepts concerning hydrogen and helium trapping by vacancy clusters. The micro-structural information is essential for understanding the helium mechanism in tungsten. The production of hydrogen or helium in the irradiated materials is dependent on the following nuclear reactions: (n, α) and (n, p).

$$ {}_1H^2 + {}_1H^3 \rightarrow He^4 + n + 17.6\ \text{MeV} \quad (1)$$

$$ {}_1H^2 + {}_1H^2 \rightarrow He^3 + n + 3.2\ \text{MeV} \quad (2)$$

$$ {}_1H^2 + {}_1H^2 \rightarrow He^3 + H^1 + 4.0\ \text{MeV} \quad (3)$$

Helium-3 and helium-4 are non-radioactive light isotopes of helium. The nucleus of helium-3 atom consists of two protons and one neutron, in contrast to two neutrons in ordinary helium. ^{3}He is a fermion while ^{4}He is a boson. The properties and behavior of ^{3}He could not be explained by the chemistry of helium, but it is explained by quantum mechanics. In addition to the fusion reactions, the reactions with neutrons are important in order to breed tritium in plasma machine. Neutron irradiation creates atomic displacements and generates helium or hydrogen in samples by atomic transmutations. Identification and behavior of defects in tungsten is important because they greatly affect its mechanical and electronic properties. The present investigation has focused on the model calculations of positron lifetime of defects in bcc structure of tungsten containing hydrogen or helium. Tungsten is considered as plasma facing materials as a first wall coating for Demo-related blankets and as amour material for gas-cooled divertors. Helium created by nuclear reaction during neutron irradiation of tungsten becomes stabilized in vacancies and changes the mechanical properties of tungsten. The retention of hydrogen or helium in the W first wall could be determined by hydrogen–defect and helium–defect interaction. At elevated temperatures, point defects by mutual recombination gather into clusters and grow into the nano-voids. Tungsten and tungsten alloys are among the structural materials to operate at high temperatures in high irradiation damage environments. Tungsten ($4f^{14}5d^{4}6s^{2}$) has bcc lattice. The interactions between hydrogen and defects in metals like Be, W, V [6] plays an important role in various areas of technology of plasma machines and in future thermonuclear reactor ITER [7]. The importance of hydrogen or helium concentrations with respect to fusion reactor more systematic studies on the nano-voids containing hydrogen and helium are needed on advanced reactor materials like W, Be, V, Fe. The major difficulties in investigating intrinsic defect properties in tungsten after neutron irradiation originate from the hydrogen or helium interactions with vacancies and vacancies clusters. Hydrogen concentration as low as 10^{-5} atomic fractions causes serious mechanical changes [8]. Degradation of the mechanical properties increases with increasing of hydrogen or helium contents. Interstitial hydrogen has a high mobility in the metals [9,10]. The solubility of hydrogen in metals is usually small and it is known to increase in the presence of nano-voids. This is due to the trapping of hydrogen at these defects. The positron density distribution in a pure vacancy and vacancy-hydrogen or vacancy-helium complexes is very different [11].

The present paper deals with model calculations of the positron lifetime parameters for defects in tungsten containing hydrogen or helium and distribution of the hydrogen and helium in nano-voids. The purpose of the study is to obtain absolute values of positron lifetime parameters. Experimental methods for investigating the properties of helium and hydrogen in tungsten are field ion microscopy (FIM), TEM transmission electron microscopy (TEM), positron annihilation (PA), perturbed $\gamma\gamma$ angular correlations (PAC), electron energy loss spectroscopy (EELS), thermal desorption (THDS), nuclear scattering methods (NSC). Direct experimental information on single hydrogen or helium atoms and nano-voids containing hydrogen or helium in tungsten lattice, could be obtain only by PA, FIM and PAC. The determination of positron lifetimes for various defects in tungsten containing hydrogen or helium has not been completely performed yet both in the experiments and in the model calculations. In this respect, the main emphasis has been put on the investigation of empty nano-voids and nano-voids with hydrogen or helium and their behavior at different gas atoms concentrations. As pointed out in paper [12] the large nano-void can be considered as an internal surface. It was found that helium bubbles had large binding energies with vacancies. In the computer calculations, a cell with 2000 atoms is typically sufficient to obtain accurate results for positron lifetime and positron ground energy for a large nano-void.

2. Positron lifetime computation method

The positron lifetime and the momentum density depend on the site where the annihilation takes place. If the positron is trapped by a vacancy-type defect in a solid its lifetime increases from that of a delocalized positron due to the reduced electron density. These changes in the positron annihilation characteristics make the positron annihilation spectroscopy powerful to study the electronic and ionic structures and the associated processes of defects in solids. The goal of our work has also been to support the experimental positron research by theoretical calculations of positron annihilation characteristics. We have developed models (within the density functional theory) to describe positron states and annihilations in solids. Calculations were done in presence of different kinds of defects. In Born-Oppenheimer approximation, electrons follow adiabatically the movements of the ions. The calculation of positron states in solids has been done by the solution of Schrödinger equation for a system containing interacting electrons and ions. The time scale of electrons is several orders of magnitude smaller than the ionic time scale and therefore it can be assumed that the electrons adapt to the current ionic positions. The Born-Oppenheimer approximation allows the ionic positions to be treated as a set of parameters. The Kohn-Sham method [13] of density functional theory [DFT] considers the problem of many interacting electrons into problem of solving one-particle Schrödinger equations. This simplification makes large-scale first-principle electronic structure calculations possible by local density approximation [LDA] [13]. The Kohn-Sham equation is an effective single-particle Schrödinger equation that has exactly the same form as the Schrödinger equation for the non-interacting electrons in an external potential. The electron density the effective potential and the corresponding Hamiltonian should be calculated until the total energy does not change more than a present limit. Plane-waves are used as basis functions. Plane-waves form a complete set that is simple and unbiased. The plane-waves do not depend on the ionic positions.

The electron dynamics in tungsten is modeled by a set of parameters and analytical functions, which in DFT depend on the mutual positions of the atoms in the calculated configuration. The parameters give information about the energy and correlation forces acting between the particles. The ground state properties of the interacting electron gas were determined by the electron density. First-principles calculations can also be used to provide reliable predictions of interstitial properties that can be used in molecular dynamics [MD] and kinetic Monte Carlo [kMC] studies of point defect evolution in irradiated materials [14]. The positron trapping is reflected by a lowered annihilation rate, due to the lower electron density. The model consisted of several thousands of W atoms in a bcc lattice. We have considered 2000 particles in a cubic cell with periodic boundary conditions. The time step was 0.0025 ps, the equilibration process was 5 ps and the calculation of positron lifetimes was made after another 40 ps. The number of electrons in tungsten is large so the calculation could be treated using statistical arguments. According to DFT, the correct ground-state energy of an interacting electron gas in an external potential can be determined by minimizing the functional of the electron density. That is why we do not need the full wave function but only the electron density. In order to understand the effect of hydrogen or helium decoration of vacancies on the positron lifetime, DFT calculations have been carried out as a function of the vacancy-cluster size in tungsten and vacancy-clusters in tungsten containing hydrogen or helium. The DF calculations for tungsten have been done with linear combination of Gaussian-type orbitals (LCGTO) based on the Kohn-Sham local-density approximation (LDA) and Honenberg-Kohn [15] DF formalism in which the ground state property is a function of the electron density in position space [13,15]. The lattice distance for tungsten is 3.163 Å [5.98 au] while the distance between the bulks nearest-neighbors is 2.74Å [5.18 au]. The basis sets LCGTO are presented in Ref. [16,17]. The hydrogen and helium electrons are considered as valence electrons. The coordinate $z = 0$ corresponds to the starting point of the positron penetration through the (111) face of the tungsten bcc structure for vacancy clusters in tungsten and for clusters containing hydrogen or helium. The ratio of hydrogen and helium atoms versus vacancy atoms in nano-voids has been varied. The positions of the atoms were allowed to relax to a minimum energy configuration. Lattice relaxations have been taken into account. The lattice relaxation around a vacancy results in about 9 % reduction of the effective vacancy volume, which decreases the positron lifetime at a vacancy. In the DF, formalism [13,15] proposed by Hohenberg, Kohn and Sham the ground state energy of an interacting electron gas is a function of the electron density $n(r)$. In the variational principle, the electron wave function satisfies the following equations:

$$\{-\nabla^2 + V_{eff}[n(r),r]\}\psi_i(r) = \varepsilon_i\psi_i(r), \tag{4}$$

where ε_i and ψ_i are the eigenvalues and wave function, respectively. Equation 4 is the effective single-particle Schrödinger equation that has exactly the same form as the Schrödinger equation for the non-interacting electrons in an external potential; $n(r)$ is the electron density at the position r.

$$n(r) = \sum_i |\psi_i(r)|^2, \tag{5}$$

The effective potential is:

$$V_{eff}[n(r),r] = \phi(r) + V_{xc}[n(r)], \tag{6}$$

We consider that the density $n(r)$ and the potential $V_{eff}(n,r)$ have the periodicity of the tungsten lattice. At first approximation the energy of the tungsten atom, at side r is assumed the same as if the tungsten atom is placed into a uniform electron gas at that density. In regions far from the nucleus, the valence electrons are most important [18]. V_{eff} is experienced by the electrons and $\phi(r)$ is the electrostatic potential derived from Poisson's equation.

$$\nabla^2\phi = -4\pi[n(r) - n_{ext}(r)], \quad \textbf{(7)}$$

where $n_{ext}(r)$ is the positive-charge density distribution. $V_{xc}[n(r)]$ in equation (6) is the exchange correlation potential, which is the derivative of the exchange correlation energy $E_{xc}[n(r)]$ with respect to the electron density $n(r)$. The magnitude of the correlation energy is important for lattice energy parameters [19]. When the changes of $n(r)$ in the volume are small enough, we get a satisfactory result using the LDA [14,16]:

$$E_{xc}[n(r)] = \int n(r)\varepsilon_{xc}[n(r)]dr, \quad \textbf{(8)}$$

where $\varepsilon_{xc}[n(r)]$ is the exchange correlation energy per particle in a homogeneous electron gas of density $n(r)$. The exact form of E_{xc} is not known. The electronic structure of the point-like defect is obtained from the self-consistent solution of equation (4). The single-electron Schrödinger equation entering in the LDA is solved in the framework of the LCGTO method [16]. We calculate the density at the position of hydrogen or helium atoms in a lattice of tungsten atoms, whose electron density is known. We consider that the size of nano-void is large enough to include the density of many neighbors. The positron wave function of the defect system is obtained by a method similar to that for electrons established by Puska and Nieminen [20], which satisfies the Schrödinger equation. Kimball and Shortly [21] introduced a method where the wave electron function is obtained through an iterative relaxation scheme. To solve the Schrödinger equation the numerical method given by Kimball and Shortley [21] was used.

$$\{-\nabla^2 + V_+[n(r), r]\}\psi_+(r) = \varepsilon_+\psi_+(r). \quad \textbf{(9)}$$

where ψ_+ is the positron wave function and V_+ is the effective positron potential.

The positron lifetime is obtained by the wave functions and energy eigenvalues of the electron and positron in the defect system. The positron annihilation rate λ or the lifetime τ of a positron in an inhomogeneous electron gas is proportional to the electron density at the positron position.

$$\frac{1}{\tau} = \lambda = \int dr|\psi_+(r)|^2\Gamma[n(r)], \quad \textbf{(10)}$$

where $\Gamma[n(r)]$ is the local annihilation rate which is a function of electron density $n(r)$ and including the enhancement. Due to the attraction between the electron and the positron the electron density around a positron will be enhanced. The electron density $n(r)$ in the imperfect metal lattice is composed of free and core electrons, they are expected to react differently to the presence of positron; therefore the local annihilation rate $\Gamma(n(r))$ is divided into the following two parts. Γ is composed of two electron densities (for valence electrons and for core electrons)

$$\Gamma[n(r)] = \Gamma_V[n_v(r)] + \Gamma_{in}[n_{in}(r)], \quad \textbf{(11)}$$

3. Results and discussion

Calculations were done in presence of different defects, vacancies, vacancy-cluster and vacancy clusters containing hydrogen or helium.

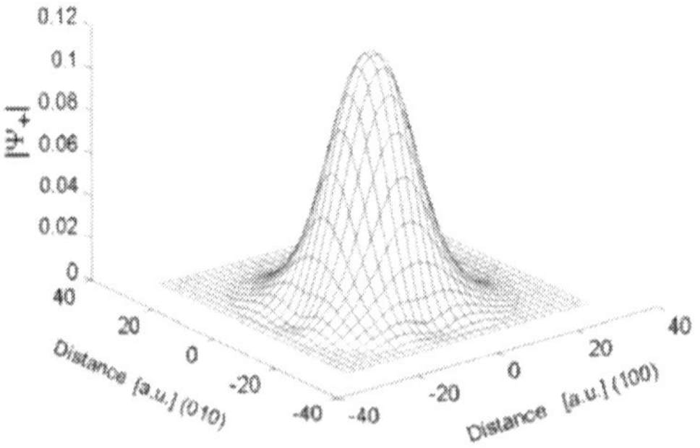

Figure 2a: Isometric plot of calculated localized wave function of positron trapped by mono-vacancy in tungsten.

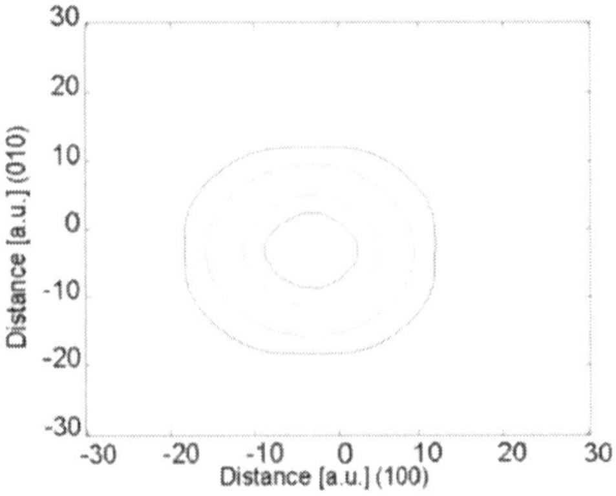

Figure 2b: Contour plot of calculated localized wave function of positron trapped by mono-vacancy in tungsten.

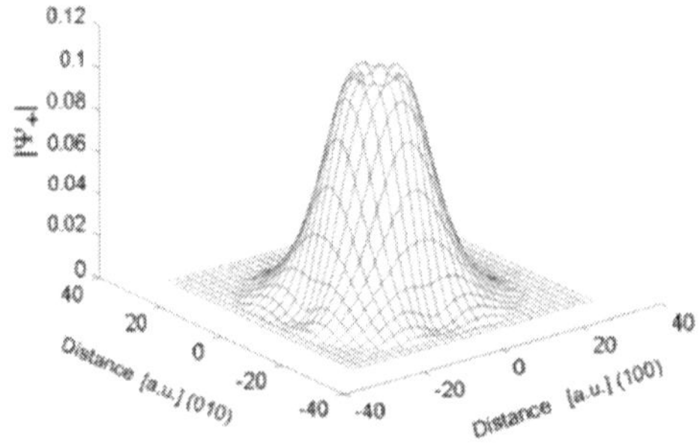

Figure 3a: Isometric plot of calculated localized wave function of a positron trapped in a mono-vacancy bound with one hydrogen atom in tungsten

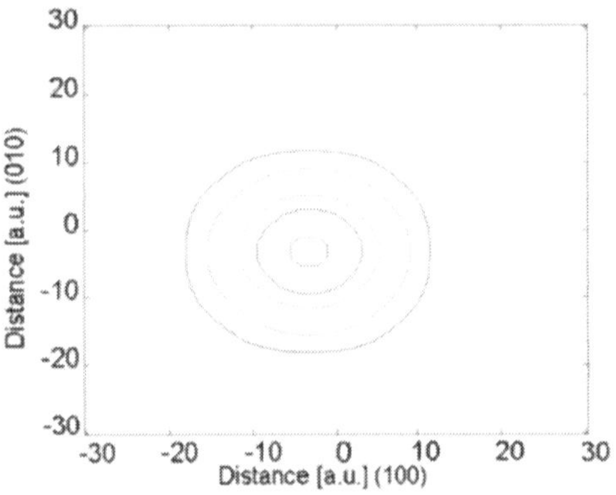

Figure 3b: Contour plot of calculated localized wave function of a positron trapped in a mono-vacancy bound with one hydrogen atom in tungsten

In figure 2a,b are shown the isometric plot and corresponding contour plot of the calculated positron density distribution [PDD] in the tungsten mono-vacancy. The contour plot of the screening potential components in the [100] plane in tungsten shows that potential is spherical close to the atomic sites. In a perfect tungsten lattice, the positron density is delocalized over the whole lattice. The calculated positron lifetime of 108 ps corresponds to those at the matrix when no defects exist in the perfect tungsten lattice. The calculated value of the positron lifetime in pure tungsten is supported by the experimental value 105 ± 3 ps reported earlier in the literature [22] both values are in good agreement. The positron lifetime calculations have been performed in bcc tungsten lattice containing vacancies, empty nano-voids and nano-voids consisting from 1 to 6 vacancies with 1 to 10 hydrogen or helium atoms. The positron lifetimes in nano-voids have been computed as a function of the number of vacancies and number of included hydrogen (nH-mV) or helium (nHe-mV) atoms, where n and m are the number of introduced hydrogen or helium atoms and vacancies number, respectively. The lifetime of trapped positron is larger than that of the positron annihilating in the free state, because the density of electrons inside a vacancy is less than that in the interstitial region. The positron lifetime was calculated as 200 ps for monovacancy in tungsten. It is evident that positron is localized at the center of the monovacancy as shown in figure 2a,b. In figure 3a,b are shown the isometric plot and the contour plot of PDD for tungsten mono-vacancy with hydrogen 1V-1H. The procedure is similar to that in paper [20]. In figure 4 are shown the results of positron lifetime calculations for different number of vacancies in tungsten lattice. Our calculations show that positron in tungsten is bound to the open defects even it contains a hydrogen or helium atom. The lifetime of positrons annihilating in voids is longer than that for mono-vacancies and gives information for presence of vacancy-cluster with two or more vacancies. Positron lifetime in tungsten has a value of 437 ps for nano-voids containing 37 vacancies. The result shows a correlation between the size of nano-voids containing hydrogen or helium and computed positron lifetime. The longer lifetime corresponds to three-dimensional vacancy-clusters, namely, nano-voids, which are observed in bcc and fcc metal lattices, like V, Fe and Mo.

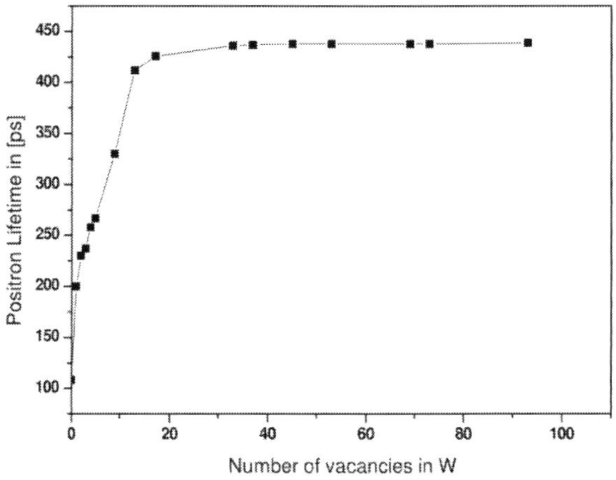

Figure 4: Correlation between positron lifetime and the number of vacancies in the nano-void

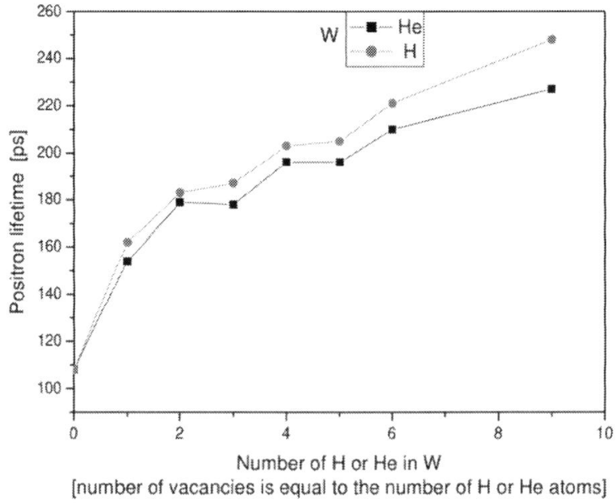

Figure 5: Correlation between positron lifetime and nano-void containing equal number of gases atoms and vacancies.

In figure 5 are shown the calculated results for vacancies containing equal number of hydrogen or helium atoms. It is seen that positron lifetime increased with the included number of hydrogen or helium atoms. Helium is mobile in tungsten below room temperature. Above 10 Å helium atoms are very stable up to temperature closed to the melting temperature of sample. Helium is practically insoluble in tungsten. Our results show that helium atoms are trapped to vacancies. A significant amount of hydrogen or helium remains in nH-mV or nHe- mV nano-voids, but these nano-voids are small and they could not be observed by TEM. The computed positron lifetime for nano-voids containing (1V + nH, 2V + nH, 6V + nH) different number of hydrogen atoms in tungsten are shown in figure 6. The obtained results present the influence of the hydrogen atoms on vacancies in the nano-voids.

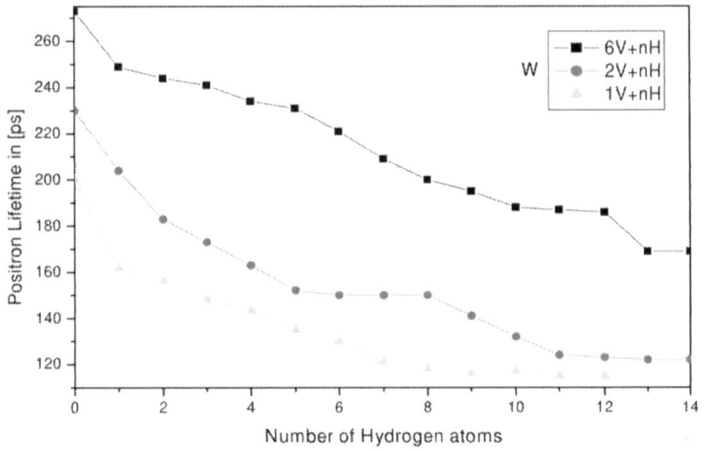

Figure 6: Correlation between positron lifetime and the number of hydrogen atoms in nano-void containing 1V+nH, 2V+nH, 6V+nH

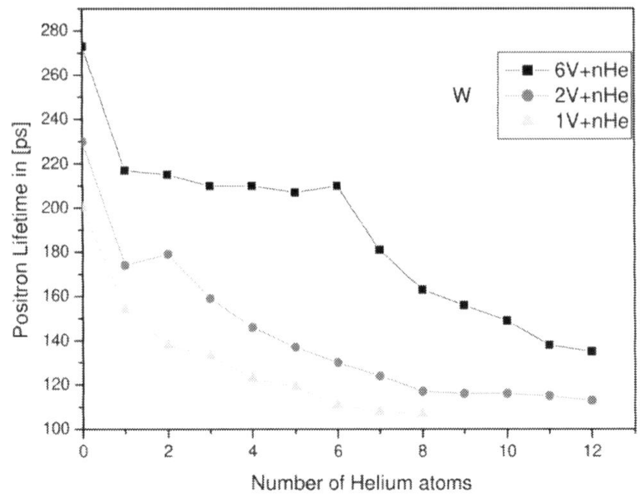

Figure 7: Calculated positron lifetime in nano-void containing 1V, 2V, 6V, and various He atoms.

Binding of hydrogen and helium atoms with vacancies in tungsten decrease positron lifetime because the introduced electron density by hydrogen or helium atoms led to increasing of the valence electron density. The positron lifetime decreases from 200 ps to 162 ps when hydrogen is bound with the vacancy (1V + 1H). It is seen that the values of calculated positron lifetimes decrease with increase of the number of hydrogen atoms. Positron lifetime decreases as 162 ps, 156 ps, 148 ps, 143 ps and 130 ps for a (1V + 1H, 1V + 2H, 1V + 3H, 1V + 4H and 1V + 6H), respectively. In figure 6, the computed value of PLT decreases with increase of hydrogen atom concentration and reaches a value of 130 ps for 1V + 6H atoms. The decrease in positron lifetime suggests an increase in hydrogen density inside of the nano-voids. The comparing of the positron lifetime for vacancy with hydrogen to vacancy in pure tungsten shows that the calculated values of the lifetimes differ. This result for first time indicates that the presence of single H atom bound with mono-vacancy affect the process of vacancies in tungsten. It is interesting to note that our results show a sensitivity of cluster for 2V containing nH atoms. The computed lifetimes for 2V + nH atoms show decrease of positron lifetime with increasing of hydrogen atoms in tungsten nano-voids. In figure 6 is observed dependence of the 6V + mH atoms from the hydrogen in the vacancy cluster in tungsten sample. Positron lifetime for vacancy-cluster 6V + 10H atoms in tungsten is 188 ps. The computer results or 1V + mH, 2V + mH and 6V + mH clearly show that increasing of hydrogen concentration in vacancy cluster leads to decrease of positron lifetime. In figure 7 are shown the results for nano-voids containing (1V + nHe, 2V + nHe, 6V + nHe) helium atoms in tungsten. The results show that binding of helium atoms with vacancies decrease positron lifetime. The obtain correlation between positron lifetime and nano-void size for first time demonstrates that positron lifetime is very sensitive method for determination of helium concentration in tungsten.

The calculated positron lifetime for various defects and voids containing hydrogen and helium in tungsten can be summarized as shown in figure 8. It is observed that with increase of nano-void size and number of hydrogen atoms positron lifetime has a higher values. We consider that positron wave function is localized at the inner space of the nano-void. When the nano-void contains hydrogen atoms, the positron wave function is localized not at the center of the nano-voids, which gives a lifetime of order 132 ps for 2V + 10H and 6V + 10H 188 ps, respectively. It is clear that PLT depends on behavior of H or He filled vacancies in these simulations.

It is observed that with increase of nano-void size and number of helium atoms positron lifetime has a higher values. We consider that positron wave function is localized at the inner space of the nano-void. When the nano-void contains helium atoms the positron wave function is localized not at the center of the nano-voids, which gives a lifetime of order [132 ps for 2V + 10H and 6V + 10H 188 ps], respectively.

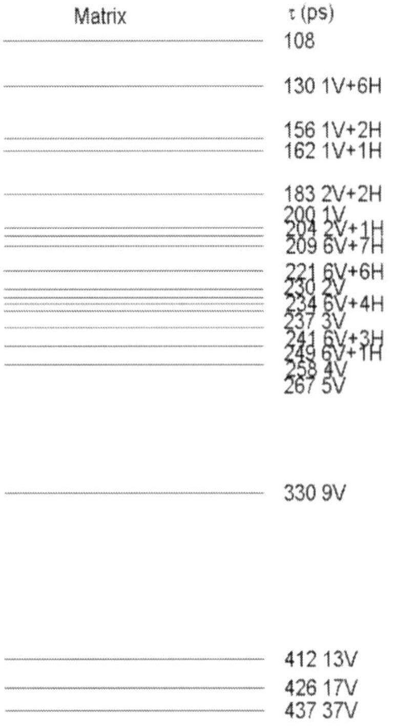

Figure 8. Calculated values of positron lifetime for various defects in W; mono-vacancy, divacancy, empty nano-voids, nano-voids containing hydrogen atoms.

4. Conclusion

This study is the first positron lifetime quantum-mechanical calculations of nano-voids and nano-voids containing hydrogen or helium in tungsten. The positron lifetime correlates with the magnitude of electron density. The study sheds light on the nano-defects formation and their evaluation in tungsten containing hydrogen or helium. It was found that a positron is bound to a metal mono-vacancy even if it contains a hydrogen or helium atoms. The model calculated positron lifetimes of nano-void containing gases atoms are shorter than in the empty void of the same size. The vacancy-clusters without hydrogen or helium are active positron traps, if once they are bound with hydrogen, they become less effective in the trapping of positrons. These results provide physical insight for positron interaction in tungsten defects and can be used for prediction of hydrogen as well as tritium-H^3 or helium-He^4 generation for the design of fusion reactors.

Acknowledgement: The EURATOM/INRNE Fusion Association supports this work.

References

[1] Schultz P J and Lynn K G 1988 *Rev Modern Phys.* **60** 71
[2] Puska M J and Nieminen R M 1994 *Rev. Modern Phys.* **66** 841
[3] Katayama K, Imaoka K, Okamura T and Nishikawa M 2007 *Fusion Engineering and Design* **82** 1645–50
[4] Lee H T, Haasz A A, Davis J W and Macaulay-Newcombe R G 2007 *JNM* **360** 196–207
[5] Xu Q, Yoshiie T and Huang H C, 2003 *NIM* B **206** 123–126
[6] Troev T, Nancov N, Petrov L and Popov E, 2008 *Reprort Letters Physics* **1** 746892
[7] Garner F A et al. 2001 *J. Nucl. Mater.* **296** 66–82
[8] Thompson A W and Bernstein I M 1986 *Effect of hydrogen behavior of materials* (AIME, New York)
[9] Sakamoto Y 2000 *Hydrogen in Metals System, Solid State Phenomena* **73-75** 137–206
[10] Ishizaki T, Xu Q, Yoshiie T, Nagata S, Troev T, 2002 *J. Nucl. Mat.* **307–311** 961

Journal of Physics: Conference Series **207** (2010) 012033 doi:10.1088/1742-6596/207/1/012033

[11] Shivachev B L, Troev T and Yoshiie T 2002 *J. Nucl. Mater.* **306** 105
[12] Puska M J 1991 *J. Phys. Condens. Matter* **3** 3455
[13] Kohn W, Sham L J 1965 *Phys. Rev.* A **140** 1133
[14] Domain C and Becquart C S 2001 *Phys. Rev.* B **65** 024103
[15] Hohenberg P, Kohn W 1964 *Phys. Rev.* B **136** 864
[16] Chiarello G, Andzelm J, Fournier R, Russo N, Salahub D R 1988 *Surf. Sci.* **202** L621
[17] Vosko H, Wilk L, Nussair M 1980 *Can. J. Phys.* **58** 1200
[18] Jansen H J F and Freeman A J 1984 *Phys. Rev.* B **30** 561
[19] Wigner E, 1934 *Phys. Rev.* **46** 1002
[20] Puska M J, Nieminen R M 1983 *J. Phys.* F **13** 333
[21] Kimball G E and Shortley G H 1934 *Phys. Rev.* **45** 815
[22] Ziegler R and Schaefer H E 1987 *Mater. Sci. Forum* **15-18** 145

Application of the Fokker Planck approach for modelling of nano-structural defects in fusion materials

Tzveta Apostolova and Troyo Troev

Institute for Nuclear Research and Nuclear Energy, 72 Tzarigradsko chaussee blvd., 1784, Sofia, Bulgaria

tzveta130468@abv.bg

Abstract. The materials used in the first wall of ITER and fusion power plants will be required to withstand neutron bombardment produced in fusion plasmas. The development of models describing the structural defects in fusion materials has become an important element in the development of a prototype of a fusion power reactor. Modeling of structural defects requires new modeling approaches giving theoretical predictions that could be experimentally validated. In this review paper, we propose the Fokker–Planck kinetic equation for modeling of nanovacancies, nanovoids and dislocation loops in fusion materials under cascade producing irradiation. It will be shown that in the simplest approximation that includes the effects of the statistical and cascade induced fluctuations, the kinetics of the evolution of the nanostructural defects can be described by the Fokker-Planck equation for the defect size distribution function. The proposed model can be applied to fusion materials under irradiation conditions that represent those of the first wall of a fusion reactor.

1. Introduction

In the Brownian motion case the complete solution of a macroscopic system would consist in solving all the microscopic equations of the system. Since this cannot be done in general a stochastic description is used instead – the system is described by macroscopic variables which fluctuate in a stochastic way.

The Fokker–Planck equation is an equation of motion for the distribution function of fluctuating microscopic variables. For a deterministic treatment the fluctuations of the macroscopic variables are neglected. By solving the Fokker–Planck equation one obtains distribution function from which any averages of microscopic variables are obtained by integration. Since the application of the Fokker–Planck equation is not restricted to systems near thermal equilibrium, we may apply it to systems far from thermal equilibrium, for instance, solids under intense laser or neutron radiation. Moreover, the Fokker–Planck equation not only describes stationary properties, but also the dynamics of systems, if the appropriate time-dependent solution is used.

The Fokker–Planck theory has been widely applied in the studies of laser interaction with dielectric and semiconducting materials. It was demonstrated [1] that when the same model system was considered in calculations using the Fokker–Planck approach (FPA) and the Boltzmann transport equation (BTE) by means of the Monte Carlo method, the results of both approaches showed good agreement in the high laser field regime. The FPA considers transport as a certain diffusive-drifting "motion" of the carriers in the energy space. An electron in the CB of a dielectric or a semiconductor can continuously gain or lose energy through interaction with phonons and the incident laser field. In this picture, the energy change of electrons is small if the laser field is not extremely strong, and the electron energy displacements are random with time. This is reminiscent of the Brownian motion of conduction electrons in energy space and Uhlenbeck and Ornstein [2] proposed a Fokker–Planck equation and applied it to the time evolution of the electron distribution function, where an assumption that electron energy does not change significantly over many collisions was introduced. The FPA method is semi-classical by nature and applicable when the average electron energy is much larger than the exchanged energy with phonons. The FPA has the advantage of being analytical, saves computational time and allows a more transparent physical interpretation.

Modeling the nanostructure evolution of materials driven far from equilibrium by intense neutron irradiation has recently become important for the development of a prototype fusion power station. Cascade irradiation produces small mobile and immobile vacancy and interstitial clusters. Recognizing the importance of stochastic fluctuations a nanostructure evolution theory can be formulated using the Fokker–Planck equation, to take into account the stochastic effects of the fluxes of mobile defects, arising from the random nature of diffusion jumps and cascade initiation.

The goal of this paper is to illustrate the successful application of the Fokker–Planck theory approach for describing the microscopic processes in dielectrics and semiconductors under laser field irradiation and to demonstrate how this theory may be used for modeling of structural defects in fusion materials under cascade producing irradiation.

2. Fokker–Planck approach for short pulse laser interaction with dielectric and semiconductor materials

Previously we have performed theoretical investigations of picosecond-to-femtosecond laser pulse interaction with selected dielectrics and semiconductors.

The Fokker–Planck quantum dynamical theory is a first-order approximation of a general semi-classical Boltzmann theory. We derive the kinetic Fokker–Planck-type equation for excited conduction electrons in bulk semiconductors systematically in the presence of a pulsed laser beyond the classical limit. We start with the total Hamiltonian of the electron–phonon system irradiated by a pulsed laser field in second quantization [3]. Initially using Fermi's Golden Rule we obtain a dynamic equation by including only the electron–phonon interaction within the first order perturbation theory. Limiting ourselves to the diffusive limit, we expand the electron distribution function and get the Fokker–Planck equation for the thermal motion of electrons in energy space. The intraband transitions of conduction electrons cannot optically respond to the spatially uniform laser field due to the impossibility of conserving energy and momentum simultaneously during the absorption of a photon by an electron, and the effect of the laser field treated classically is included in the equation using the following considerations: (a) after the laser field is applied there is no net momentum change of electrons within an interval of collision time τ_p (many time periods of the oscillating laser field) over which two successive collisions with phonons occur, and this is different from the case with a dc electric field; (b) electrons will gain net power from the laser field. In the presence of the laser field, a drift in energy space is obtained which is described by

$$\frac{\partial f^e{}_k}{\partial t} = \lim_{\Delta t \to 0}\left(\frac{1}{\Delta t}\right)\left[f^e\left(E^e{}_k - \left(dE^e{}_k/dt\right)\Delta t, t\right) - f^e\left(E^e{}_k, t\right)\right] = -\left(\frac{dE^e{}_k}{dt}\right)\frac{\partial f^e{}_k}{\partial E^e{}_k}.$$

Since the time period of the laser field $2\pi/\Omega_L$ is much shorter than τ_p, a time average is taken of the Fokker–Planck-type equation over many periods of the laser field, an averaged fluctuation which represents the average energy gained by electrons by absorbing the laser power $E_{tr} = \ll \Delta E^e{}_k \gg = \int dE^e{}_k \Delta E^e{}_k(t) f^e{}_k$ is introduced and for $E_{tr} \ll E^e{}_k$ the expansion

$f^e_k\left(E^e_k - \Delta E^e_k\right) \approx f^e_k\left(E^e_k\right) - E_{tr}\dfrac{\partial f^e_k}{\partial E^e_k}$ is used. Finally we obtain the generalized Fokker–Planck-type equation [3]:

$$\frac{\partial f^e_k}{\partial t} + V^*_k \frac{\partial f^e_k}{\partial E^e_k} - D^*_k \frac{\partial^2 f^e_k}{\partial \left(E^e_k\right)^2} = A_k f^e_k + S_k \tag{1}$$

In equation (1) we have included the coupling of the energy drift of electrons in the presence of the laser field to the phonon-assisted intraband electron transitions which gives an additional anti-diffusion current and have given analytical expressions for all the source terms up to second order perturbation theory, including single-photon absorption, impact ionization due to Coulomb interaction between electrons and holes, and non-radiative recombination due to a phonon-mediated interaction. We calculate numerically the hot electron distribution function which can be used to evaluate the electron number density, the average electron energy, and the non-equilibrium electron temperature. The magnitude of the distribution function becomes very large at high intensity I_L of the incident laser field because of the large optical absorption proportional to I_L. The increased absorption directly leads to a larger electron number density (solid curve) in figure 1. The shifted up peak of the electron distribution function in energy produces a larger average electron kinetic energy when the laser intensity becomes strong as can be seen in figure 2.

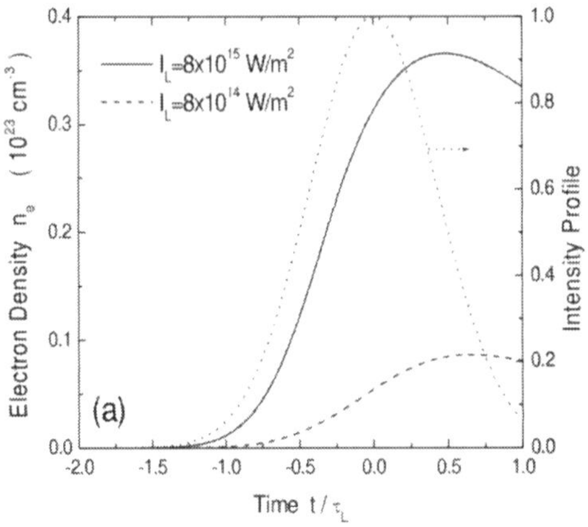

Figure 1. Comparison of the electron number density as a function of the scaled time with higher or lower laser intensity.

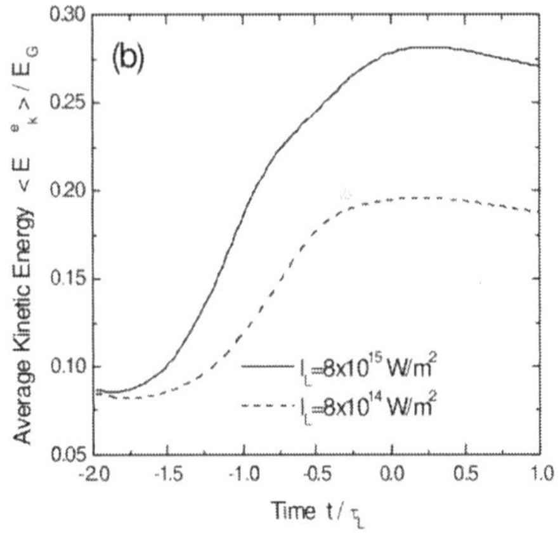

Figure 2. Comparison of the scaled average electron energy as a function of the scaled time with higher or lower laser intensity.

In addition, we have developed a model for electrons in a bulk semiconductor material interacting with an infrared laser field and an applied strong dc electric field. In the model we couple the first order force balance equation for the slow center-of-mass motion of electrons and the Fokker–Planck-type equation for the ultrafast relative scattering motion of the excited electrons beyond the relaxation-time approximation. The Fokker–Planck type equation obtained is an approximation to the Boltzmann scattering equation which we systematically derive when the Fermi energy of the electrons is much larger than the phonon and photon energies and the drift energy of the electrons [4].

We find that the drift velocity of electrons increases with the amplitude of the infrared laser field when its polarization is parallel to the dc field due to a suppression of momentum relaxation as seen in figure 3 but decreases with the infrared field amplitude for the perpendicular polarization (figure 4) due to an enhancement of momentum relaxation. The heating of electrons is described beyond the first order energy balance equation. For both cases the non-equilibrium electron temperature increases with increasing of the infrared field amplitude due to the increase of the average kinetic energy of electrons by power absorption from the applied infrared laser field [4].

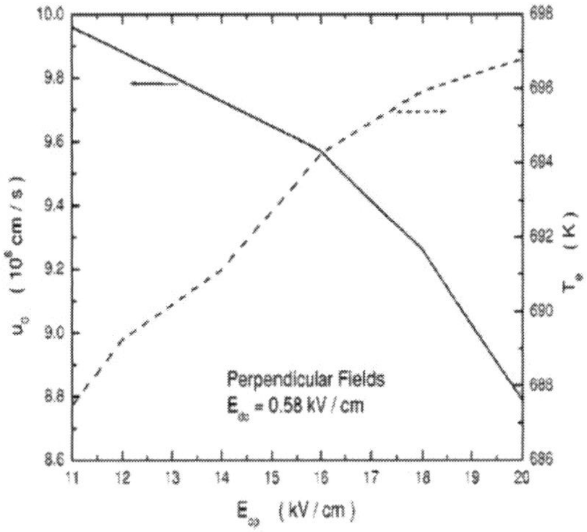

Figure 3. Calculated drift velocity and nonequilibrium electron temperature as functions of the amplitude of the incident infrared field when it is parallel to the dc field.

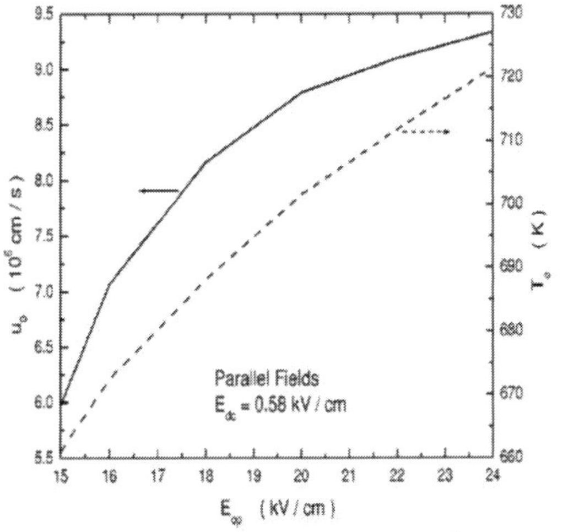

Figure 4. Calculated drift velocity and nonequilibrium electron temperature as functions of the amplitude of the incident infrared field when it is perpendicular to the dc field.

3. Nanostructure defect evolution under cascade-damage irradiation

A basic assumption implicit in the conventional theory of nanostructure evolution in irradiated materials based on the chemical reaction rate theory is that the defects (vacancies and interstitials, etc.) are produced continuously and uniformly in time and space at some appropriate mean value. In contrast, under cascade damage conditions, point defects are produced in discrete packages, randomly in time and space so the conventional picture of point-defect production is not appropriate for irradiation by neutrons and heavy ions, where point defects are produced in cascades in a highly localized and segregated fashion. Because of the random nature of both the diffusing jumps and the cascade occurrence, point defect arrival rates at sinks are also probabilistic and fluctuate all the time.

The nanostructure evolution under nonequilibrium irradiation involves random aggregation or absorption of mobile point defects plus the formation of vacancies. It can be described by the theory of stochastic processes which starting point is a master equation (e.g. Smoluchowsky–Chapman –Kolmogorov equation for a Markovian process) for the distribution of defect sizes. The master equation can be reduced to a deterministic rate equation for the concentration of specific defects if the transition probabilities in the equation are replaced by reaction rates. This mean-field approximation does not take into account the statistical nature of defect production, cascade effects, and the arrival and absorption of defects at defect clusters. One unique feature of nanostructure evolution under

Journal of Physics: Conference Series **207** (2010) 012034 doi:10.1088/1742-6596/207/1/012034

irradiation is that the size of atomic transitions is generally smaller the defect cluster size and the mobility of large defect clusters is negligible, and, hence, one can derive a Fokker–Planck equation as an approximation to the master equation.

In the simplest approximation, within which the stochastic effects due to the random nature of both the point-defect migratory jumps and cascade initiation can be included, the kinetic equation for the void evolution takes the form of the Fokker–Planck equation [5]:

$$\frac{\partial f(n,t)}{\partial t}+\frac{\partial}{\partial n}\left[V(n,t)-\frac{\partial D(n,t)}{\partial n}\right]f(n,t)=j_0. \tag{2}$$

Here $f(n,t)$ is the defect density distribution function in the space of defect sizes at time t. The drift term on the right hand side the Fokker–Planck equation describes the mean defect growth rate defined by the time-averaged fluxes of free vacancies and interstitials towards the defect cluster. The diffusivity term governs the stochastic spread of the defect size distribution function by taking into account the stochastic fluctuations in defect cluster sizes initiated by the cascades. It is related to the average point-defect fluxes and cascade properties. The last two terms in the equation are source and sink terms. They describe, respectively, the production rate of small interstitial defect clusters by cascades or by random jumps, and the rate of their removal by absorption at dislocation network and interstitial loops by sweeping.

$$V(n)=\frac{3n^{1/3}}{a^2}\left[D_v C_v - D_i C_i - D_v C^e{}_s(n)\right]$$

is the defect growth rate defined by the mean field theory. The parameters are as follows: $a=[3\Omega/4\pi]^{1/3}$ where Ω is the atomic volume, C_j is the concentration of defects, D_j is the defects diffusion coefficient $(j=i,v)$ being the indices for voids and interstitials. $C^e{}_s(n)=C_\infty\exp\left(\frac{2\gamma_s\Omega}{kTR_c}\right)\approx C_\infty\left(1+\frac{2\gamma_s\Omega}{kTR_c}\right)$ is the equilibrium concentration of defects in a neighborhood of a void with radius $R_c(n)=an^{1/3}$, C_∞ is the equilibrium vacancy concentration; γ_s is the surface tension coefficient, k and T are the Boltzmann constant and absolute temperature respectively.

In the equation (2)

$$D(n)=D^s(n)+D^c(n)+D^e(n)$$

is the "diffusivity" that describes the diffusive spread of the defect size distribution in time due to the stochastic fluctuations of point defect fluxes.

$D^s(n)=\frac{3n^{1/3}}{2a^2}\left\{D_v\left[C_v-C^e{}_s(n)\right]+D_i C_i\right\}$ is the contribution to the diffusivity due to the random migratory jumps;

$D^c(n)=\frac{3n^{2/3}}{4a}\left[\frac{G_v\langle N^2{}_{dv}\rangle}{k_v N_{dv}}+\frac{G_i\langle N^2{}_{di}\rangle}{k_i N_{di}}\right]$ is the contribution to the diffusivity due to random cascade initiation;

$D^e(n)=\frac{9n^{2/3}D_v C_s(n)}{2a^2}$ is the contribution to the diffusivity due to random emission of point defects. In the expressions above N_{dj} is the average number of point defects, $\langle N^2{}_{dj}\rangle$ is the average square number of point defects and k_j^2 is the total sink strength for point defects of type j [6].

The term on the right hand side of the Fokker–Planck equation describes the rate of production of small interstitial clusters by cascades and the rate of their removal by absorption at dislocation network and interstitial loops by sweeping. For simplicity it is assumed that the initial sizes of the defects are the same (n_0 being the initial number of defects) which is reflected in the δ function.

$$j_0 = \frac{\varepsilon_i G}{n_0}\delta(n - n_0) - (\rho_N + \rho_l)Wf(n,t),$$

where ε_i is the fraction of point defects immobilized in primary clusters, $G = Q\Omega N_d$ is effective generation rate of free point defects (Q – mean number of cascades per unit volume per unit time, N_d – average number of free point defects), ρ_N and ρ_l are the dislocation densities for the network and the interstitial loops, respectively, and W is the reaction constant between the primary clusters and the sweeping dislocations [7].

Vacancy clusters consisting of several vacancies are mobile and a void shrinking below a minimum size $n_{\min}$ can no longer be treated as a void. In this case, the left boundary condition of the Fokker–Planck equation should be a vanishing size distribution function at any time and for defect size equal to the minimum one: $f(n = n_{\min}, t) = 0$.

To ensure a finite total defect number density the right boundary condition of the Fokker–Planck kinetic equation should be a vanishing size distribution function at any time and for defect size equal to infinity [8]: $f(n \to \infty, t) \to 0$.

4. Conclusions

On the basis of the Fokker–Planck approach (FPA) used for describing the microscopic processes in dielectric and semiconductor materials irradiated by short pulsed laser we demonstrate that FPA can be applied for studying nanostructural defect evolution under cascade-damage irradiation in different ITER fusion materials.

The FPA can be used as an alternative method to Density functional (DFT) method for describing the defect evolution in fusion materials. The numerical results of the two methods can be compared.

Acknowledgments

One of the authors T. A. was supported by the National Research Council of the USA while doing part of this research. The author also wishes to especially acknowledge her co-authors in the referenced papers Dr. Danhong Huang, Dr. David Cardimona and Dr. Paul Alsing from the United States Air Force Research Laboratory for their contribution in the research. This work was supported partly by the Fusion Association EURATOM/INRNE.

References

[1] Bringuier E 1997 *Phys. Rev. B* **56** 5328; Bringuier E 1998 *Phys. Rev. B* **57** 2280
[2] Uhlenbeck G E and Ornstein L S 1930 *Phys. Rev.* **36**, 923
[3] Apostolova T, Huang D, Alsing P, McIver J, Cardimona D A 2002 *Phys. Rev. B* **66** 075208
[4] Huang D, Apostolova T, Alsing P, Cardimona D A 2004 *Phys. Rev. B* **69** 75214
[5] Semenov A A, Woo C H 1999 *Appl. Phys. A* **69** 445
[6] Semenov A A, Woo C H 2002 *Phys. Rev. B* **66** 024118
[7] Semenov A A, Woo C H 1993 *J. Nucl. Mater.* **205** 74
[8] Semenov A A, Woo C H 2006 *Phys. Rev. B* **74** 024108

Numerical Simulations of Displacement Cascades in Irradiated with Intensive Neutron Flux Iron

N Nankov, L Petrov, E Popov and T Troev

Institute for Nuclear Research and Nuclear Energy, Bulgarian Academy of Science, 72 Tzarigradsko Chaussee Blvd., 1784 Sofia, Bulgaria

nnankov@inrne.bas.bg

Abstract. Numerical calculations of the radiation defects created in iron irradiated by fast neutrons (E > 0.1 MeV) were carried out. The total atom damages in iron at a neutron flux of 1.4×10^{15} n/cm^{2}s^{-1} were determined to be 37.2 displacements per atom (dpa) per full power year (fpy). The helium concentration of 375 appm/fpy was obtained in irradiated iron for the same total neutron flux. A multi-scale numerical model has been developed in order to obtain better values for the neutron damages and the concentration of impurities. Using this approach the concentration of Fe, Mn and Cr isotopes as well as the concentration of helium and tritium has been determined as a result of the fast neutron irradiation.

1. Introduction

Iron is considered as one of the major materials for the first wall of the upcoming new generation fusion prototype reactors ITER and DEMO aimed at demonstrating the technical feasibility of electrical power production by means of (D,T) fusion reaction:

$$^{2}\mathrm{H} + {}^{3}\mathrm{H} \rightarrow {}^{4}\mathrm{He}\,(3.5\,\mathrm{MeV}) \; + \; {}^{0}\mathrm{n}\,(14.1\,\mathrm{MeV}).$$

Iron has high radiation resistance, high chemical stability, low hydrogen isotope retention and good high temperature mechanical strength especially for fusion applications. Its function would be the protection of actively cooled wall structures from high heat fluxes in order to sustain the component life time and the plasma compatibility. Another reason for the selection of Fe would be its ability to produce only stable or very short lived Mn and Cr isotopes due to nuclear transmutations caused by fusion neutrons. Additionally iron and the stainless steel are widely used low cost materials which would significantly lower the costs of the first wall (FW) blanket. In order to test the iron properties under intensive fast (E > 0.1 MeV) neutron irradiation numerical as well as experimental studies are necessary. Several nuclear facilities are proposed [1] for experimental studies of irradiation effects on material due to the neutron flux from ITER and DEMO. Another approach relays on numerical simulations of the iron response to the intensive high-energy neutron irradiation. The purpose of our study was the numerical calculation of the damage of Fe sample caused by intensive fusion neutron irradiation. In order to achieve our goal we have used the multi-scale approach [2]. First we have simulated the interaction of neutrons with macroscopic iron target at high energy (E > 0.1 MeV). For this purpose the fusion neutron spectrum data was taken from the experiment as an input for the simulation. We have recorded the initial energy and position of iron primary knock-on atoms (PKA)

Journal of Physics: Conference Series **207** (2010) 012035 doi:10.1088/1742-6596/207/1/012035

and of all isotopes created in the nuclear reactions at the output of the calculation. The obtained results allowed us to calculate the concentration of Fe, Mn and Cr isotopes as well as the concentration of helium and deuterium and used it as an input for the next step in our simulation. Using Monte-Carlo (MC) simulation we have calculated the atom displacements caused by iron PKA. At the end of the simulation the final position of the atoms involved in damage cascades have been recorded. Finally, statistical methods were applied to calculate the atom displacements and the number of the produced vacancies and interstitials as a function of the PKA energy for the created isotopes.

2. Methodology of the simulation

Different approaches are necessary for the simulation of the beryllium behavior under irradiation with high energy (E > 0.1 MeV) neutrons and high fluence up to 8×10^{19} n/cm^2 for various energies and time scales. ^{4}He, ^{2}H and isotopes like ^{55}Mn and ^{53}Cr created by inelastic neutron capture have to be precisely determined. In the present work the initial neutron energy distribution for the simulation was taken from [3]. The typical neutron spectrum of D-T reactor for high energies (E > 0.1 MeV) is shown in figure 1.

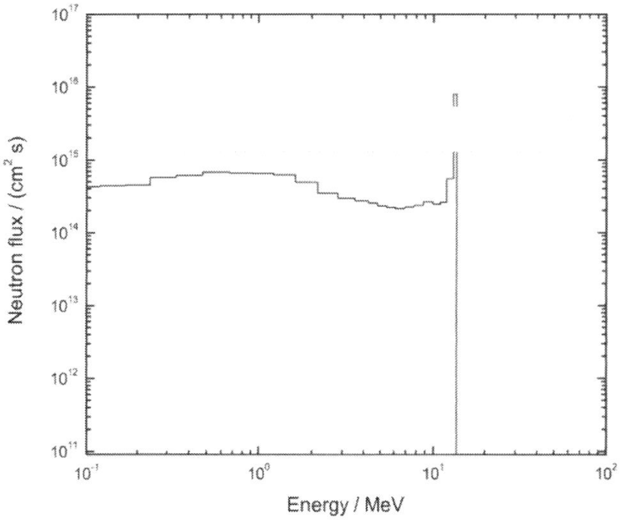

Figure 1. Total neutron flux from plasma in D-T reactor.

One can see that the sharp neutron peak of the fusion neutrons is located at 14.1 MeV and its intensity is more than one order of magnitude higher than the intensity of the low energy neutrons. One can see on the other hand that the variation of intensity of the neutrons with lower energies is small and they can be considered as a background. One can expect from the spectrum that most of the radiation damages are due to fusion neutrons at 14.1 MeV. That is why in our calculations we have taken into consideration only neutrons with energies up to 14 MeV. We created a computer Monte-Carlo (MC) code based on the Geant4 CERN package [4] to simulate ion and neutron interactions in Fe homogeneous target at high energies (E > 0.1 MeV). The processes like neutron elastic and inelastic scattering, neutron capture with and without emission of charged particles were taken into consideration in the Be target. The low energy processes like gamma ray and secondary electron emissions as well as all types of electromagnetic interaction of charged particles and photons with matter were included in the present calculations. The high precision database for low-energy neutron cross section NDL ver. 3.20 has been used with the purpose to achieve high accuracy of the neutron interaction in Fe. Our computer simulations were carried out for macroscopic cylindrical Be target with diameter of 60 mm and length of 50 mm. The Fe PKA for the damage cascades due to the neutron scattering and capture were considered for small cylindrical volume with diameter of 30 nm and length of 30 nm. It has been aligned with the beam axes and positioned at 1 cm behind the target surface. The total neutron fluence used in the simulation was of about 8×10^{19} n/cm^2 ($E > 0.1$ MeV). Nuclear transmutations which play an important role in helium, deuterium and tritium formation were also taken into account. The interaction of fast neutrons in the Fe target which were taken into consideration in our code can be expressed by following reactions:

$$^{56}\mathrm{Fe} + {}^{0}\mathrm{n} \rightarrow {}^{56-xn}\mathrm{Fe} + x{}^{0}\mathrm{n}$$
$$^{56}\mathrm{Fe} + {}^{0}\mathrm{n} \rightarrow {}^{55}\mathrm{Mn} + {}^{2}\mathrm{H}$$
$$^{56}\mathrm{Fe} + {}^{0}\mathrm{n} \rightarrow {}^{53}\mathrm{Cr} + {}^{4}\mathrm{He}$$

In figure 2a is shown the total neutron interaction cross section in ^{56}Fe taken from the ENDF/B database and used in our simulation. The resonance structure is dominant for the neutron interactions at energies lower than 3 MeV. At neutron energies higher then 5 MeV up to 14.1 MeV the neutron interaction cross section is almost constant and has the value of about 5 barn. In figure 2b are shown the production cross sections of ^{4}He and ^{3}H used in our numerical modeling. One could see that the ^{4}He production has a maximum at the energy of the fusion neutrons which is 14 MeV. The ^{3}H production has a threshold of 12 MeV and therefore it takes place in the fusion reactor. As a result the concentration of ^{3}H in the fusion neutron irradiated ^{56}Fe is expected to be several orders of magnitudes lower than the concentration of ^{4}He. The integral cross section for ^{4}He production is about 200 times lower than the total neutron interaction cross section in ^{56}Fe. Therefore one could expect that the most of the created defects are due to Fe recoils as a result of neutron elastic and inelastic collisions.

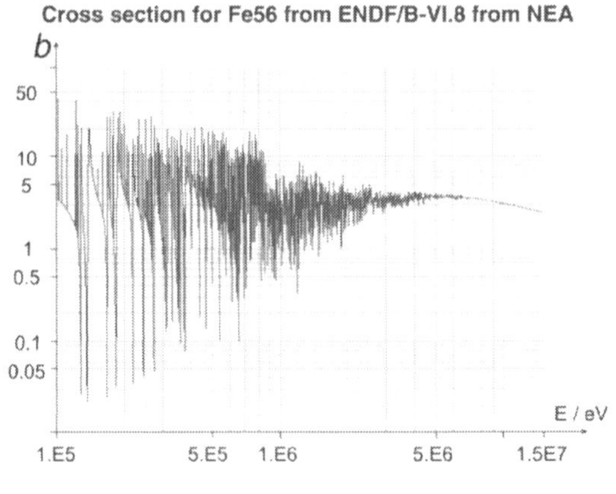

Figure 2a. Total neutron interaction cross section.

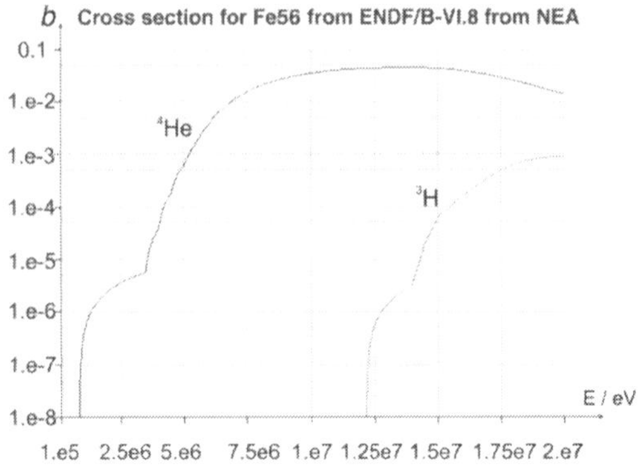

Figure 2b. Production cross sections of ^{4}He and ^{3}H for neutrons at different energies.

The next step in our computer simulations was to carry out structural damage calculations using the results from the described high-energy neutron interactions (figure 3). For this purpose we used the binary collision approximation (BCA) and the SRIM MC code [5] because it relays on the universal pair potential of Ziegler, Biersack and Littmark [6,7]. The SRIM code allows us to calculate the stopping of the PKA at energies higher than 10 keV/u with accuracy of 10%. In this approach the crystal structure of Fe has been neglected. Iron has been considered as a homogeneous continuous medium. Every PKA from the high energy neutron interaction with Fe atoms was tracked down to its stop in the medium. According to the model, the target atom which is hit by PKA and starts a recoil cascade is identified with its recoiling energy. For each cascade we consider the displacement collisions, vacancy production, replacement collisions and interstitial atoms, as described below. The number of displacement collisions records how many target atoms were set in motion in the cascade with energies above their displacement energy which is specified at the input. The Fe target vacancies are next items which were recorded.

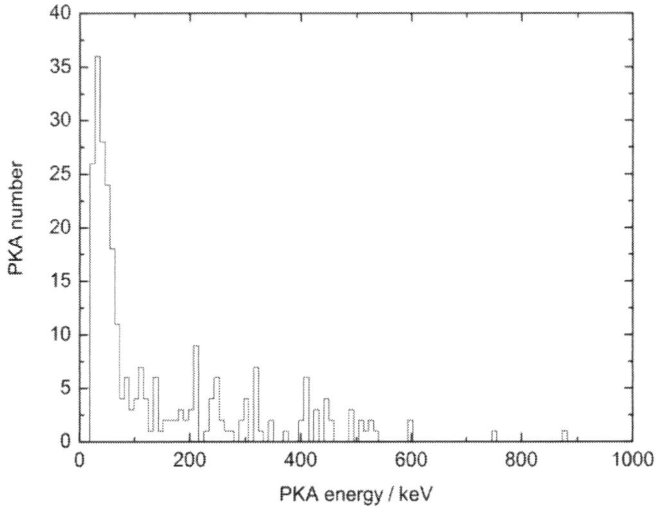

Figure 3. PKA energy distribution in 30x30x30 nm^3 simulation domain after the neutron irradiation.

The number of Frenkel pairs N_{NRT} produced per cascade can be estimated by the NRT formula [8]:

$$N_{\text{NRT}} = 0.8E_{\text{D}} / 2E_{\text{d}},$$

where E_d is the average displacement energy for all crystallographic directions and E_D is the damage energy which is the fraction of recoil energy that goes into displacement damage, after subtracting the portion dissipated for electron excitation. In our simulation the average displacement energy E_d has been chosen to be 40 eV. The lattice binding energy was taken to be 3 eV equal to the cohesive energy of the Fe crystal. At the end of every cascade the ion type, the PKA energy, the recoil energy and the atom positions have been recorded for further processing.

3. Results and discussion

Our SRIM-based calculations show that the damages, produced by fast neutron fluence up to 8×10^{19} n/cm^2, correspond to 0.151 dpa. The total damages of 37.2 dpa/fpy were calculated using the NPRIM code [3]. ^{4}He and ^{53}Cr concentration was calculated to be about 0.4 appm. According to calculations using the NPRIM code [3] it reaches 375 appm/fpy. Those values are in a good agreement with the results, obtained in [1]. The ^{2}H and ^{55}Mn concentrations were determined to be about 7% of the He concentration.

In order to investigate the influence of the energy distribution of the ^{56}Fe PKAs, created as a result of fusion neutron induced nuclear reactions, we have calculated the total number of defects in displacement cascades as a function of the PKAs energy. The result is shown in figure 4.

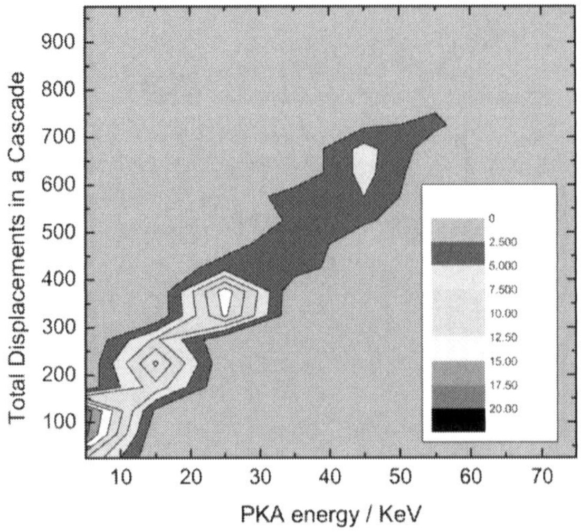

Figure 4. Correlation between PKA energy distribution and the total displacements in a cascade after the neutron irradiation.

One could see that most damages are created by PKAs with energy up to 100 keV. Most probable cascades have between 50 and 150 displacements. We have found that they are well concentrated in the PKA energy range up to 10 keV. The maximal damage is caused by PKAs with energy around 30 keV. PKAs with higher energies could produce more displacements but the number of such cascades is small probably due to the fact that the most efficient energy transfer in ^{56}Fe occurs at lower energies.

If the PKA has enough energy it can produce secondary recoils which could initiate additional cascades. In our simulation we obtained values for the number of such sub-cascades as a function of the PKA energy. The results indicate that ^{56}Fe PKA energies up to 50 keV have higher probability to induce sub-cascades. We determined that the most probable number of sub-cascades is around 150. Similar results are discussed in [1,9]. The authors pointed out that for PKA with energy higher than some critical value in the range of 10–40 keV the formation of sub-cascades is more probable. This result is in agreement with our numerical calculations. We established that low energy cascades ($E <$ 100 keV) are responsible for about 90% of the defects production.

In our simulation we investigated the vacancy and self-interstitial recombination due replacement collisions. We found that the number of recombination processes was about 2.5 % of the total number of displacements.

In the last step in our multi-scale model calculation we determined the depth profile of the vacancy distribution inside the ^{56}Fe sample. The result is shown in figure 5. It indicates that although the damages are created in a volume of $30\times30\times30$ nm^3 the damage extends up to 400 nm. This is due to the large number of displacements in the cascades as well as to the large number of sub-cascades. The distribution has a sharp maximum in the area where PKAs are created as a result of the neutron interactions. As we already shown this fact can be explained with the large probability of the damage creation due low energy PKAs. The small maximums in the distribution indicate the presence of sub-cascades. All defects are attractive trapping centers for He and H atoms and their properties could be studied by means of precise non-destructive experimental techniques like positron annihilation spectroscopy [10] or transmission electron microscopy (TEM) for point defects larger than 2 nm.

4. Conclusions

Numerical simulations of creation of point defects in ^{56}Fe, irradiated by intensive fast ($E > 0.1$ MeV) neutron fluence up to 8×10^{19} n/cm^2 has been performed. The neutron spectrum represents the energy

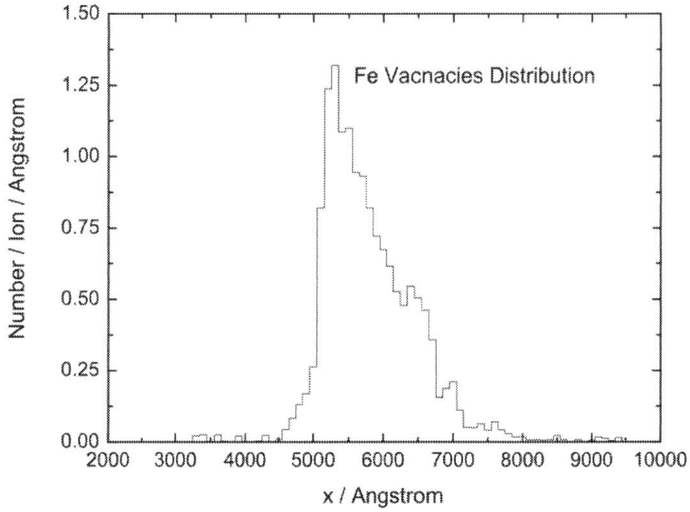

Figure 5. Vacancy distribution in ^{56}Fe sample after the fusion neutron irradiation

structure of the fusion neutrons and the 14.1 MeV peak. Neutron interactions with ^{56}Fe like elastic and inelastic scattering, gamma ray emission and nuclear transmutations have been considered. Our multi-scale numerical model includes two steps. In the first step precise calculations of the PKA energies and positions for Fe recoils have been made. The calculated damage of 0.151 dpa and 37 dpa/fpy was compared to the available data in the literature and a good agreement was found. He content was determined to be 0.4 appm which is in agreement with the available experimental measurements and calculations. ^{2}H content was estimated to be 7% of the He content. The correlation between the PKAs energy and the damages was established. It was shown that the most vacancies are created by PKAa with energy below 50 keV with maximum around 30 keV. In the second step of our numerical modeling the depth profile of the vacancy distribution in the radiated sample was determined. It shows sharp maximum at the depth where PKAs were created as a result of the neutron irradiation.

5. Acknowledgments

This work was supported by European Fusion Development Agreement (EFDA) collaboration and the European Commission. The authors gratefully acknowledge K. Berovski and S. Peneva for their participation in discussions and their contribution to the technical preparation of the manuscript.

References

[1] Vladimirov P, Möslang A 2004 *J. Nucl. Mat.* **329-333** 233–237
[2] Troev T, Nankov N, Petrov L and Popov E 2008 *Research Letters in Physics* **2008** 746892 (doi:10.1155/2008/746892)
[3] Shimakawa S et al. 2002 *Proc. of the 11th International Symposium on Reactor Dosimetry,* (Brussels, Belgium, REACTOR DOSIMETRY IN THE 21ST CENTURY)
[4] Inguimbert C and Gigante R 2006 *RIEEE Transactions on Nuclear Science* **53** 270
[5] *SRIM-2006*, www.srim.org
[6] Ziegler J, Biersack J P, Littmark U 1985 *The Stopping and Range of Ions* (Vol. 1, Pergamon, New York)
[7] Robinson T, 1992 *Nucl. Instr. Meth. Phys. Res. B* **67** 396
[8] Norgett M J, Robinson M T, Torrens I M 1975 *Nucl. Eng. Design* **33** 50
[9] *Materials Assessment Report* (G 74 MA 10 01-07-11 W0.2 Ch 2.1 pp 29)
[10] Troev T et al. 2006 *J. Nucl. Mater* **359** 93–101

Third International Workshop & Summer School on Plasma Physics 2008 IOP Publishing
Journal of Physics: Conference Series **207** (2010) 012036 doi:10.1088/1742-6596/207/1/012036

The distributions of plasma parameters in the "Magnetor" device with hot cathode discharge

I A Shchelkanov[1,3], D S Naumov[1], G V Krashevskaya[1], M M Tsventoukh[1,2], D E Vovchenko[1], G V Khodachenko[1], V A Kurnaev[1]

[1]Moscow Engineering and Physics Institute (State University), 115409 Kashirskoe shosse 31, Moscow, The Russian Federation

[2]P N Lebedev Physical Institute RAS, 119991 Leninsky ave. 53, Moscow, the Russian Federation

E-mail: adwier@rambler.ru

Abstract. The dipole magnetic configuration as realized in planet magnetospheres is of interest for controlled fusion plasma confinement. This type of magnetic trap is free of convective instability. Distributions of plasma parameters in "Magnetor" device with double dipole magnetic configuration that allows restrict considerable volume of confined plasma are investigated. The comparison of plasma parameters distribution inside the trap for two plasma generation methods: microwave discharge at 2,45GHz and hot cathode discharge in crossed E and B fields are made. It is shown that plasma is localized inside the magnetic separatrix for both types of discharges. To determine possible additional plasma movements caused by the presence of the electric field in the magnetic configuration measurements of plasma flows in discharge with the hot cathode in crossed E and B fields are carried out. Two directions of plasma flows with flux density $10^{16}\div10^{17}cm^{-2}s^{-1}$ inside the magnetic trap are determined: the flow along the axis leading particles leaving from a trap through weak magnetic field region and the flow caused by plasma azimuth drift.

1. Introduction

Investigation of new magnetic configurations for thermonuclear plasma confinement is still an open problem. The main idea of such investigations is to develop magnetic configurations suppressing plasma instabilities. One of them is a dipole magnetic configuration realized in the planet magnetosphere.

A dipole magnetic configuration created by DC coil current is stable against the most hazard instability - convective instability. The criteria of convective stability was proposed in [1,2]. According to this criteria $\nabla P \nabla U < \gamma P(\nabla U)^2 / U$, where $U = -\oint dL / B$ integral along magnetic field line, P – plasma pressure, γ – adiabatic index, B – magnetic field, plasma is stable against interconvertible – convective disturbance, if the plasma pressure gradient lower than magnetic field pressure gradient.

Investigation of dipole magnetic configuration is carried out in the USA (CTX [3], LDX [4]) and the Japan (RT devices [5]).

A new plasma device "Magnetor" [6] with the dipole magnetic field restricted by an additional DC coil current has been created in Moscow Engineering and Physics Institute (State University) in 2003. In all these installations the plasma generated by the electron-cyclotron resonance (ECR) discharge at low pressure is investigated. Here the fast electron fraction (with energy up to hundreds keV) creates the most of plasma pressure comparable with magnetic field pressure what allows to work in high β modes. The opportunity of obtaining the plasma configuration, in which additional density is concentrated alone magnetic surface, is of interest. The present work is devoted to the realization of plasma filling of the allocated magnetic surface via discharge with the hot cathode and to research of the key characteristics of a plasma configuration in comparison with the case of ECR plasma generation.

2. Description of the installation and plasma generation methods

The scheme of the installation is presented in the Figure 1.

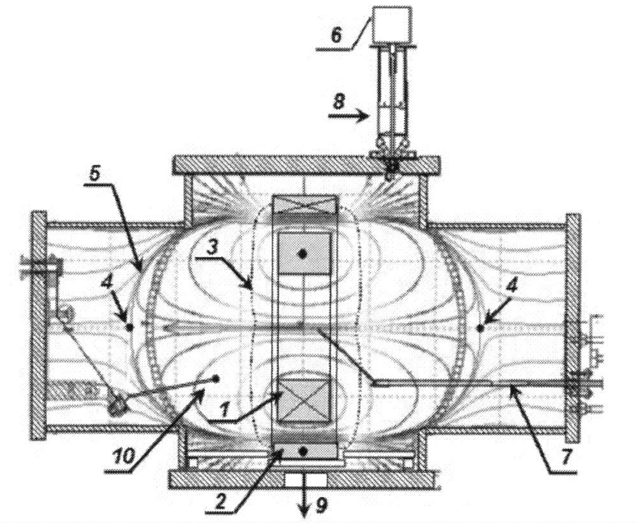

Figure 1. The scheme of "Magnetor" device: 1- the main coil, 2- the additional coil, 3- ECR surface (fundamental frequency B_{res}=875Gs), 4 – magnetic fields zero, 5- separatrix, 6 – magnetron, 7 – movable probe, 8 - microwave input, 9 – vacuum pumping, 10 – cathode-emitter

The magnetic system is located inside stainless steel vacuum chamber with volume of 0.12 m^3. It consists of the two DC current coils: the main one and the additional. These coils have the same axis and placed in one plane. The additional coil is oppositely directed to the main coil current. The magnetic field of the additional coil restricts the dipole magnetic field, created by the main DC current coil. DC currents are limited to 40 kA and -20 kA in the main and the additional coils correspondingly. There are two magnetic plugs: the first one is in the center of magnetic system (~1,3 kGs) the second is in the region between coils and its maximal magnetic field intensity (~1.8 kGs). Here, in ring region between coils with strongest magnetic field, the main coil supports are placed. Magnetic coils are isolated from the vacuum chamber.

The magnetic field has separatrix – the magnetic field surface separating magnetic field lines closed around the main coil from those being closed around additional coil. Separatrix passes through zero magnetic field points. Current values in each magnetic coil are set to create the magnetic field with the sepratrix inside the vacuum chamber.

Plasma-forming gases are air and argon, working pressures range is 10^{-5} - 10^{-3} Torr.

There are two methods of plasma generation.

Primarily, plasma is created by microwave low pressure discharge using the electron-cyclotron resonance (ECR) condition. The microwave source is the magnetron with power 900W, at frequency 2.45GHz. The microwave pulse length is 10ms with 50Hz repetition frequency. The antenna loop is used for the microwave power input into the vacuum chamber. The plasma source is the resonance surfaces crossed all magnetic lines inside the separatrix. The fundamental frequency 2.45GHz corresponds to the field strength of 875Gs. Fundamental resonant field surfaces (B_{res}=875Gs) have the shape similar to the disks placed at both sides of the coils plane. Higher harmonics resonant field surfaces ($B_2=B_{res}/2=437.5$Gs, $B_3=B_{res}/3=291.7$Gs etc.) have the similar shape and are placed separately from the coils plane. As it was estimated experimentally the absorbed microwave power is about 200W (~20% of total) per each 10ms pulse.

Additionally to the microwave ECR plasma source the system of discharge with hot cathode was developed for abilities to an additional filling of one narrow magnetic surface and also for the investigation of ExB effects on the plasma confinement. The hot cathode is tungsten spiral filament of 1cm longways. It is located inside the separatrix, the anode is the vacuum chamber. The position of the hot cathode is set to get the largest plasma generation region. The power supply provides negative bias on the cathode up to the -1000 V, and current up to 5 A.

Diagnostics of plasma is carried out by the movable probes. Probes allow obtaining information on local plasma parameter and its distribution and provide adequate accuracy of relative measurements.

3. Experimental results

3.1. The electron temperature distribution and plasma concentration distribution inside the separatrix

Experiments on research of microwave discharge plasma parameters were carried out under the following conditions. The currents were in the main coil 33 kA and in the additional one 17 kA, thus the maximal value of a magnetic field in a ring plug between the coils was 1.5kGs. The plasma-forming gas was air, the working pressure was maintained at $5\cdot10^{-5}$Torr.

The plasma ignition experiments in crossed ExB fields with the heated cathode currents in coils had values 17kA for the main and 9kA for the additional, so the maximal value of the magnetic field in the trap was 780 Gs. The region of magnetic field and electric field intensity values for an opportunity of discharge ignition in crossed ExB fields was defined according to theory Haefer [7] and Redhead [8]. The voltage -780V was supplied on the cathode relative the vacuum chamber wall, the discharge current of has been limited by 300mA. The hot cathode was situated in coordinate z=-130mm, r=50mm in the different part of magnetic trap relative to the probe system. Plasma-forming gas was air, working pressure was maintained $7\cdot10^{-5}$Torr. Attempts to ignite discharge at low pressure with the cold cathode or without imposing an electric field have come to the end with failure.

The working pressure and coils temperature were controlled during the experiments.

Double cylindrical and plate probes were used for diagnostics plasma parameters. The length of the probe was 5 mm, diameter of cylindrical probe was 0.4 mm, the width of the plate probe was 5 mm, thickness was 0,1 mm. The distance between probe electrodes was 5 mm. The criterion of double probe application possibility is an absence of probe electrode shielding one by another. This criterion is satisfied if the distance between probe electrodes more than 10 Debie radiuses.

Magnetic field inside the "Magnetor" device is strongly not uniform. It varies from from ~1 Gs to ~1.5 kGs. Therefore the ratio of the electron Larmour radius to the probe radius is changed from 0.3 to 280 (expected electrons temperature is about ~10eV). So in the region of the magnetic field value higher than 700 Gs electrons become magnetized for the probe, therefore it is possible mistake in measurement of temperature here. Ions are not magnetized in any point inside the magnetic configuration.

According to the experimental data the electron temperature is constant inside the separatrix. In the region outside of the separatrix measurements were not carried because of the low plasma concentration value. Low plasma density is the reason of overlapping electrode sheath and it is impossible to define the temperature. The value of electron temperature in the case of ECR discharge is 10 eV and in the case of the hot cathode discharge is 8 eV.

The density of plasma has been calculated from the ion saturation current measurements according to the Bohm formula in the assumption of the electronic temperature constancy. Measurements of an ion current were carried out at negative shift of -100V on a probe relative to reference electrode.

The results of plasma density distribution measurements inside the separatrix are in the figure 2.

The obtained density distributions of the microwave plasma and plasma discharge in crossed ExB fields with the hot cathode are shown on figure 2 and figure 3 accordingly.

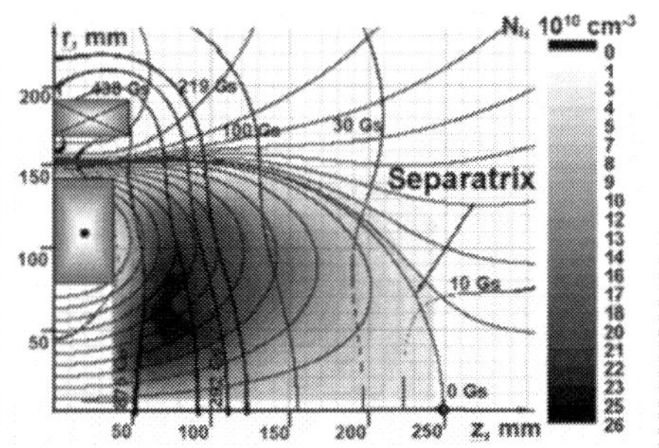

Figure 2. Density distribution of the microwave plasma inside separatrix

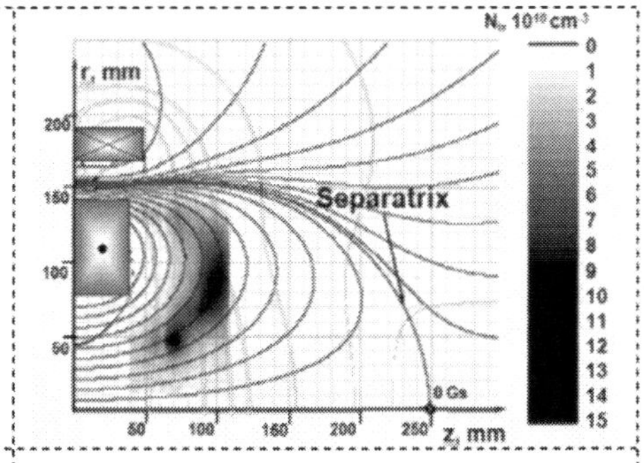

Figure 3. Density distribution of the plasma discharge in crossed ExB fields

It can be seen that the plasmas of these discharges are localized inside the separatrix. The reducing of the plasma density in the direction of the magnetic system axis and in a direction of magnetic field coils is observed in both cases. The reducing of microwave discharge density in the direction of magnetic field coils is connected with the presence of magnetic field lines closed on the main coil bandage. Here, at high field side, the low mirror ratio (< 3) lead to the particle losses on a surface of the main coil. The particles diffusion towards the region of high magnetic field at the hot cathode discharge is highly difficult; therefore plasma filling of this region does not occur. Also one can see that the magnetic surfaces located closely to the cathode are mainly filled.

3.2. Plasma flow investigation inside the magnetic configuration

The presence of the electric field in the magnetic configuration can lead to the additional to the diffusion directed particles and plasma movement in ExB fields (in particular – the plasma rotation). For the experimental plasma flux characterization the Mach probe technique was used. Two plane probes separated by the insulator (figure 3) must be directed normally to the plasma flux direction, so one of probes is placed at the upstream side and another one is placed at the downstream side. As the plasma source the hot cathode discharge was used.

From each probe the ion saturation currents are measured. The ratio of the upstream and the downstream currents to the probes allows to define the plasma flow velocity via formula follows from the numerical investigations [9]:

$$\frac{J_{up}}{J_{down}} = \exp(K\frac{v_d}{c_s}) = \exp(KM) \tag{1}$$

J is the current density, K is the calibration factor, depending on plasma conditions, v_d is the plasma flow velocity, c_s is the ion sonic speed.

The formula for K factor estimation has been evaluated by Hutchinson calculations [10]:

$$K = \frac{1}{(0,43\cdot\sqrt{1+\frac{Ti}{Te}})}, \tag{2}$$

where T_i and T_e are the ion and electron temperatures.

The ion temperature was not measured. It is assumed Ti = 0,025эВ. Therefore for parameters plasma of hot cathode discharge according to (2) the factor K is about 2.34.

For the investigation of the plasma fluxes with different directions the Mach-probe system was oriented by different angles to the device axis and to the magnetic field lines.

Figure 4 represents the distribution of the ion currents ratio from probe plates inside the separatrix.

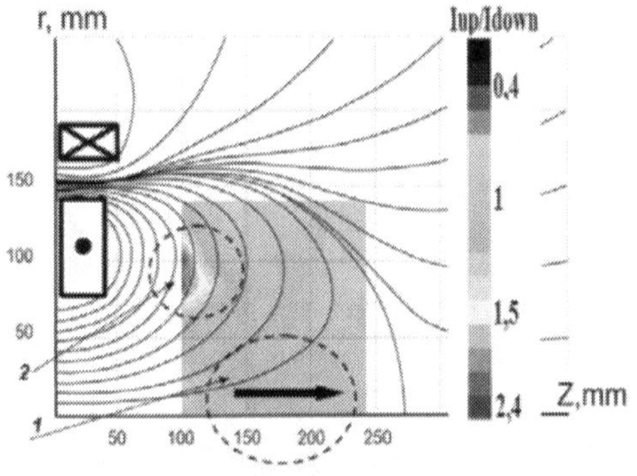

Figure 4. Areas of measured non-zero plasma flows inside separatrix

The ion currents ratio inside the separatrix mostly is about one. That indicates the absence of essential plasma flow. Only two small regions have high ion current ratio. They are numbered by 1 and 2 in the figure 4.

The first region has been found near to the magnetic system axis. The plasma flow is directed along this axis out from the magnetic trap. Magnetic field lines of paraxial region pass through the area of a weak magnetic field. The velocity of charged particles flow in this direction is about $\sim 10^5$ cm/s (M >1), the corresponding plasma flux can be estimated as $\sim 10^{16}$ cm^{-2} s^{-1}.

The second region consists of a flow in azimuth φ-direction, i.e. around «Magnetor» axis. This flow ca be driven both by particles drift due to the magnetic field lines curvature and by the crossed ExB fields azimuth motion. The plasma velocity of this flow is about $\sim 3 \cdot 10^5$ cm/s (M <1), and the corresponding plasma flux can be estimated as $\sim 10^{17}$ cm^{-2} s^{-1}.

4. Conclusions

Density distribution of the plasma produced by both the ECR microwave discharge and the hot cathode discharge is localized inside the magnetic separatrix and isolated from the vacuum vessel walls.

The distribution of plasma density depends on a position and extent of a charged particles source.

The possibility of filling with plasmas the chosen magnetic field surface of bi-dipole magnetic configuration was demonstrated during the experiments with the discharge in crossed E×B fields. Comparison of plasma density distribution for two types of discharges allows concluding that plasma distribution depends on an arrangement and extent of a charged particles source.

There are two directions of plasma movement inside the magnetic trap were observed. The first one is the axial plasma flux, which is driven by the particles losses though the weak field regions. The second one is the azimuth flow, which can be related to the both magnetic field lines curvature particle drift and to the ExB drift. The corresponding plasma fluxes can be estimated in the range $10^{16} \div 10^{17}$ cm^{-2} s^{-1}.

Acknowledgements

The work was supported by ROSATOM contract № H.4д.47.03.08.153.

References

[1] Kadomtsev B B 1958 *Plasma Physics and the Problem of Controlled Thermonuclear Reactions*

Journal of Physics: Conference Series **207** (2010) 012036 doi:10.1088/1742-6596/207/1/012036

ed. M A Leontovich (Moscow: Publ. Comp. Acad.Sci. USSR) **4** 370 (in Russian)

[2] Rosenbluth M N, Longmire C L 1957 *Ann. Phys* **1** 120–140.

[3] Levitt B, Maslovsky D and Mauel M E 2002 Measurement of the global structure of interchange modes driven by energetic electrons trapped in a magnetic dipole *Phys. Plasmas.* **1 9** 2507

[4] Garnier D T et al 2006 *Phys. Plasmas* **13** 056111

[5] Goto T, Yatsuka E, Morikawa J and Ogawa Y 2006 Plasma Production by Electron Cyclotron Heating on the Internal Coil Device Mini-RT *Japanese Journal of Applied Physics* **45 6A** 5197-5202

[6] Berdnikova M M, Vaitonene A M, Vaitonis V V et al 2003 *Vopr. At. Nauki Tekh., Ser. Termoyadernyi Sintez* **1** 22-28 (in Russian)

[7] Haefer R 1953 Acta *Physica Austriaca* **7** 251-278

[8] Redhead P A 1958 *Canadian J. of Phys.* **36** 255-270

[9] Chung K S and Hutchinson I H 1998 *Phys. Rev. A* **38** 4721

Model Calculations of Edge Dislocation Defects and Vacancies in α-Iron Lattice

L.Petrov, T.Troev, N.Nankov, E.Popov

Institute for Nuclear Research and Nuclear Energy, Bulgarian Academy of Sciences, 72 Tzarigradsko Chaussee Blvd., 1784 Sofia, Bulgaria

Email: lpetrov@inrne.bas.bg

Abstract. Two models of defects in perfect α-iron lattice were discussed. In the perfect bcc iron lattice 42×42×42 a_0 (a_0 = 2,87 Å) an edge dislocation was created, moving the second half of the bulk on one a_0 distance. This action generates a little volume in the middle of the bulk witch increases of the positron lifetime (PLT) calculated using the superimposed-atom method of Puska and Nieminen [1]. The result of 118 ps PLT in simple edge dislocation's model is in a good concurrence with earlier publications and experimental data [2]. Through the dislocation line one, two and three vacancies were localized. These models give the results for PLT of 146, 157 and 167 ps respectively. The computer simulations were performed using Finnis–Sinclair (FS) N-body potential.

1. Introduction

The knowledge of the behaviour of the basic metals as iron, chromium, beryllium and so on used in different scientific and industry branches like nuclear physics investigations and nuclear reactors through the process of exploitation is very important. There are very well known papers [1–2] that examine theoretical models of defects in metal lattices. This paper continues the consideration of models closer to the practice [3].

The basic theory of model calculating of defects in perfect lattices was given in [1,4,5]. Positron spectroscopy is capable of monitoring of micro-voids even at the earliest stages of their creation. The method is based on positron trapping at open volume defects and it was also used for the characterization of mono-vacancies and vacancy clusters [3,6].

In a perfect α-iron lattice matrix with dimension 42×42×42 (total 18,522 atoms) a perfect edge dislocation was made by the next procedure. The coordinates $x_{i,j,k}$ and $z_{i,jk,}$ of the atoms are translated according to elasticity theory of dislocations before calculating the total energy of the model crystal:

$$x_{i,j,k} = x_{i,j,k} + U(x,z) \qquad z_{i,j,k} = z_{i,j,k} + V(x,z) \qquad (1)$$

where $U(x,z)$ and $V(x,z)$ depend on the parameters of bcc iron lattice.

Modeling the vacancies in the lattice and relaxation of the transformed matrix are made according to the procedures described in [3].

2. Calculation method

The PLT in defects is calculated according to the method of Puska and Nieminen [1]. The electron density distribution $n(r)$ is approximated by superimposing the spherical averaged charge densities $n_{at}(r)$ (r is the distance from the atomic center) of an isolated atom:

$$n(r) = \sum_i n_{at}(|r - R_i|) \tag{2}$$

where R_i is the position of the ith atom.

The Coulomb potential $V_C(\mathbf{r})$ is represented by superimposing the function $V_{at}(\mathbf{r})$ of distance $\mathbf{r}$ from given atom:

$$V_C(\mathbf{r}) = \sum_i V_{at}(|r - R_i|) \tag{3}$$

The correlation potential $V_{corr}(n(r))$ practically was determined by Boroński and Nieminen [4]. The total potential felt by a positron is:

$$V_{tot}(r) = V_C(r) + V_{corr}(n\_(r)) \tag{4}$$

The Kimball–Shortly method is used for obtaining the positron wave function $\psi_+(\mathbf{r})$ by numeric solution of the Schrödinger equation. The energy eigenvalue is obtained in the every interaction while a saturated minimum value is reached.

The positron annihilation rate is expressed by:

$$\lambda = \pi r_o^2 c \int \mathrm{d}r n_+(r) x [n_v \Gamma_v(n_v(r)) + n_c(r)\Gamma_c + n_d(r)\Gamma_d], \tag{5}$$

where $n_+ = \psi_+^2$ is the density of positron, and n_v, n_c and n_d represents the density of valence, core and d-electrons respectively; r_o is the classical electron radius and c is the speed of light. Γ_v, Γ_c and Γ_d are the corresponding enhancement factors due to the correlation between positrons and electrons. For core electrons $\Gamma_c = 1,5$. According [7]:

$$\Gamma_v(n_v) = 8/3 + 1/6\, r_v^3 \tag{6}$$

where $r_v = (3/4\pi\, n_v)^{1/3}$ is the density parameter of the valence electrons. The value of $\Gamma_d = 2,42$ corresponding to PLT = 108 ps in the perfect bcc iron lattice was obtained (nearer to experimental data).

PLT [τ] is given by:

$$\tau = 1/\lambda \tag{7}$$

3. Results and discussions

In Table 1 are given the main results of the model calculations made in this work. In figure 1 are shown graphically the same data.

Journal of Physics: Conference Series **207** (2010) 012037 doi:10.1088/1742-6596/207/1/012037

Table 1: Main results of calculations

No	Lattice state	Calculated PLT [ps]	PLT according to earlier lit. data in [3] [ps]	
			Theoretical	Experimental
1	Perfect lattice	108	110	110
2	One vacancy in the middle of bulk	179	181	175
3	Edge dislocation	118	117	
4	Edge dislocation + 1 vac. in the dislocation line	146	140	150
5	Edge dislocation + 2 vac. in the dislocation line	157		
6	Edge dislocation + 3 vac. in the dislocation line	167		

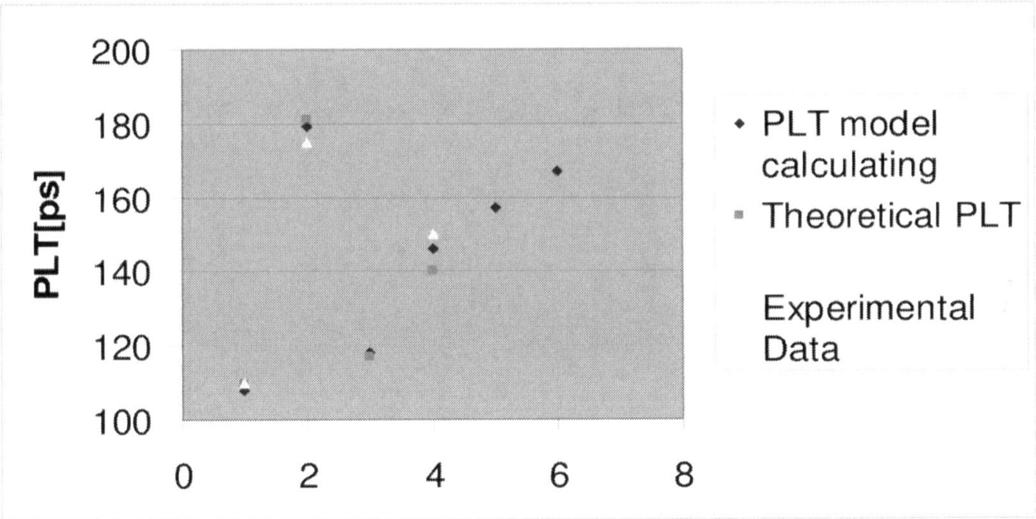

Figure 1. Comparison of the PLT from model calculating and the experimental data.

In figures 2, 3 are shown the positron wave functions in simple vacancy in the perfect lattice, edge dislocation.

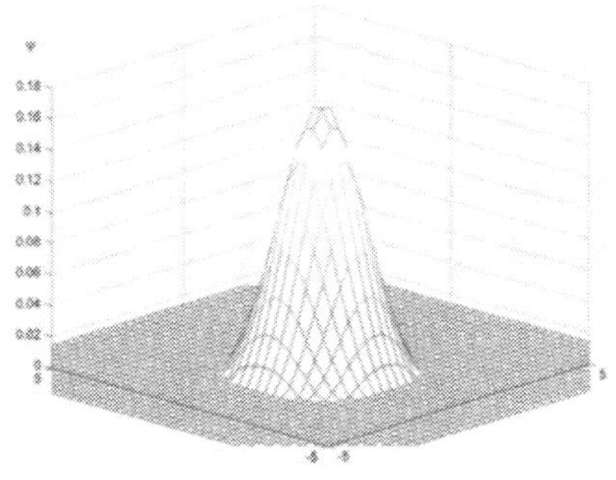

Figure 2. Simple vacancy in the middle of the lattice

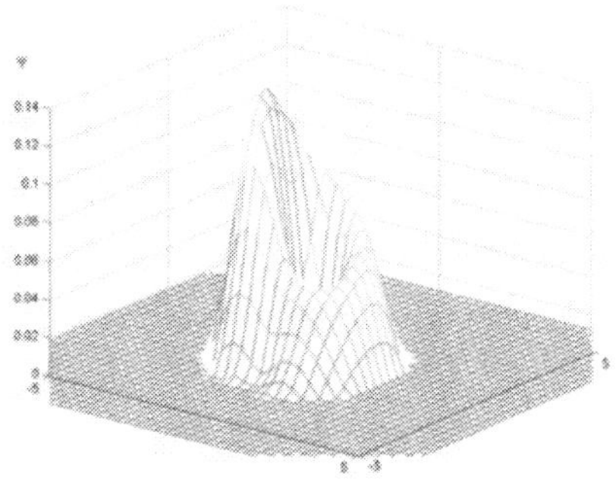

Figure 3. Perfect edge dislocation

4. Conclusions

The PLT of 118 ps for pure edge dislocation in α-iron has been obtained. For the model with edge dislocation and vacancies in the dislocation line PLTs of 146, 157 and 167 ps respectively have been calculated. Our calculated results show deviations of about 20 % from the PLT in pure vacancies. However, the vacancy and vacancy clusters contribute to the PLT much more than the pure dislocation. Our results give information about the type of defects in α-iron lattice in good agreement with the experiment. There are differences in the PLT for various vacancy localizations outside the dislocation and along the dislocation line. The PLT in the case of compact three-vacancy cluster is about 10 ps higher than the three-vacancy nano-void along the dislocation line.

Acknowledgement: The EURATOM/INRNE Fusion Association supports this work.

References

[1] Puska M J, Nieminen R M 1994 *Rev. Modern Phys.* **66** № 3 841–897
[2] Kamimura Y, Tsutsumi T, Kuramoto E 1995 *Phys. Review B* **52** 879–886
[3] Petrov L, Nanakov N, Popov E, Troev T 2007 *17th IAEA Technical Meeting on Research Using Small Fusion Devices* 22–24 October, Lisbon, Portugal
[4] Bronski E and Nieminen R M 1986 *Phys. Review B* **34** 3820
[5] Kuramoto E, Tsutsumi T, Ueno K, Ohmura M, Kamimura Y 1999 *Computational Materials Science* **14** 28–35
[6] Ishizaki T, Xu Q, Yoshiie T, Nagata S, Troev T 2002 *J. Nucl. Mater.* **307–311** 961–965
[7] Brandt W, Reinheimer J 1971 *Phys. Lett.* **35**A 109

Positron Lifetime Calculation of Defect in Palladium Containing Hydrogen and Helium

T Troev, E Popov, P Staykov, L Petrov, K Berovski, S Peneva and N Nankov

Institute for Nuclear Research and Nuclear Energy, Bulgarian Academy of Sciences, Sofia, Bulgaria

E-mail: troev@inrne.bas.bg

Abstract. Lattice defects containing hydrogen and helium in Pd have been studied by positron model calculations using Density Functional Theory [DFT] in its local density approximation [LDA] [1] implementation. A correlation between the positron lifetime, positron ground stage energy and number atoms of H and He has been established. It was found that the value of positron lifetime for perfect Pd lattice is 106 ps and for mono vacancy is 184 ps which corresponds to the experimental values.

1. Introduction

Positron quantum mechanical calculations were carried out for investigation of defects in palladium containing hydrogen and helium atoms. Recently the convergence of electronic structure calculations for super cell size has been studied for vacancy, vacancy clusters and vacancy clusters contacting hydrogen and helium atoms [2,3]. The electron wave function in palladium has been obtained in LDA to the DFT. The determination of electron and positron states in condensed matter has been done on the basis of a generalization of DFT known as two component DTF (TCDFT). The ground state electron and positron densities are calculated solving a set of one particle Kohn-Sham equation for electrons and positrons. The positron model calculations have been performed in fcc palladium lattice containing vacancies, empty nano-voids and nano-voids consisting from various number of vacancies and gases atoms.

On the basis of the calculated results the behaviors of the empty nano-voids and nano-voids with hydrogen were discussed. The lifetime of trapped positron is sensitive to the vacancies defect as well as to the nano-voids with different ratio of hydrogen and helium atoms to the number of vacancies [4,5].

2. Theory

For more exact treatment the reader is referred to [2,6].

2.1. Positron lifetime

The positron annihilation rate is proportional to the electronic density at positron position and can be calculated from the overlap integral as:

$$\lambda = \frac{1}{\tau} = \pi r_0^2 c \int d\,\mathrm{r}\, n_+(\mathrm{r}) n_-(\mathrm{r}) g(0;n_+;n_-), \quad (1)$$

where r_0 is the classical electron radius, c is the speed of light, n_+ and n_- are the positron and electron densities and $g(0;n_+;n_-)$ the electron–positron pair correlation functional evaluated at the positron [2]. The zero positron density limit of the pair correlation function $g(0;n_+;n_-)$ is denoted by

γ and called the enhancement factor of the electron density at the positron. The most trivial form is simply $\gamma = 1$, which is called the independent particle model (IPM) approximation.

2.2. Two component density functional theory

To solve the needed electron and positron densities a two component generalization of the density functional theory (TCDFT) can be used. This means writing the total energy functional $E[n_-, n_+]$ for the electron-positron system in external potential V_{ext} in terms of average electron and positron densities as

$$E[n_-, n_+] = F[n_-] + F[n_+] + \int d\,\mathrm{r} V_{ext}(\mathrm{r})[n_-(\mathrm{r}) - n_+(\mathrm{r})] - \\ - \int d\,\mathrm{r} \int d\,\mathrm{r}' \frac{n_-(\mathrm{r}) n_+(\mathrm{r}')}{|\mathrm{r} - \mathrm{r}'|} + E_c^{e-p}[n_+, n_-] \tag{2}$$

with

$$F[n] = T[n] + \frac{1}{2} \int d\,\mathrm{r} \int d\,\mathrm{r}' \frac{n_-(\mathrm{r}) n_+(\mathrm{r}')}{|\mathrm{r} - \mathrm{r}'|} + E_{xc}[n] \tag{3}$$

denoting the one component functional for electrons and positrons [3]. Here $T[n]$ is the kinetic energy functional of no interacting electrons or positrons, E_{xc} is the exchange correlation energy and $E_c^{e-p}[n_+, n_-]$ the positron–electron correlation energy functional.

From this one can proceed with a generalized Kohn-Sham variation method to obtain ground state electron and positron densities. This requires solving one particle Schrödinger equations for electron and positron wave functions. For positron this reads

$$-\frac{1}{2} \nabla^2 \varphi_i^+(\mathbf{r}) + V_{eff}(\mathbf{r}) \varphi_i^+ = \varepsilon_i \varphi_i^+(\mathbf{r}) \tag{4}$$

with effective potential written as

$$V_{eff}(\mathrm{r}) \varphi_i^+ = \phi(\mathrm{r}) + \frac{\delta E_{xc}[n_+]}{\delta n_+(\mathrm{r})} + \frac{\delta E_c^{e-p}[n_+, n_-]}{\delta n_+(\mathrm{r})}. \tag{5}$$

Here we have

$$\phi(\mathrm{r}) = \int d\,\mathrm{r}' \frac{-n_-(\mathrm{r}') + n_+(\mathrm{r}') + n_0(\mathrm{r}')}{|\mathrm{r} - \mathrm{r}'|}, \tag{6}$$

for the total Coulomb potential and n_0 the charge density from which the external potential V_{ext} arises. For electron the effective potential is the same as (5) with n_+ and n_- swapped and the sign of the Coulomb potential reversed.

From the obtained wave functions, the electron and positron densities defined as

$$n_-(\mathrm{r}) = \sum_{\varepsilon_i \le \varepsilon_f} |\varphi_i(\mathrm{r})|^2, n_+(\mathrm{r}) = \sum_i^{N_+} |\varphi_i^+(\mathrm{r})|^2 \tag{7}$$

can be calculated. The summation on electron states is up to the Fermi level but usually only one positron ($n_+ = 1$) needs to be considered. The Kohn–Sham equations must be solved self-consistently using iterative scheme simultaneously for electron and positron states.

2.3. Local density approximation

Unfortunately, in equation (5) we still have unknown terms E_{xc} and E_c^{e-p} that calls for approximations and the results may strongly depend on them. The usual way to continue is to employ the local density approximation (LDA) [1] for exchange-correlation energy E_{xc} in which one writes

$$E_{xc}^{LDA}[n] = \int d\,\mathrm{r}\, n(\mathrm{r}) e_{xc}^{hom}(n(\mathrm{r})). \quad \textbf{(8)}$$

The exchange correlation energy E_{xc} is thus approximated by using the exchange correlation energy of the homogeneous electron gas e_{xc}^{hom} per electron with the replacement of the constant density n_0 by the local density $n(\mathrm{r})$ of the inhomogeneous system, known from quantum Monte Carlo simulations.

From this the annihilation rate λ is calculated within the LDA as

$$\lambda = \pi r_0^2 c \int d\,\mathrm{r}\, \gamma^{LDA} |\varphi_+(\mathrm{r})|^2 |\varphi_-(\mathrm{r})|^2 \quad \textbf{(9)}$$

where φ_+ is the positron wave function [7].

To solve is used a parameterized positron wave function as

$$|\varphi_+\rangle \approx a_0 + a_1\left(\mathrm{erf}\left(\frac{r}{a_2}\right)\right)^{a_3} = a_0 + a_1\left(\frac{1}{\sqrt{2\pi}} \int_{-\infty}^{r} e^{-\left(\frac{r}{\sqrt{2a_2}}\right)^2} dr\right)^{a_3}, \quad \textbf{(10)}$$

where r is the distance from the center of the atomic sphere and a_i are fitted parameters, $i \in \{1,\ldots,3\}$. The parameters have been calculated using the LMTO-ASA method [8].

3. Result

The positron lifetime in vacancy clusters increases from 184 to 417 ps with the cluster size, where the cluster size grows up from 1 to 38 vacancies. These calculations are shown in figure 1 with graphic correlations between positron lifetime and positron ground stage energy as a function of number of vacancies. Here we observe that positron lifetime has correlation from positron ground state energy like conversely proportion.

These results are well compatible with experimental results and theoretical morphology for this problem. With other words the positron lifetime grows up with decrease in the electron density and also the positron ground stage energy decreases in parallel with the same decrease of the electron density.

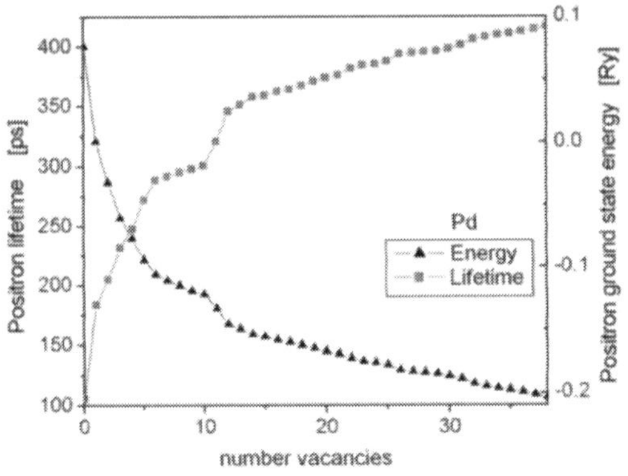

Figure 1. Correlations between positron lifetime and positron ground stage energy as a function number of vacancies in the nano-void

In figure 2a are shown the correlations between positron lifetime and positron ground stage energy as a function of number atoms of H and He in mono vacancy. We see that positron lifetime has again conversely proportion to positron ground stage energy but with growth of number of atoms H and He the positron lifetime decreases, when in the same time positron ground stage grows up. In other words the positron lifetime decreases with the growth of the electron density and the positron ground stage energy also grows up with the same increase of the electron density. The electron density from H and He atoms becomes equal to the electron density from perfect lattice when the positron lifetime becomes equal to the positron lifetime of perfect lattice. This view is confirmed in figure 2b where the correlations between lifetime and positron ground stage energy are shown as a function of number of atoms H and He in two vacancies.

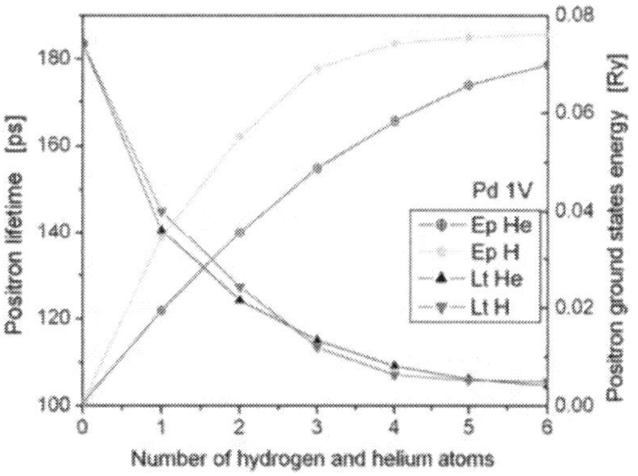

Figure 2a. Correlations between positron lifetime and positron ground stage energy as a function of number atoms H and He in mono vacancy.

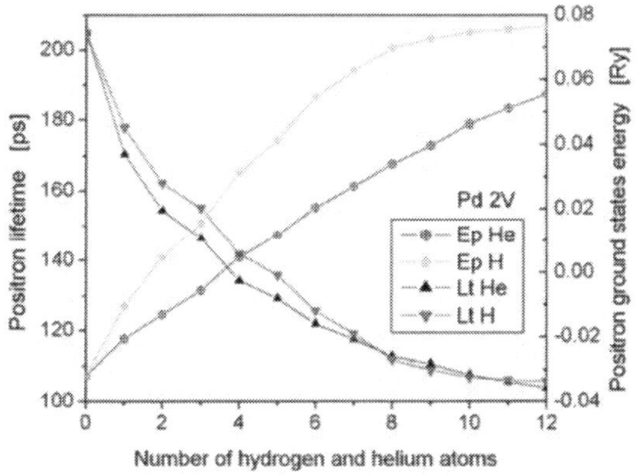

Figure 2b. Correlations between positron lifetime and positron ground stage energy as a function of number atoms H and He in two vacancies.

4. Conclusions

This study is the first positron model calculation for defects and nano-voids containing hydrogen and helium in Pd lattice. A correlation between the positron lifetime and number of atoms H and He is established. We found that the value of positron lifetime for perfect Pd lattice is 106 ps, for mono vacancy – 184 ps, for 1V1H – 145 ps, where for 1V1He the value is 141 ps.

Acknowledgement: The EURATOM/INRNE Fusion Association supports this work.

References

[1] Dreizler R M and Gross E K U 1990 *Density Functional Theory* (Springer-Verlag)
[2] Schultz P J and Lynn K G 1988 *Rev. Mod. Phys.* **60** 701–779
[3] M. J. Puska and R. M. Nieminen 1994 *Rev. Mod. Phys.* **66** 841–897
[4] Ishizaki T, Yoshiie Q Xu T, Nagata S and Troev T 2002 *J Nucl. Mater.* **961** 307–311
[5] Shivachev B L, Troev T and Yoshiie T 2002 *J. Nucl. Mater.* **306** 101
[6] Saarinen K, Hautojärvi P and Corbel C 1997 *Identification of Defects in Semiconductors* (Academic Press) A **388** 434
[7] Alatalo M, Barbiellini B, Hakala M, Kauppinen H, Korhonen T, Puska M J, Saarinen K, Hautojärvi P and R. M. Nieminen 1996 *Phys. Rev.* B **54** 2397
[8] Valminen P, 2002 *Special Assignment* (Helsinki University of Technology) Espoo

AUTHOR INDEX

Álvarez, R 158
Alves, L L 158
Apostolova, Tzveta 201
Badulescu, M 116
Bardamid, A F 107
Belyaeva, A I 107
Benova, E 136, 180
Berovski, K 223
Blagoev, A 112
Bogdanov, T 180
Bozhinova, I 112
Braginsky, O V 95
Calixto-Rodriguez, M 120
Calzada, M D 170
Canal, C 47
Castillo, F 120, 124
Christova, M 142, 148
Clément, F 56, 62
Cvelbar, U 69
Damyanova, M 184
Dankov, Plamen 101
Dias, F M 21
Dimitrijevic, M S 142, 148
Drenik, A 69
Epstein, I L 1, 11
Faulques, E 132
Fetisov, I K 128
Gamero, A 89
Gegiuc, A 40
Gigosos, M A 89
Gueorguiev, V 101, 132
Hadjichristov, G B 132
Halamova, Ivana 85
Havlícková, E 174
Held, B 56
Hrach, R 174
Iacob, C 116
Ivanov, I V 128
Ivanov, T. 101, 132
Ivanov, V G 132
Khodachenko, G V 128, 213
Konovalov, V G 107
Kossyi, I A 30
Kovacevic, A 142
Kovalev, A S 95
Kovtun, K V 107
Krashevskaya, G V 213
Krcma, Frantisek 75, 81, 85
Kurnaev, V A 213
Lebedev, Yu A 1, 11
Leclaire, C 62
Lecoq, E 56, 62
Loiseau, J-F 56
Lopaev, D V 95
Lungu, C P 116
Mankelevich, Yu A 164
Marinov, Y G 132
Martínez, H 120, 124
Mendez, E 124
Mihaila, I 40
Mozetic, M 69
Mozgrin, D V 128
Muñoz, J 170
Naidenkova, D I 107
Nankov, N 190, 207, 219, 223
Naumov, D S 213
Orial, G 62
Osorio-González, D 124
Palomares, J M 89
Panousis, E 56, 62
Pashov, A 112
Peña, Y 120
Pencheva, M 136
Peneva, S 223
Petrov, D B 128
Petrov, L 207, 219, 223
Pointu, A M 47
Popa, G 40
Popov, E 190, 207, 219, 223
Porosnicu, C 116
Prochazkova, Jana 75
Proshina, O 152
Rakhimov, A T 95
Rakhimova, T 95, 152, 164
Reyes, P G 124
Ricard, A 47, 56, 62
Rusinov, I 112
Ryzhkov, V I 107
Sabchevski, S 184
Sahal-Bréchot, S 148
Sanchez-Juarez, A 120
Sazavska, Vera 81
Shakhatov, V A 1, 11
Shchelkanov, I A 213
Shkurenkov, I A 164
Shtan', A F 107
Shukshina, T V 128
Simic, Z 148
Simsova, Tereza 81
Slatin, K A 107
Snegirev, A M 128
Sola, A 89
Solodovchenko, S I 107
Stanev, A 112
Stara, Zdenka 75, 85
Staykov, P 223
Stefanov, Peter 101
Surdu-Bob, C C 116
Tatarinov, A V 11
Torres, J 89

AUTHOR INDEX

Trembach, O V..107
Troev, T 190, 201, 207, 219, 223
Tsventoukh, M M..213
Ursu, C 40
Van Der Mullen, J J A M 89
Vasil'Ev, A A 107
Vasilieva, A N...95
Vesel, A 69
Villeger, S..47
Voitsenya, V S 107
Voloshin, D 152
Vovchenko, D E...213
Yoshiie, T 190
Zemanek, Nikola 81
Zhelyazkov, I 184
Zyryanov, S M...95

Institute of Physics
Dirac House, Temple Back
Bristol BS1 6BE UK

proceedings
.com